MOLECULAR BIOLOGY
INTELLIGENCE
UNIT

Tight Junctions

Lorenza Gonzalez-Mariscal, Ph.D.
Department of Physiology, Biophysics and Neuroscience
Center for Research and Advanced Studies (CINVESTAV)
México D.F., México

LANDES BIOSCIENCE / EUREKAH.COM
GEORGETOWN, TEXAS
U.S.A.

SPRINGER SCIENCE+BUSINESS MEDIA
NEW YORK, NEW YORK
U.S.A.

TIGHT JUNCTIONS

Molecular Biology Intelligence Unit

Landes Bioscience
Springer Science+Business Media, LLC

ISBN: 0-387-33201-4   Printed on acid-free paper.

Springer Science+Business Media, LLC, 233 Spring Street, New York, New York 10013, U.S.A.
http://www.springer.com

Please address all inquiries to the Publishers:
Landes Bioscience / Eurekah.com, 810 South Church Street, Georgetown, Texas 78626, U.S.A.
Phone: 512/ 863 7762; FAX: 512/ 863 0081
http://www.eurekah.com
http://www.landesbioscience.com

Printed in the United States of America.

9 8 7 6 5 4 3 2 1

Library of Congress Cataloging-in-Publication Data

Tight junctions / [edited by] Lorenza González-Mariscal.
    p. ; cm. -- (Molecular biology intelligence unit)
  Includes bibliographical references and index.
  ISBN 0-387-33201-4
  1. Tight junctions (Cell biology) I. González-Mariscal, Lorenza. II. Title. III. Series: Molecular biology intelligence unit (Unnumbered)
  [DNLM: 1. Tight Junctions. QU 350 T566 2006]
QH603.C4T56 2006
571.6--dc22
                         2006007715

CONTENTS

EDITOR

Lorenza Gonzalez-Mariscal
Department of Physiology, Biophysics and Neuroscience
Center for Research and Advanced Studies (CINVESTAV)
México D.F., México
Email: lorenza@fisio.cinvestav.mx
Chapters 7, 11

CONTRIBUTORS

James Melvin Anderson
Departments of Cell and Molecular
    Physiology and Medicine
University of North Carolina
    at Chapel Hill
Chapel Hill, North Carolina, U.S.A.
Email: jandersn@med.unc.edu
Chapter 3

Maria S. Balda
Division of Cell Biology
Institute of Ophthalmology
University College London
London, U.K.
Email: m.balda@ucl.ac.uk
Chapter 8

Abigail Betanzos
Department of Pathology
Center for Research and Advanced
    Studies (CINVESTAV)
México D.F., México
Email: betanzos@yahoo.com.mx
Chapter 7

Jeff M. Bronstein
Department of Neurology, RNRC
David Geffen School of Medicine
    at UCLA
Los Angeles, California, U.S.A.
Email: jbronste@ucla.edu
Chapter 14

Matthias Bruewer
Department of General Surgery
University of Muenster
Muenster, Germany
Email: bruwer@uni-muenster.de
Chapter 10

Marcelino Cereijido
Department of Physiology,
    Biophysics and Neuroscience
Center for Research and Advanced
    Studies (CINVESTAV)
México D.F., México
Email: cereijido@fisio.cinvestav.mx
Chapter 1

Yan-Hua Chen
Department of Anatomy and Cell Biology
East Carolina University Brody
    School of Medicine
Greenville, North Carolina, U.S.A.
Email: cheny@mail.ecu.edu
Chapter 2

Sandra Citi
Department of Molecular Biology
University of Geneva
Geneva, Switzerland
Email: sandra.citi@molbio.unige.ch
Chapter 5

Rubén Gerardo Contreras
Department of Physiology,
    Biophysics and Neuroscience
Center for Research and Advanced
    Studies (CINVESTAV)
México D.F., México
Email: rcontrer@fisio.cinvestav.mx
Chapters 1, 11

María del Refugio García-Villegas
Department of Physiology,
    Biophysics and Neuroscience
Center for Research and Advanced
    Studies (CINVESTAV)
México D.F., México
Email: rgarciav@fisio.cinvestav.mx
Chapter 1

Klaus Ebnet
Institute of Cell Biology
ZMBE
University of Münster
Münster, Germany
Chapter 13

Judith Eckert
Developmental Origins of Health
    and Disease
School of Medicine
University of Southampton
Southampton, U.K.
Email: J.J.Eckert@soton.ac.uk
Chapter 12

Natalie D. Eddington
Department of Pharmaceutical Sciences
Pharmacokinetics-Biopharmaceutics
    Laboratory
School of Pharmacy
University of Maryland at Baltimore
Baltimore, Maryland, U.S.A.
Email: neddingt@rx.umaryland.edu
Chapter 15

Alan S. Fanning
Department of Cell
    and Molecular Physiology
University of North Carolina
    at Chapel Hill
Chapel Hill, North Carolina, U.S.A.
Email: alan_fanning@med.unc.edu
Chapter 6

Alessio Fasano
Mucosal Biology Research Center
University of Maryland
    School of Medicine
Baltimore, Maryland, U.S.A.
Email: AFasano@Peds.umaryland.edu
Chapter 15

Tom P. Fleming
School of Biological Sciences
University of Southampton
Southampton, U.K.
Email: tpf@soton.ac.uk
Chapter 12

David Flores-Benítez
Department of Physiology,
    Biophysics and Neuroscience
Center for Research and Advanced
    Studies (CINVESTAV)
México D.F., México
Email: dflores@fisio.cinvestav.mx
Chapter 11

Kristopher K. Frese
Abramson Family Center
Research Institute
Philadelphia, Pennsylvania, U.S.A.
Email: kfrese@mail.med.upenn.edu
Chapter 9

Daniel A. Goodenough
Department of Cell Biology
Harvard Medical School
Boston, Massachusetts, U.S.A.
Email:
    daniel_goodenough@hms.harvard.edu
Chapter 2

Laurent Guillemot
Department of Molecular Biology
University of Geneva
Geneva, Switzerland
Email:
    laurent.guillemot@molbio.unige.ch
Chapter 5

Yutaka Hata
Department of Medical Biochemistry
Graduate School of Medicine
Tokyo Medical and Dental University
Tokyo, Japan
Email: yuhammch@med.tmd.ac.jp
Chapter 4

Susumu Hirabayashi
Department of Medical Biochemistry
Graduate School of Medicine
Tokyo Medical and Dental University
Tokyo, Japan
Email: shirmmch@tmd.ac.jp
Chapter 4

Miriam Huerta
Department of Genetics
    and Molecular Biology
Center for Research and Advanced
    Studies (CINVESTAV)
México D.F., México
Email: huertam@cinvestav.mx
Chapter 7

Blanca Estela Jaramillo
Department of Physiology,
    Biophysics and Neuroscience
Center for Research and Advanced
    Studies (CINVESTAV)
México D. F., México
Email: bejarami@fisio.cinvestav.mx
Chapter 7

Ronald T. Javier
Department of Molecular Virology
    and Microbiology
Baylor College of Medicine
Houston, Texas, U.S.A.
Email: rjavier@bcm.tmc.edu
Chapter 9

Emma Kavanagh
Division of Cell Biology
Institute of Ophthalmology
University College London
London, U.K.
Email: e.kavanagh@ucl.ac.uk
Chapter 8

Isabel J. Latorre
Abramson Family Center
Research Institute
Philadelphia, Pennsylvania, U.S.A.
Email: ilatorre@mail.med.upenn.edu
Chapter 9

Andrea Lippoldt
Max Delbrück-Center Berlin-Buch
Department of Neurology
Schering AG Berlin
Berlin, Germany
Email:
    Andrea.Lippoldt@SCHERING.DE
Chapter 13

Esther Lopez-Bayghen
Department of Genetics
    and Molecular Biology
Center for Research and Advanced
    Studies (CINVESTAV)
México D. F., México
Email: ebayghen@prodigy.net.mx
Chapter 7

Qun Lu
Department of Anatomy
    and Cell Biology
East Carolina University
    Brody School of Medicine
Greenville, North Carolina, U.S.A.
Email: luq@mail.ecu.edu
Chapter 2

Karl Matter
Division of Cell Biology
Institute of Ophthalmology
University College London
London, U.K.
Email: k.matter@ucl.ac.uk
Chapter 8

Asma Nusrat
Department of Pathology
    and Laboratory Medicine
Emory University
Whitehead Research Building
Atlanta, Georgia, U.S.A.
Email: anusrat@emory.edu
Chapter 10

Noha N. Salama
Department of Pharmaceutical Sciences
Pharmacokinetics-Biopharmaceutics
    Laboratory
School of Pharmacy
University of Maryland at Baltimore
Baltimore, Maryland, U.S.A.
Email: nohasalama@gmail.com
    or nsala002@umaryland.edu
Chapter 15

Bhavwanti Sheth
School of Biological Sciences
University of Southampton
Southampton, U.K.
Email: B.Sheth@soton.ac.uk
Chapter 12

Liora Shoshani
Center for Research and Advanced
    Studies (CINVESTAV)
México, D.F, México
Email: shoshani@fisio.cinvestav.mx
Chapters 1, 11

Fay Thomas
School of Biological Sciences
University of Southampton
Southampton, U.K.
Email: F.C.Thomas@soton.ac.uk
Chapter 12

Seema Tiwari-Woodruff
Department of Neurology
Geffen School of Medicine at UCLA
Los Angeles, California, U.S.A.
Email: seemaw@ucla.edu
Chapter 14

Anna Tsapara
Division of Cell Biology
Institute of Ophthalmology
University College London
London, U.K.
Email: a.tsapara@ucl.ac.uk
Chapter 8

Christina M. Van Itallie
Department of Medicine
University of North Carolina
    at Chapel Hill
Chapel Hill, North Carolina, U.S.A.
Email: christina van-itallie@med.unc.edu
Chapter 3

Hartwig Wolburg
Institute of Pathology
University of Tübingen
Tübingen, Germany
Email:
    hartwig.wolburg@med.uni-tuebingen.de
Chapter 13

PREFACE

Tight junctions (TJs) are cell-cell adhesion belts that encircle epithelial and endothelial cells at the limit between the apical and the lateral membrane. These junctions are crucial for the establishment of separate compartments in multicellular organisms and for the exchange of substances between the internal milieu and the external environment. The perception of TJs has changed over the years. From being regarded as static paracellular seals, they have come to be perceived as dynamic structures that adjust their morphology and function in response to physiological, pharmacological and pathological challenges. The roles that TJs play in epithelial and endothelial cells has also widened, and nowadays this structure is regarded not only as a fence that limits within the membrane, the movement of proteins and lipids between the apical and basolateral membranes, or as a gate that regulates in a size and charge selective manner the transit of ion and molecules through the paracellular pathway, but also as a structure integrated by molecules that participate in the control of cell proliferation. These observations highlight the importance of understanding TJ physiology in order to develop effective strategies for the treatment of pathological conditions such as cancer and autoimmune diseases.

This broader perception of TJs is reflected in all the chapters of the book and has been attained thanks to the identification in recent years of a wide array of proteins that constitute TJs in epithelial and endothelial cells as well as in central nervous system myelin.

The first chapter of this book describes the evolution of the transporting epithelium from primitive metazoans to mammalian epithelia. Then, Chapters 2, 3, 4, 5, 6 and 7 review the characteristics and functionality of different proteins of the TJ studied in cell culture models, natural epithelia, mutant mice and heritable human diseases. Chapters 7, 8 and 9 analyze the presence of TJ proteins at the nucleus, as well as their association with nuclear proteins and their role in the regulation of cell proliferation and cancer. The interaction and regulation of TJs by the actin cytoskeleton is described in Chapter 10, while the regulatory role that lipids and some proteins exert on TJs is described in Chapter 11. Chapter 12 reviews the biogenesis of TJs during development in the mouse embryo and includes emerging data from other mammalian species. The particular characteristics of endothelial TJs are discussed in Chapter 13, giving particular emphasis to those found in the blood brain barrier. The structure and molecular composition of TJs present in central nervous system myelin is discussed in Chapter 14. The book ends with a chapter dedicated to the novel and promising strategy of enhancing drug delivery via modulation of TJs.

All these topics have been addressed by a group of leading experts from around the world. I thank them for their effort and hope that this book will be helpful to a broad audience of readers interested in the biology of epithelial and endothelial cells.

Lorenza Gonzalez-Mariscal, Ph.D.

CHAPTER 1

Evolution of the Transporting Epithelium Phenotype

Marcelino Cereijido, María del Refugio García-Villegas, Liora Shoshani and Ruben Gerardo Contreras

Abstract

Metazoans and transporting epithelia (TE) kept a strict correlation throughout evolution because a cell lodged in an intimate tissue and surrounded by an extracellular space less than a micron thick would quickly perish were it not for the intense and highly selective exchange of substances across TE. The main cellular features of TE are **tight junctions** and apical/basolateral **polarity** involving close to a hundred molecular species exquisitely assembled. Even when at the dawn of metazoan, junctions and polarity must have been much simpler, it is hard to imagine how the molecules that are involved might have coincided in the same organism and within a few minutes. The present chapter attempts to solve this conundrum by discussing several clues.

1. Polarity, as well as certain molecules involved in its generation and maintenance, are even present in unicellulars.
2. Molecular species belonging to septate and occluding junctions can be found in unicellulars, albeit fulfilling different roles.[3] Primitive metazoan might have had very simple epithelia of a transient nature, that helped to retain nutrients and signal molecules for short periods, then opened to allow the whole mass of cells to be flushed by the environment (*"Thrifty sponge"*).
4. Early metazoan might have compensated the inefficiency of their primitive epithelia with a large surface-to-volume ratio.
5. Finally, the possibility exists that cells might have proliferated without completely detaching from each other, and preserving the orientation of their mitotic spindle, thereby generating an ample overall polarized epithelium that would create an internal environment even before an internal body of somatic cells would grow inside (*"the mare nostrum metazoan"*).

Introduction

A unicellular organism in the sea exchanges substances with its environment, and this environment behaves as an infinite and constant reservoir that is not exhausted by the removal of nutrients nor spoiled by the excreted wastes. On the contrary, when a cell is lodged in a recondite fold of the brain, the liver or any other organ of a mammal, its surrounding is reduced to an extremely thin film, that nevertheless behaves as an infinite and constant reservoir as if it were an immense ocean. This is due to the fact that this milieu is in quick equilibrium with blood, which is carried to ample areas of transporting epithelia (intestinal, renal, gills, etc.) (Fig. 1) that act as interfaces with the external environment that extrudes metabolic wastes and takes up nutrients. Therefore, the life of higher metazoan is unavoidably dependent on transporting epithelia.

Tight Junctions, edited by Lorenza Gonzalez-Mariscal. ©2006 Landes Bioscience and Springer Science+Business Media.

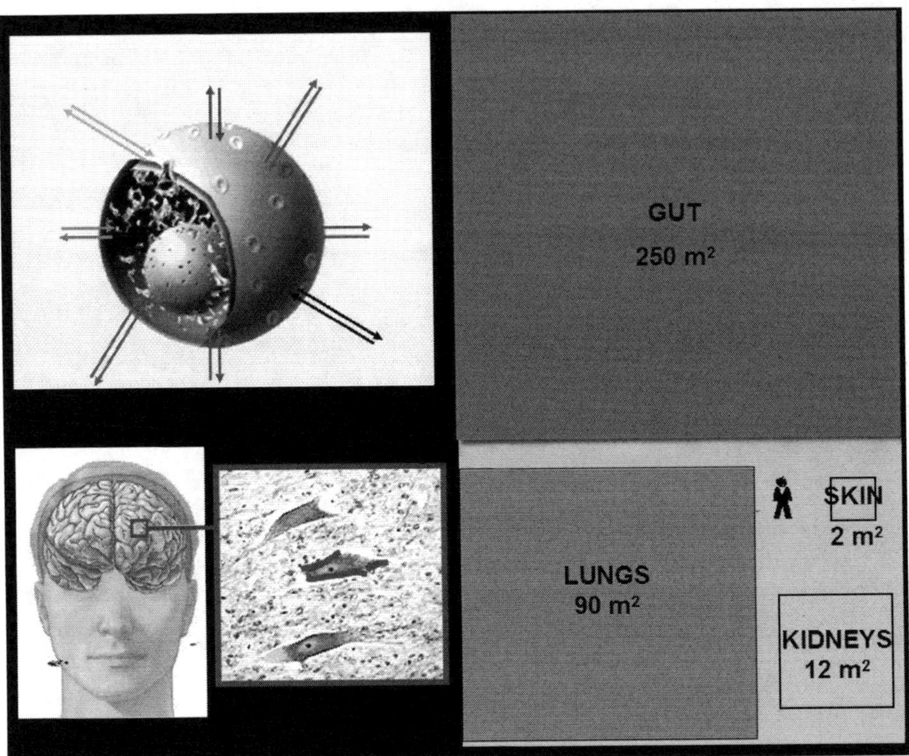

Figure 1. Regardless of how recondite its position in the body of a metazoan, cells always exchange with an environment that behaves as an infinite reservoir. Top left) A single cell in the sea exchanges substances freely, without exhausting nor contaminating its milieu. A cell lodged in a tissue of a metazoan instead (bottom left) exchanges with a thin layer of extracellular fluid (bottom center: red) that would be quickly spoiled, were it not for a circulatory apparatus that takes substances and nutrients to and from enormous areas of epithelia. Therefore, the existence of metazoan is possible because of their exchange of substances with the environment across huge epithelial layers whose cells have two basic properties: tight junctions and polarity. To give a visual idea of the size of these epithelia, we compare (right) the area of some of the main ones with a human figure. A color version of this figure is available online at www.Eurekah.com.

We will not discuss in this chapter metazoan like the sponge, where the environment can circulate through the body of the animal, and thus does not depend on the exchange through epithelia. For the same reason, we would not deal with "multicellulars" like *Dictiostelium discoideum*, where in a given moment cells clump together and form a transient slug that enables the cells to migrate. Metazoan somewhat more elaborated than sponges, like *Cnidaria*, whose most well known representative is perhaps the hydra, have an enormous surface-to-volume ratio, and this large area compensates for the fact that the properties of their epithelium might not be as elaborated as, say, the intestinal mucosa or the renal tube of mammals. It should be pointed out though, that epithelia of *Cnidaria* and other lower metazoan present today may not be "primitive" as they had the opportunity to adapt through hundreds of million years, and are in fact highly elaborated and efficiently adapted.

The so called "transporting epithelia phenotype" is constituted by cells that form continuous layers whose outermost cell layer is surrounded by occluding junctions (TJs) that transform the epithelium into an effective permeability barrier, and are polarized: their apical domain has structural, biochemical and physiological properties different from those of the basolateral region.

Since the multitude of molecules that constitute the TJ and other cell-cell and cell-substrate junctions, the large number of membrane molecules that have a polarized distribution, as well as the highly complex mechanisms responsible for junctions and polarization, were perfected along ages of evolution, we face a conundrum for explaining the relationship between transporting epithelia and multicellularity. Thus, it is obvious that this large number and variety of molecular species involved, that required millions of years of evolution, might not have coincided within minutes in the same multicellular organism. This very short period ranging in minutes was required by the fact that the cellular mass of primitive metazoan would perish, unless its internal milieu was restored by exchange with the external surrounding. We have discussed elsewhere[1] how this enigma might have been solved, and Figure 2 depicts some plausible alternatives, that might have participated in the dawn of transporting epithelia and metazoans.

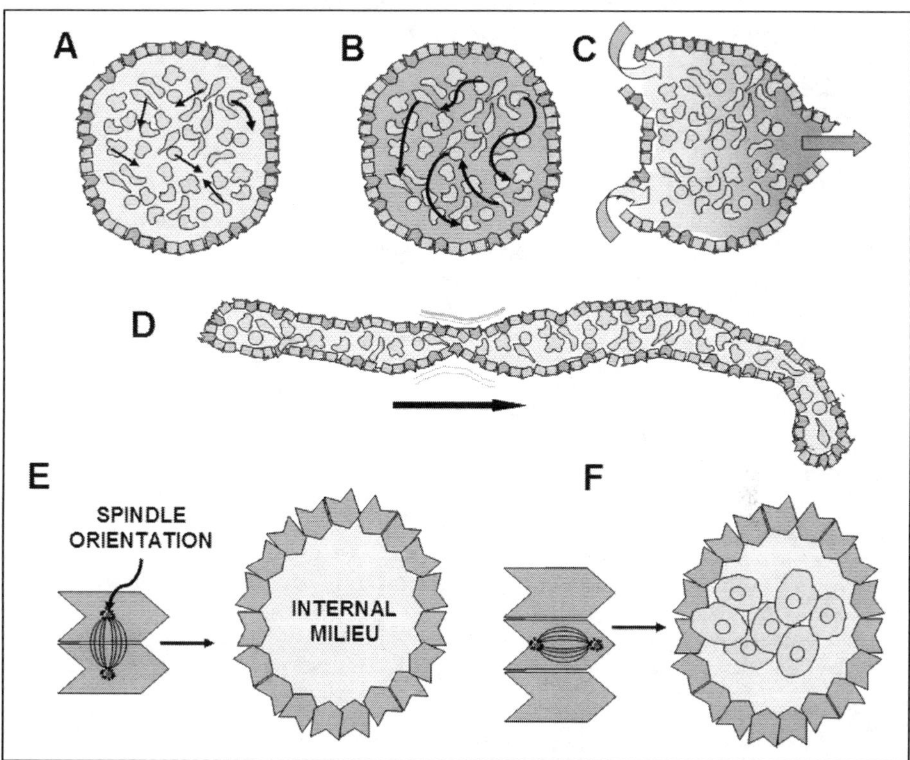

Figure 2. Hypothetical steps towards the emergence of metazoan surrounded by a transporting epithelium. Top: "thrifty sponge": A) In a given moment, cells are surrounded by a primitive epithelial layer of non-polarized cells, and secret a valuable substance (a nutrient? a signal?). Surrounding cells not only lack polarity, but do not have a selective permeability either. B) Although cells profit from the valuable substance that they secrete, the internal milieu is exhausted of nutrients and polluted by wastes. C) As a consequence of intoxication, or because of an increase in osmolarity of the internal milieu, the "thrifty sponge" bursts, and the organism is flushed by the outer solution. D) A conglomerate of cells adopts a very flat structure, and is surrounded by a primitive epithelium without specific permeability, yet the area-to-volume ratio is large enough to insure a satisfactory exchange with the environment. Peristaltic movements (arrow) might help stir the internal milieu. E) "Mare nostrum". A polarized cell divides, its descendants stay attached to each other, generate an internal milieu and, because of the orientation of the mitotic spindle, all descendant cells have the same polarization. F) Perpendicular orientation of the mitotic spindle provokes the development of an internal body of cells. Taken with kind permission from Cereijido et al, Physiol Revs 2004; in press.[1]

Experimental Approaches

After several decades of characterizing junctions and polarity in mature tissues, attempts were made for understanding how are these established. To the best of our knowledge, Oxender and Christensen[2] were the first to attempt to assemble an artificial epithelium by sandwiching single Ehrlich ascites cells between Millipore filters separating two chambers. Yet these cells lacked an intrinsic asymmetry, remained oriented at random, did not attach, and no asymmetry of the whole preparation was detected. Further attempts used epithelial cells dislodged from epithelia through treatment with Ca^{2+} chelants and hydrolyzing enzymes. A crucial step was given when the epithelium of a frog skin[3] was separated and it proved to preserve the overall properties of a transporting epithelia. Therefore, the second step was to mince the epithelium, dislodge the cells, and investigate whether these maintained a satisfactory ionic steady state, responded to ouabain, amiloride, changes in temperature, and displayed other basic properties of natural epithelia.[4,5] Finally, the strategy was to seed the cells on glass or Millipore filters to form a sort of artificial epithelium. Unfortunately, cells performed poorly in culture, detached easily, died soon, and prompted us to resort to established cell lines. We chose the MDCK line derived by Madin and Darby[6] from canine kidney, which was generally used to grow viruses, because it retained sufficient differentiation as to secrete fluid.[7] We cultured these cells on translucent supports (a nylon cloth coated with collagen) on which cells can easily be observed[8-12] or nitrocellulose filters. A similar preparation was developed by Misfeldt et al.[13] This approach gave us an opportunity to study the development of the transporting epithelia phenotype. Thus we used it to study the synthesis, assembly and sealing of TJs,[9-11,14-17] as well as the polarity of Na^+,K^+-ATPase;[15,18-21] Enrique Rodriguez-Boulan introduced the use of viruses that bud either apically (e.g., Flu) or basolateraly (e.g., vesicular stomatitis) to track the fate of specific proteins during apical or basolateral polarity,[22-27] and Carlos A. Rabito analyzed the onset of polarized co-transporters.[28-30] Mary Taub perfected the approach by developing totally defined culture media,[31-33] and complementary procedures were refined to open and reseal the TJs by removing and restoring Ca^{2+},[11,34-37] follow the cascades of phosphorylation involved,[38-40] the role of the cytoskeleton,[41,42] the participation of protein synthesis and sorting,[24,26,43,44] the polarized distribution of ion channels,[45-50] the involvement of E-cadherin,[51-53] the effect of agents that induce differentiation,[54] and the vectorial movement of receptors.[55] A distinct advantage of cultured monolayers is derived from the possibility of labeling beforehand a given cell type, and mixing them with other type from different epithelia and even different animal species,[56-58] and other characteristics illustrated below, and reviewed in Cereijido,[59] Cereijido and Anderson (2001)[a].[60]

Besides of the preparation just described, the information discussed in this review was obtained with suspended clumps and cysts of epithelial cells,[61] yeasts, *Caenorhabditis elegans* and *Drosophila*,[62] fertilized eggs, and the first stages of development from egg to embryo.[63]

The Development of Tight Junctions

Most of the cellular mechanisms and steps involved in the making of a TJ started by cultivating adult cells that had lost junctions and polarity through harvesting with trypsin and EDTA. The process was usually triggered by Ca^{2+} that activates a cascade of reactions detailed in Shoshani and Contreras.[64] It must be taken into account that this synthesis of junctions and polarity takes place in cells that already have all the mechanisms and molecules involved, or can synthesize them de novo if these were destroyed by trypsin. However it is taken for granted that the processes would mimic the normal synthesis and assembly, an assumption justified by a host of studies on natural preparations, such as the synthesis of TJs in the villi of the intestinal mucosa, and other instances where cells migrate from the depth of a crypt to the apex of the villum, or that can be observed in the steps from morula to embryo.[63]

---

[a] For a vignette on the development of this model system see Current Contents 1989; 32(46):1; and The Physiologist 2003; 46:114-115.

The Development of Polarity: The Role of the β-Subunit of the Na⁺,K⁺-ATPase

Until recently there was a great deal of misunderstanding about the relationship between TJs and polarity: (1) TJs were regarded as responsible for the apical/basolateral polarization because polarization coincides with the assembly and sealing of the TJ. Furthermore, when TJs are opened by the use of chelants such as EGTA or EDTA, apical and basolateral markers mix and polarization vanishes. It was soon realized that chelants affect both TJs and polarity, and the development of the transporting epithelium phenotype seems to be an all-or-nothing process. In fact, it was recently shown that it depends on the activation of a single gene, LKB1, an homologous of PAR-4 gene in *C. elegans* and *D. melanogaster*.[65] Interestingly, the activation of gene LKB1 can trigger the development of the transporting epithelium phenotype even in the absence of cell-cell contacts. (2) Secondly, it was gradually realized that it may be the other way around, because the TJ itself results from a delicate mechanism of polarization, reflected not only in its precise position at the apical/basolateral boundary, but in the fact that the TJ is not a conglomerate of static molecules, but a vertiginous steady state assembly of molecules that arrive and depart from other organelles, notably the nucleus (see the chapter by Lopez-Bayguen et al). (3) In keeping with this dynamic picture, Mauchamp et al[66] have shown that pig thyroid cells in suspensions tend to form clumps with the apical side outwards and the junction toward the outermost end of the intercellular space, yet the addition of serum causes a reversal of both, polarity and TJs. (4) Contreras et al[20] have shown that MDCK cells maintained in monolayers in absence or sufficiently low Ca^{2+} have no TJs, and Na⁺,K⁺-ATPase is distributed at random all over the plasma membrane. In this situation, the addition of 1.8 mM Ca^{2+} triggers junction formation so fast, that a large number of enzymes are trapped in the "wrong", apical side. Yet these misplaced Na⁺,K⁺-ATPases are subsequently removed in spite of the presence of the TJ, and new enzyme is inserted correctly in the basolateral domain.

But the main purpose of the present chapter is not to review the current status of the relationship of junctions and polarity in mammalian epithelial cells, but to imagine how they may have occurred in ancestral epithelia at the dawn of metazoan. In this respect we have been investigating the polarized distribution of Na⁺,K⁺-ATPase, an enzyme that is not only recognized as the star of vectorial transport across epithelia from the seminal model of Koefoed-Johnson and Ussing[67] model, that was the blueprint for further models for almost all epithelia[12,19,68,69] but, due to the particular balance of ions it produces, is the driving force for other vectorial transports, such as sugars, aminoacids, phosphate, K^+, H^+, Ca^{2+}, Mn^{2+}, Cu^{2+}, Fe^{3+} and Zn, etc. Na⁺,K⁺-ATPase is polarizedly distributed in MDCK cells.[19,20,29,70] It occupies the lateral plasma membrane, a position that favors the transport of ions towards the intercellular space, and may therefore drive the net movement of solutes towards the inner bathing solution, as this space is opened toward this solution and closed on the outer end by the TJ.[1,71]

The Na⁺,K⁺-ATPase is a member of the family of P-type ATPases. ATP hydrolysis by these ATPases includes a step of transfer of the terminal phosphoryl group of ATP onto the carboxyl group of an aspartic acid residue that is located in the active site of the enzyme.[72] The formation of a phosphorylated intermediate during the catalytic cycle is a characteristic of P-type ATPases, that distinguishes them from V-ATPases and F-ATPases.[73,74] The Na⁺,K⁺-ATPase is a heteromultimeric membrane enzyme constituted by α, β, and γ subunits. The 112 kDa α-subunit, often called the catalytic subunit, bears the site for ATP hydrolysis, and the binding sites for Na⁺, K⁺ and cardiac glycosides. The β-subunit is part of the Na⁺,K⁺- and H⁺,K⁺-ATPases but is not included in other P-type ATPases, for example Ca^{2+}-ATPase of both plasma membrane and sarcoplasmic reticulum do not have it. It is composed of about 370 amino acids, with the first 30 ones exposed to the cytosol, and 300 fold to form the extracellular portion, that has 3 disulphide bonds. There are 3 N-glycosylation consensus sequences (NXS or NXT) in the extracellular domain.[75,76] The polypeptide chain of the β-subunit weights 32-35 kDa, and when fully glycosylated can reach an overall apparent molecular mass of 55-60 kDa. Analysis of β-subunits in which the disulphide bonds were removed through substitution in baculovirus-infected insect cells, reveals that the S-S bridges are not important for assembling the heterodimer, yet they are required for membrane targeting.[77,78] The β-subunit has been identified as a factor responsible for cell adhesion in nervous tissue.[163] The activity of Na[+],K[+]-ATPase can be influenced by lectins, in particular, concanavalin A,[79] and galectins (various lectins of animal origin), some of which affect cell adhesion[80] and provide in this way signal transmission to the catalytic subunit. Three β-isoforms can be attributed to Na[+],K[+]-ATPase in mammals (β$_1$, β$_2$ and β$_3$). Significantly, for the nongastric H[+],K[+]-ATPase, no specific β-subunit has been identified. X,K-ATPase β isoforms exhibit only about 20-30% overall sequence identity, but share several structural features. β-isoforms exhibit a tissue specific distribution with β$_1$ of Na[+],K[+]-ATPase being expressed ubiquitously, β$_2$ mainly in the heart, skeletal muscles, and glial cells, and β$_3$ in many tissues.

Based on the evidence that the polarized distribution of membrane molecules results from an addressed targeting plus a selective retention in submembrane scaffolds[45,71] we have investigated the role of the β-subunit.

α/β-Subunit Interactions

The subunits of Na[+],K[+]-ATPase are synthesized independently in the endoplasmic reticulum and assembled in this organelle. The expression of α and β-subunit isoforms follows a tissue-specific pattern.[81-83] Newly synthesized Na[+],K[+]-ATPase is directly addressed to the basolateral membrane domain in MDCK cells.[84-87] This targeting seems to be determined by the impossibility of this enzyme to board the glycosphingolipid (GSL)-rich rafts that assemble in the Golgi complex, and form vesicles that carry proteins towards the apical domain. This exclusion may be overcome by endowing the Na[+],K[+]-ATPase with a sequence signal from the 4th transmembrane segment of the α-subunit of H[+],K[+]-ATPase (TM4 signal) that suffices to readdress the Na[+],K[+]-ATPase towards the apical domain.[88,89]

Role of the cytoskeleton. The activity and polarized distribution of renal Na[+],K[+]-ATPase appears to depend on a connection of ankyrin to the spectrin-based membrane cytoskeleton, as well as on association with actin filaments. Na[+],K[+]-ATPase not only co-purifies with ankyrin, spectrin and actin, but also with three further peripheral membrane proteins, pasin 1 and pasin 2,[90] and moesin.[91] A specific binding site for ankyrin has been localized on the α-subunit.[92] Interestingly, the *Drosophila* Na[+],K[+]-ATPase α-subunit conserves an ankyrin binding site,[93] and co-distributes with ankyrin and spectrin in polarized fly cells.[94-97] Yet, its polarized distribution was not altered in spectrin-null mutants.[98,99] Furthermore, Dubreuil et al[97] using mutations in the *Drosophila* β-spectrin gene, provided the first direct evidence that β-spectrin determines the subcellular distribution of the Na[+], K[+]-ATPase and that this role is independent of α-spectrin.

Unlike most epithelia, the retinal pigment epithelium (RPE) distributes the Na[+],K[+]-ATPase to the apical membrane.[100-103] Early studies on rat RPE monolayers Gundersen et al[102] suggest that an entire membrane-cytoskeleton complex is assembled with opposite polarity. Recent studies have shown that other polarized membrane proteins such as viral envelope membrane proteins do preserve their polarity in RPE cells.[100,104] Furthermore, Rizzolo and Zhou[105] used the RPE of chicken embryos, and observed that the Na[+],K[+]-ATPase and α-spectrin segregate into different regions of the cell.

Na[+],K[+]-ATPase, Cell Attachment and the Position of the TJ

The complex mechanisms described above are the product of a long evolution, and might not have been present in primeval metazoans of pre-Cambrian times, more than 600 millions years ago. The first metazoans exchanging through epithelia must have had much simpler mechanisms. One of the simplest occurring in an epithelium, would be a membrane molecule with the capacity to attach to a similar one in the neighboring cell, so the pair would get stuck at the contacting border (Fig. 3). The Na[+],K[+]-ATPase might constitute a case in point. Axelsen and Palmgren[106] propose that the third subunit of bacterial K[+]-ATPase (KdpC) may be the

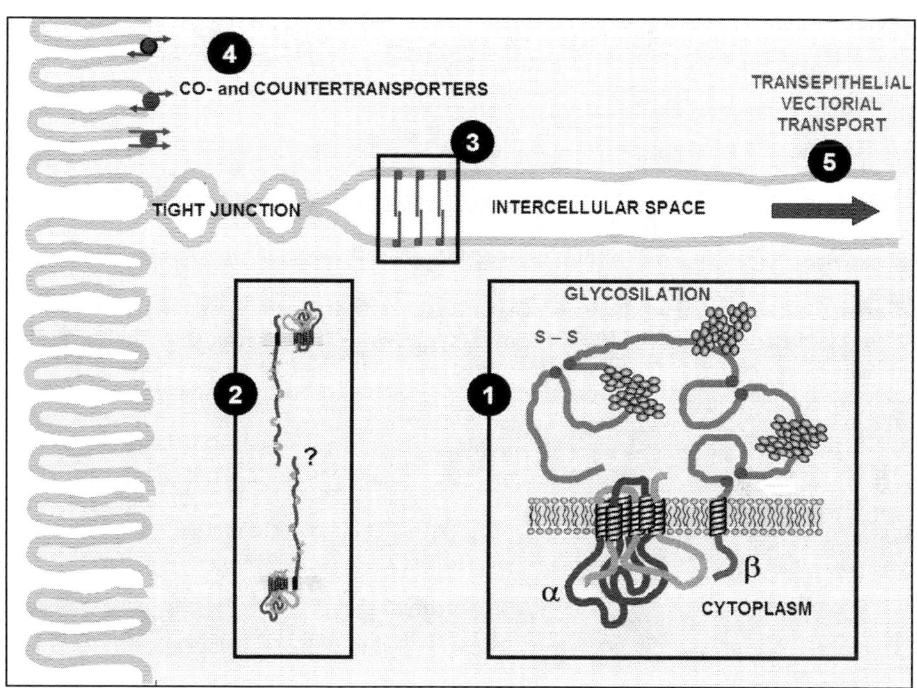

Figure 3. Evolution of transepithelial transport. Hypothetical molecular and cellular events that may have intervened in the generation of a primeval transporting epithelium. Two neighboring epithelial cells have their apical domain toward the left. 1) The α-subunit of Na⁺,K⁺-ATPase is represented as a membrane protein with several transmembrane domains, and a β one with the characteristics of a cell-attaching molecule: a short cytoplasmic domain, a single transmembrane domain, and a long extracellular fragment with S-S links (red) and glycosylation (green). Although both subunits are intimately associated, they are represented as separated from each other to avoid crowding. For the same reason the γ-subunit is omitted. 2) The complex of α plus β-subunits becomes stabilized at the lateral border of the cell, because it is the only position in the plasma membrane where the β subunits of the two neighboring cell can interact. 3) Interacting Na⁺,K⁺-ATPases from neighboring cells are represented in a much simplified way. Because of the TJ, ions pumped into the intercellular space can only diffuse inward. 4) the polarized expression of co- and countertransporters of sugars, aminoacids, and a host of other ions driven by the asymmetries built up through the work of the Na⁺,K⁺-ATPase, originate a vectorial transport across the whole epithelium. In principle, the TJ may occupy any position between the two cells. However, selective pressure must have pushed the TJ toward the outermost end of the intercellular space, because in this configuration the intercellular space is open towards the inner side of the body, and none of the absorbed nutrients would leak back toward the outer solution (5). As discussed in the text, the β-subunit of the Na⁺,K⁺-ATPases confers some attaching capability to the cell.[71] Yet the main attaching molecule triggering the synthesis and assembly of TJs and Na⁺,K⁺-ATPase polarization seems to be the E-cadherin, which is not represented in the drawing. A color version of this figure is available online at www.Eurekah.com.

ancestor of the X,K⁺-ATPase β-subunit, due to the similar assembly functions they both perform. As discussed above, in higher species the β-subunit is tightly and specifically bound to the α-subunit,[72] and the association begins as soon as they are synthesized in the endoplasmic reticulum[84,85] so that both reach the plasma membrane forming already a hard-to-dissociate dimer, in which β helps α to trap K⁺ in a pocket during the pumping cycle.[107] Yet it is not expected that they would be firmly bound in early organisms like *Caenorhabditis elegans*, because sequence analyses show no association domains.[108,109] Therefore it is conceivable that the α-subunit became a stable resident of the lateral space of transporting epithelia by virtue of being bound to a β-one, which in turn has the ability to dwell in this position because of a linkage it makes with another β-subunit placed in the plasma membrane of a neighboring cell (Fig. 3).

However, a Na^+,K^+-ATPase located in a primitive epithelium as depicted in Figure 3 would be inefficient, as roughly one half of the pumping fluid would leak back to the outer environment. Combinations of Na^+,K^+-ATPase at the cell-cell border with TJs would probably select epithelia in which TJs would be pushed toward the outermost end of the intercellular space, because it would insure a maximal yield of fluid transported vectorialy towards the inside (Fig. 3).

Evolution of the Epithelial Vectorially Transporting Phenotype

Evidence that mollusks can absorb ions from dilute media was already presented by August Krogh[110] (see also ref. 111). Water, ions, lipids, aminoacids, and sugars were found to be transported by similar mechanisms in the intestine in *Ascaris*, earthworms, fresh water clams, insects as well as in the mammalian intestine.[112-116]

Evolutionists refer to **homology** when two organs have the same role and evolutionary origin, but reserve **analogy** (an anatomical term) and **homoplasy** (its molecular counterpart) for organs that, in spite of playing the same physiological role, result from different phylogenetic processes. A familiar example of the latter are the wings of the butterfly and those of the bat. In this respect, the relationship between different occluding junctions (e.g., TJ and septate junction) is still debatable. Thus they play the role of occluding junctions,[117-119] and share considerable homology between some of their molecules, such as claudins, ZO-1 and dlg.[120,121] Nevertheless these types of junctions do show considerable differences in composition as well as in their evolution.[122,123]

The next step in Evolution is constituted by *Cnidaria*, whose most conspicuous representative is the hydra, a fresh water coelenterate first described by Abraham Tembley in 1744 (as cited by Gierer and Meinhardt,[124]). It reproduces through stem cells, and although its cells are not known to have apical/basolateral polarity, the organism does show regional differentiation, attributed to gradients of substances secreted by some of its cells (Gierer and Meinhardt).[124] Fei et al[125] have found that *Hydra vulgaris* has an homologue of ZO-1, termed HZO-1, that is a MAGUK protein, a family with members such as ZO-2, dlg-A and TamA, which are involved in a wide variety of cellular functions such as TJ formation, cell proliferation, differentiation, and synapse formation.

Studies on the in vivo distribution, stability and function of β-catenin in embryos of the sea anemone *Nematostella vectensis* (Cnidaria, Anthozoa) have shown that in *N. vectensis* β-catenin is differentially stabilized along the oral-aboral axis. Thus, this molecule is accumulated and translocated into nuclei in cells at the site of gastrulation and used to specify entoderm.[126] Wikramanayake et al[126] results suggest a key role for the Wnt/β-catenin pathway in the evolution of axial asymmetries and entoderm formation in an ancestor that existed before the evolution of the mesodermal germ layer. Until recently, it was assumed that polarity and axis formation have evolved only in metazoan phyla higher than *Cnidaria*. One key molecule involved in the signal transduction causing tissue polarity is Frizzled, a seven-transmembrane receptor that is activated by the Wnt secreted glycoproteins. Adell et al[127] reported the isolation of the Frizzled gene from the demosponge *Suberites domuncula* (*Sd-Fz*), expressed in cells close to the surface of the sponges and in the pinacocytes of some canals, and its mRNA is upregulated during the formation of three-dimensional sponge cell aggregates in culture. These results provide the first experimental evidence of expression of genes probably involved in polarity already in the lowest metazoan phylum (*Porifera*).

Evolution of the Na⁺,K⁺-ATPase

Okamura et al[108,109] correlated the molecular evolution of the 11 subfamilies of the P-type ATPases within the establishment of different kingdoms. Interestingly, heavy metal transporters

(Type 1B) and intracellular Ca-ATPases (type 2A) are probably the most fundamental for life since these two types are found in every kingdom. They appeared very early in evolution, and played critical roles in ion homeostasis. On the other hand, ancestral animals probably found the Na$^+$,K$^+$-ATPase and H$^+$,K$^+$-ATPase (type 2C) and got rid of the proton pump (type 3A) present in the lineage of Plants and Fungi. While the divergence of substrate-specificity (e.g., ion selectivity) occurred early in the evolution of P-type ATPases and has been conserved ever since,[106] the divergence of the P-type ATPase isoforms occurred after vertebrates and invertebrates separated.[128] The study of evolution of P-type ATPases has been recently enriched with new information from the database for the nematode *Caenorhabditis elegans* and the fly *Drosophila melanogaster*. Focusing on key domains of both α- and β-subunits, such as the ouabain binding site and the α/β assembly site, a set of novel isoforms that retain an ancestral characteristic of the Na$^+$,K$^+$- and H$^+$,K$^+$-ATPase have been identified.[108,109,129] In the phylogenetic analysis, the β-subunits of *C. elegans* and *D. melanogaster* form a unique single cluster. Interestingly, *D. melanogaster* possesses two distinct types of β-subunits, one type closer to the vertebrate β-subunits, and the other sharing more homology with the *C. elegans* β-subunits. As mentioned before, the functional expression of the Na$^+$,K$^+$-ATPase requires the assembly of the α- and β-subunits. In this regard, the newly identified invertebrate subunit isoforms (Ce2C3 and Ce2C4 and Dmβ4-6) are of particular interest because they lack the characteristic domains that have been demonstrated to be critical for α/β assembly.[130-132] Therefore, Takeyasu et al[129] and Okamura et al[108,109] suggested that these non-assembling subunits may exist as lonely subunits and may play not-yet identified function(s) other than in ATP-driven ion transport. Taken together, their studies suggest that the P-type ancestor first lacked the inhibitor binding site and the assembly domain in the α-subunit, and therefore existed as a single subunit in lower invertebrate organisms. In its evolution, this ancestral form of α-subunit acquired the abilities to bind ouabain and to assemble with β-subunit, becoming the immediate ancestor of the currently known Na$^+$,K$^+$- and H$^+$,K$^+$-ATPase family. Although the information on the molecular evolution of the β-subunit is comparatively scarce, it has been suggested that it derives from the third subunit of bacterial K$^+$-ATPase (KdpC). Alternatively, the β$_m$ protein of sarcoplasmic reticulum could also represent a primitive form of the β-subunit family of the X$^+$,K$^+$-ATPases.

Evolution of Junction Proteins

When *Dictyostelium* cells run out of nutrients, they polarize and adhere to form a multicellular structure that supports the spore head.[133] During this process, the cells at the top of the stalk tube form a constriction with rings of cells that express adherent junctions with the β-catenin analogue *aardvark* associated to the actin cytoskeleton. *aardvark* is also independently required for cell signaling,[134] and shares significant homology with *Saccharomyces cereviseae* protein Vac8, *Saccharomyces pombe* SPBC354, and plant *Arabidopsis thaliana* AB016888 sequences. This demonstrates that in spite of lacking stable junctions, protozoa appear to have molecules that coordinate cytoskeletal dynamics, the position of the mitotic spindle and cell polarity, and that may have been precursors of molecules that make AJs and TJs in metazoans.[134,135] β-catenin that has a dual role in cell-cell adhesion and cell signaling, is present in the non-metazoan amoeba *Dictyostelium discoideum*, indicating that it evolved before the origins of metazoan.[134] It seems plausible that β-catenin may have been a requisite for all multicellular development.[134] In this sense, *Dictyostelium* has β-catenin that plays signaling and adhesion roles as in vertebrate. *C. elegans* instead diverged, because it has two catenins: HMP-2 for adhesion, and Bar-1 for signaling to the nucleus.[136] It has been reported that β-catenin is involved in vital biological processes beyond cell adhesion and Wnt signaling. Thus, Kaplan et al[137] demonstrated that β-catenin is a component of the mammalian mitotic spindle that functions to ensure the proper centrosome separation, as well as a subsequent establishment of a bipolar spindle. This role of β-catenin might have been a function required in unicellulars, before the molecule became involved in adhesion processes.

Apical junctions in *Caenorhabditis elegans* have tetraspan VAB-9, that is a claudin-like membrane protein, that colocalizes with E-cadherin (HMR-1), and whose localization depends on HMR-1, α (HMP-1) and β-catenin (HMP-2).[138]

Hua et al,[139] analyzed the phylogenetic trees of connexins, claudins and occludins, and found no sequence or motif similarity between the different families studied, indicating that, if they did evolve from a common ancestral gene, they have diverged considerably to fulfill separate and novel functions.

It has been recently demonstrated that Sinuous, an homolog of claudin in *Drosophila*, not only localizes to the septate junctions, but also confers to this structure its barrier function.[121] Kollmar et al,[140] only found a single claudin gene in the urochordate *Halocynthia roretzi*, and concluded that claudins emerged when TJs replaced septate junctions. However, Asano et al,[120] found that *Caenorhabditis elegans* has no less than 4 different forms of claudins participating in the epithelial barrier.

Cadherins form a superfamily with at least six distinct subfamilies: classical or type-I cadherins, atypical or type-II cadherins, desmocollins, desmogleins, protocadherins, and Flamingo cadherins. In addition, several cadherins clearly occupy isolated positions in the cadherin superfamily[141,142] suggesting a different evolutionary origin of the protocadherin and Flamingo cadherin genes versus the genes encoding desmogleins, desmocollins, classical cadherins, and atypical cadherins. In contrast to classic cadherins, that bind catenins by the cytoplasmic domains and have genes with as much as 12 introns, nonclassic cadherins do not interact with catenins, and genes have fewer introns.[143] Phylogenetic analyses suggest that there are four paralogous subfamilies (E-, N-, P-, and R-cadherins) of vertebrate classic cadherin proteins[144] and that these genes duplicated early in evolution.[145]

Amphioxus cadherin BbC, localizes to adherens junctions in the ectodermal epithelia of embryos and confer homotypic adhesive properties. It has a cytoplasmic domain whose sequence is highly related to the cytoplasmic sequences of all known classic cadherins, but its extracellular domain lacks the classical five extracellular cadherin repeats and is similar to the extracellular domain of nonchordate cadherins.[146]

Choanoflagellates, a group of unicellular and colonial flagellates that resemble cells found only in metazoan, express sequences of proteins that encode cadherin repeats. Phylogenetic analyses reveal that choanoflagellate cadherins are most similar to protocadherins and to the Flamingo class of cadherins, demonstrating that these proteins evolved before the origin of animals and were later co-opted for development.[147]

In seven different species of ascidians, including *Ciona intestinalis*, tight junctions have been observed by electron microscopy.[148] Correspondingly the *C. intestinalis* genome contain genes that code for all the essential protein components of tight junctions: 3 claudin genes (Ci-Claudin-a,-b and –c), a single JAM gene (Ci-JAM) orthologe to human JAM3, one ZO gene (Ci-ZO) with a ZU5 domain (a domain analogous to human ZO-1 and to protein UNC-5-like netrin receptor) suggesting that the last common ancestor of ascidians and vertebrates had a single ZO gene and that human ZO-2 and ZO-3 but not ZO-1 appear to have lost a ZU5 domain after ascidians and vertebrates diverged.[149] Although an occludin gene could not be identified and may not be present in the *Ciona* genome, it has been demonstrated that occludin is not actually essential for the formation of a tight junction.[150,151]

The genome of the ascidian *Ciona intestinalis* contains genes that encode protein components of tight junctions, hemidesmosomes and connexin-based gap junctions, as well as of adherent junctions and focal adhesions, but it does not have those for desmosomes.[149] In addition, extracellular matrix proteins genes like β-netrin that have been regarded as vertebrate-specific, were also found in the *Ciona* genome. Its orthologes have not been found in either fly or nematode genomes. These results suggest that the last common ancestor of ascidian and vertebrates (i.e, the ancestor of the entire chordate clade), had essentially the same system of cell junctions as those in extant vertebrates. However, the number of genes for each family in the *Ciona* genome is far smaller than that in vertebrate genomes. In vertebrates these ancestral cell junctions then appear to have evolved into more diverse and more complex forms, compared with those in their urocordate siblings.[149]

The Dawn of Metazoans and Transporting Epithelia

The co-relationship between multicellularity and transporting epithelia is stronger than the considerable variety of adaptive structures known to exist in many organisms. But a mere accumulation of cells might not qualify as metazoan. Thus, when nutrients are exhausted or wastes accumulate, cells may temporarily shut off their living processes, or form a transient multicellular structure (e.g., *Dyctiostelium discoideum*) and migrate somewhere else. Likewise, organisms such as *Porifera*, without real tissues nor organs, have no internal milieu because the environment can circulate through the body, all cells can exchange directly with the environment, and should not concern us here. We refer instead to stable metazoans that survive thanks to an internal milieu whose stable composition depends on transporting epithelia. Although the fossil record starts some 560-600 million years ago, there is evidence that they were already existing as early as 800-1000 million years ago.[152,153] Ideas about metazoan phylogeny are numerous and somewhat conflicting.[152,154-158] Furthermore, our search for an ancestor of metazoans with true epithelia may start with a plausible idea of how such organism should have been, regardless of whether it is still surviving or has been long extinct.

Some "Theoretically" Plausible Organisms

The elaboration on a priori possible organisms might at least afford a useful working hypothesis to solve the conundrum mentioned at the Introduction on the origins of metazoans and transporting epithelia. Let us depict a few suggested elsewhere.[1]

The "Thrifty Sponge"

It is conceivable that in a given moment a group of cells became surrounded by a primitive epithelium, i.e. one whose cells did not exhibit vectorial transport, but that attach and become an impermeable barrier (Fig. 2, top). Secreted molecules such as nutrients or cAMP may thereby be momentarily conserved and used as food stuff or signals. Yet the entrapped milieu would gradually spoil, and force a transient breakage of the epithelium, thus allowing the external medium to gain access to the cells and flush the organism. One may even conceive a certain synchrony between periods of secretion and epithelial tightness. In this respect, Green and Bergquist[159-161] have found that in sponges, cell junctions are apparently formed only when required for a specific purpose.

Very Flat Organisms

Another possibility would be a very flat organism with large surface-to-volume ratio, sufficient to allow exchange with the surroundings (Fig. 2, middle). Although these putative organisms might not have had circulatory apparatuses, peristaltic movements might have helped to stir the internal milieu and favor exchange. Actually, Nature seems to have resorted to large surface-to-volume body plans to ease the survival of primitive metazoans. Thus *Cnidaria*, just one evolutionary step above sponges, have wide and flat structures.

The "Mare Nostrum" Metazoans

Up to this point we have been asking how did primeval metazoans develop their first epithelium. We cannot discard the opposite situation though. Since polarization is already observed in unicellulars, the possibility exists that an already polarized cell would proliferate without completing the separation of its descendants, and somehow profit from preserving an internal milieu (Fig. 2, bottom). Furthermore, proliferating epithelial cells usually have their mitotic spindle parallel to the surface of the epithelium, so that proliferation expands the area of the organ.[135] Yet when the spindle is perpendicular to the epithelium, it provokes the formation of an outgrowth similar in many respects to the body mass of a higher organism.

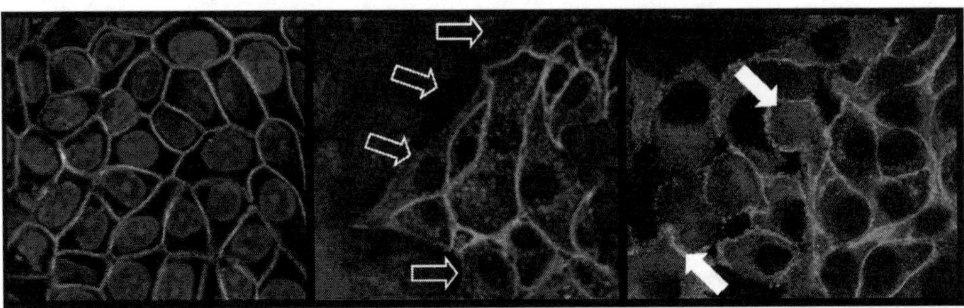

Figure 4. Patterns in the expression of Na$^+$,K$^+$-ATPase in epithelial cells. Left) In MDCK cells cultured as monolayers, Na$^+$,K$^+$-ATPase is expressed at the lateral borders (green), forming a chicken fence pattern. Center) Monolayer prepared with a mixed population of MDCK cells (derived from dog kidney) and CHO cells (from Chinese Hamster Ovary), which were previously stained with the red dye CMTMR, were treated with a first antibody raised against the β-subunit of the dog that, therefore, only detects the β-subunit of MDCK but not the one in CHO cells. The picture shows that this subunit occupies the cellular border of an MDCK cell provided the neighboring cell is another MDCK one (green), but not when it contacts CHO cells. Right) Monolayer prepared with MDCK cells and CHO cells transfected with β-subunit from the dog. Green staining reveals that CHO cells now express dog β-subunit in their plasma membrane (arrows) and, in keeping with drawings 3 and 4 in Figure 3, MDCK cells now do express their Na$^+$,K$^+$-ATPases in homotypic as well as heterotypic MDCK/CHO borders.[71] A color version of this figure is available online at www.Eurekah.com.

Na$^+$,K$^+$-ATPase and Cell Adhesion: Role of the β-Subunit in Cell Attachment

Na$^+$,K$^+$-ATPase arrives to the lateral membrane, interacts with ankyrin, and becomes anchored to the cytoskeleton that stabilizes the enzyme in this position.[162] Yet we still ignore why this enzyme binds to the cytoskeleton at the lateral borders, but does not bind to this structure when expressed at other cell borders. Furthermore Na$^+$,K$^+$-ATPase in *Drosophila* epithelia does not bind to ankyrin but is nevertheless polarized.[95] In order to develop a more plausible explanation we took as hints that:

i. When a monolayer of cultured MDCK cells is treated with EGTA, the chicken fence image of Na$^+$,K$^+$-ATPase splits into two moieties and each neighboring cell retrieves its own, indicating that each cell contributes the pumps, and opening the possibility that these interact across the intercellular space.

ii. In keeping with such conjunction, MDCK cells express Na$^+$,K$^+$-ATPase at the lateral but not at the basal borders, as if only at the intercellular space the enzyme would find an attaching partner on an opposite cell.[56,69]

iii. When co-cultured with other epithelial or fibroblastic cell, an MDCK cell only expresses Na$^+$,K$^+$-ATPase in a given border provided its neighbor is another MDCK cell (Fig. 4C), suggesting that the putative Na$^+$,K$^+$-ATPase/Na$^+$,K$^+$-ATPase interaction is a specific one.[71]

iv. As mentioned above, Gloor et al,[163] have shown that the β$_2$-subunit of glial cells acts as a cell adhesion molecule and has the corresponding structure: a short cytoplasmic tail, a single transmembrane domain, as well as long and glycosylated extracellular fragment. MDCK cells have instead the β$_1$ isoform, but its structure is almost identical to that of β$_2$. This suggests that the β-subunit of Na$^+$,K$^+$-ATPase may establish a cell-cell contact as suggested in Figure 4. (v) Furthermore, this subunit is required for the formation of the septate junction of Drosophila.[121]

vi. The β-subunit of Na$^+$,K$^+$-ATPase does participate in cell attachment.

On this basis, we have proposed that the polarized position of Na$^+$,K$^+$-ATPase at the lateral borders of epithelial cells depends upon the binding ability of its β-subunit. The expression of this enzyme at the plasma membrane facing the intercellular space would be the only place where the β-subunits of neighboring cells can interact.[15,18,69] This view has recently received experimental support from the studies illustrated in Figure 4D,E.[71] Briefly, it is observed that MDCK cells only express their β-subunits at the point of contact with another MDCK cell, but not in contacts with CHO (Chinese Hamster Ovary). However, when CHO cells are transfected beforehand with β-subunit from dog (i.e. same species as MDCK cells), MDCK cells now do express their β-subunits in heterotypic MDCK/CHO contacts. This model had recently been subject to experimental tests by Shoshani et al,[71] who have found that, in fact, transfection of β-subunits obtained from dog actually confers to CHO cells ("CHO-β") the ability to form mixed monolayers with MDCK (dog), and express not only these subunit at heterotypic (MDCK/CHO-β) borders, but force CHO-β cells to express their endogenous α subunit, in keeping with the idea that the machinery of the cell can only send to the membrane the previously associated α/β complex.

Summary

We cannot provide an answer to the question raised in the Introduction about the origin of transporting epithelia and metazoans. However, after systematizing the information obtained from widely heterogeneous fields, it seems that ancestral unicellulars already possessed most of the precursor molecules to build up transporting epithelia, and form a perhaps transient metazoan that evolution might subject to selective pressures. Attachment, both at the molecular and the cellular level, seems to have played a paramount role.

Acknowledgements

We thank Dr. Catalina Flores-Maldonado, and Lic. Isabel Larre for helpful discussion. We also acknowledge the efficient and pleasant help of Elizabeth del Oso, and the economic support of the National Research Council of México (CONACYT).

References

1. Cereijido M, Contreras RG, Shoshani L. Cell adhesion, polarity, and epithelia in the dawn of metazoan. Physiol Revs 2004; in press.
2. Oxender DL, Christensen HN. Transcellular concentration as a consequence of intracellular accumulation. J Biol Chem 1959; 234:2321-2324.
3. Rotunno CA, Zylber EA, Cereijido M. Ion and water balance in the epithelium of the abdominal skin of the frog Leptodactylus ocellatus. J Membr Biol 1973; 13:217-232.
4. Zylber EA, Rotunno CA, Cereijido M. Ion and water balance in isolated epithelial cells of the abdominal skin of the frog Leptodactylus ocellatus. J Membr Biol 1973; 13:199-216.
5. Zylber EA, Rotunno CA, Cereijido M. Ionic fluxes in isolated epithelial cells of the abdominal skin of the frog Leptodactylus ocellatus. J Membr Biol 1975; 22:265-284.
6. Madin SH, Darby NB. As cataloged in the American Type Culture Collection Catalog of Strains. 1958.
7. Leighton J, Brada Z, Estes LW et al. Secretory activity and oncogenicity of a cell line (MDCK) derived from canine kidney. Science 1969; 163:472-473.
8. Cereijido M, Robbins ES, Dolan DF et al. Membrane properties of cell of renal origin (MDCK) cultured in monolayers on a permeable and translucent support. Proc Int Union Physiol 1977; 12:250a.
9. Cereijido M, Rotunno CA, Robbins ES et al. In vitro formation of cell layers with properties of epithelial membranes. Biophys J 1977; 17:22a.
10. Cereijido M, Robbins ES, Dolan WJ et al. Polarized monolayers formed by epithelial cells on a permeable and translucent support. J Cell Biol 1978; 77:853-880.
11. Cereijido M, Rotunno CA, Robbins ES et al. Polarized epithelial membranes produced in vitro. In: Hoffman JF, ed. Membrane Transport Processes. New York: Raven Press, 1978:433-43.
12. Cereijido M, Rotunno CA. Introduction to the Study of Biological Membranes. New York, NY: Gordon and Breach, Science Publishers; 1971.
13. Misfeldt DS, Hamamoto ST, Pitelka DR. Transepithelial transport in cell culture. Proc Natl Acad Sci USA 1976; 73:1212-1216.
14. Fernandez-Castelo S, Bolivar JJ, Lopez-Vancell R et al. Ion transport in MDCK cells. In: Taub M, ed. Tissue Culture of Epithelial Cells. New York: Plenum Press, 1985:37-50.
15. Cereijido M, Ehrenfeld J, Meza I et al. Structural and functional membrane polarity in cultured monolayers of MDCK cells. J Membr Biol 1980; 52:147-159.

16. Gonzalez-Mariscal L, Chavez DR, Cereijido M. Effect of temperature on the occluding junctions of monolayers of epithelioid cells (MDCK). J Membr Biol 1984; 79:175-184.

17. Martinez-Palomo A, Meza I, Beaty G et al. Experimental modulation of occluding junctions in a cultured transporting epithelium. J Cell Biol 1980; 87:736-745.

18. Cereijido M, Stefani E, Palomo AM. Occluding junctions in a cultured transporting epithelium: structural and functional heterogeneity. J Membr Biol 1980; 53:19-32.

19. Cereijido M, Ehrenfeld J, Fernandez-Castelo S et al. Fluxes, junctions, and blisters in cultured monolayers of epithelioid cells (MDCK). Ann N Y Acad Sci 1981; 372:422-441.

20. Contreras RG, Avila G, Gutierrez C et al. Repolarization of Na+-K+ pumps during establishment of epithelial monolayers. Am J Physiol 1989; 257:C896-C905.

21. Contreras RG, Shoshani L, Flores-Maldonado C et al. Relationship between Na(+),K(+)-ATPase and cell attachment. J Cell Sci 1999; 112 (Pt 23):4223-4232.

22. Lisanti MP, Sargiacomo M, Graeve L et al. Polarized apical distribution of glycosyl-phosphatidylinositol-anchored proteins in a renal epithelial cell line. Proc Natl Acad Sci USA 1988; 85:9557-9561.

23. Lisanti MP, Caras IW, Davitz MA et al. A glycophospholipid membrane anchor acts as an apical targeting signal in polarized epithelial cells. J Cell Biol 1989; 109:2145-2156.

24. Rodriguez-Boulan E, Paskiet KT, Sabatini DD. Assembly of enveloped viruses in Madin-Darby canine kidney cells: polarized budding from single attached cells and from clusters of cells in suspension. J Cell Biol 1983; 96:866-874.

25. Rodriguez-Boulan E, Paskiet KT, Salas PJ et al. Intracellular transport of influenza virus hemagglutinin to the apical surface of Madin-Darby canine kidney cells. J Cell Biol 1984; 98:308-319.

26. Rodriguez-Boulan E, Pendergast M. Polarized distribution of viral envelope proteins in the plasma membrane of infected epithelial cells. Cell 1980; 20:45-54.

27. Rodriguez-Boulan E, Sabatini DD. Asymmetric budding of viruses in epithelial monlayers: a model system for study of epithelial polarity. Proc Natl Acad Sci USA 1978; 75:5071-5075.

28. Rabito CA, Karish MV. Polarized amino acid transport by an epithelial cell line of renal origin (LLC-PK1). The apical systems. J Biol Chem 1983; 258:2543-2547.

29. Rabito CA, Tchao R. [3H]ouabain binding during the monolayer organization and cell cycle in MDCK cells. Am J Physiol 1980; 238:C43-C48.

30. Rabito CA, Tchao R, Valentich J et al. Effect of cell-substratum interaction on hemicyst formation by MDCK cells. In Vitro 1980; 16:461-468.

31. Taub M. Tissue Cultured in Epithelial Cells. New York, NY: Plenum Press; 1985.

32. Taub M, Saier MH, Jr. Regulation of 22Na+ transport by calcium in an established kidney epithelial cell line. J Biol Chem 1979; 254:11440-11444.

33. Taub M, UB Chuman L et al. Alterations in growth requirements of kidney epithelial cells in defined medium associated with malignant transformation. J Supramol Struct Cell Biochem 1981; 15:63-72.

34. Contreras RG, Gonzalez-Mariscal L, Balda MS et al. The role of calcium in the making of a transporting epithelium. NIPS 1992; 7:105-108.

35. Contreras RG, Ponce A, Bolivar JJ. Calcium and tight junctions. In: Cereijido M, ed. Tight Junctions. 1st ed. Boca Raton: CRC Press, 1992:139-49.

36. Gonzalez-Mariscal L, Chavez DR, Cereijido M. Tight junction formation in cultured epithelial cells (MDCK). J Membr Biol 1985; 86:113-125.

37. Gonzalez-Mariscal L, Contreras RG, Bolivar JJ et al. Role of calcium in tight junction formation between epithelial cells. Am J Physiol 1990; 259:C978-C986.

38. Balda MS, Gonzalez-Mariscal L, Contreras RG et al. Assembly and sealing of tight junctions: possible participation of G-proteins, phospholipase C, protein kinase C and calmodulin. J Membr Biol 1991; 122:193-202.

39. Balda MS, Gonzalez-Mariscal L, Matter K et al. Assembly of the tight junction: the role of diacylglycerol. J Cell Biol 1993; 123:293-302.

40. Citi S. Protein kinase inhibitors prevent junction dissociation induced by low extracellular calcium in MDCK epithelial cells. J Cell Biol 1992; 117:169-178.

41. Meza I, Ibarra G, Sabanero M et al. Occluding junctions and cytoskeletal components in a cultured transporting epithelium. J Cell Biol 1980; 87:746-754.

42. Meza I, Sabanero M, Stefani E et al. Occluding junctions in MDCK cells: modulation of transepithelial permeability by the cytoskeleton. J Cell Biochem 1982; 18:407-421.

43. Griepp EB, Dolan WJ, Robbins ES et al. Participation of plasma membrane proteins in the formation of tight junctions by cultured epithelial cells. J Cell Biol 1983; 96:693-702.

44. Rindler MJ, Ivanov IE, Plesken H et al. Viral glycoproteins destined for apical or basolateral plasma membrane domains traverse the same Golgi apparatus during their intracellular transport in doubly infected Madin-Darby canine kidney cells. J Cell Biol 1984; 98:1304-1319.

45. Moreno J, Cruz-Vera LR, Garcia-Villegas MR et al. Polarized expression of Shaker channels in epithelial cells. J Membr Biol 2002; 190:175-187.
46. Ponce A, Cereijido M. Polarized distribution of cation channels in epithelial cells. Cell Physiol Biochem 1991; 1:13-23.
47. Ponce A, Bolivar JJ, Vega J et al. Synthesis of plasma membrane and potassium channels in epithelia. Cell Physiol Biochem 1991; 1:195-204.
48. Ponce A, Contreras RG, Cereijido M. Polarized distribution of chloride channels in epithelial cells. Cell Physiol Biochem 1991; 1:160-169.
49. Stefani E, Cereijido M. Electrical properties of cultured epithelioid cells (MDCK). J Membr Biol 1983; 73:177-184.
50. Talavera D, Ponce A, Fiorentino R et al. Expression of potassium channels in epithelial cells depends on calcium-activated cell-cell contacts. J Membr Biol 1995; 143:219-226.
51. Gumbiner B, Simons K. A functional assay for proteins involved in establishing an epithelial occluding barrier: identification of a uvomorulin-like polypeptide. J Cell Biol 1986; 102:457-468.
52. Gumbiner B, Simons K. The role of uvomorulin in the formation of epithelial occluding junctions. Ciba Found Symp 1987; 125:168-186.
53. Gumbiner B, Stevenson B, Grimaldi A. The role of the cell adhesion molecule uvomorulin in the formation and maintenance of the epithelial junctional complex. J Cell Biol 1988; 107:1575-1587.
54. Lever JE. Inducers of mammalian cell differentiation stimulate dome formation in a differentiated kidney epithelial cell line (MDCK). Proc Natl Acad Sci USA 1979; 76:1323-1327.
55. Matter K, Mellman I. Mechanisms of cell polarity: sorting and transport in epithelial cells. Curr Opin Cell Biol 1994; 6:545-554.
56. Contreras RG, Lazaro A, Bolivar JJ et al. A novel type of cell-cell cooperation between epithelial cells. J Membr Biol 1995; 145:305-310.
57. Contreras RG, Shoshani L, Flores-Maldonado C et al. E-Cadherin and tight junctions between epithelial cells of different animal species. Pflugers Arch 2002; 444:467-475.
58. Gonzalez-Mariscal L, Chavez DR, Lazaro A et al. Establishment of tight junctions between cells from different animal species and different sealing capacities. J Membr Biol 1989; 107:43-56.
59. Cereijido M. Tight Junctions. In: Cereijido M, ed. Boca Raton: CRC Press, 1992.
60. Cereijido M. Tight Junctions. In: Cereijido M, ed. Tight Junctions. 2nd ed. Boca Raton: CRC Press, 2001.
61. Wang AZ, Wang JC, Ojakian GK et al. Determinants of apical membrane formation and distribution in multicellular epithelial MDCK cysts. Am J Physiol 1994; 267:C473-C481.
62. Wodarz A. Establishing cell polarity in development. Nat Cell Biol 2002; 4:E39-E44.
63. Fleming TP, Ghassemifar MR, Sheth B. Junctional complexes in the early mammalian embryo. Semin Reprod Med 2000; 18:185-193.
64. Shoshani L, Contreras RG. Biogenesis of epithelial polarity and tight junctions. In: Cereijido M, ed. Tight Junctions. 2nd ed. Boca Raton: CRC Press, 2001:165-197.
65. Baas AF, Kuipers J, van der Wel NN et al. Complete polarization of single intestinal epithelial cells upon activation of LKB1 by STRAD. Cell 2004; 116:457-466.
66. Mauchamp J, Chambard M, Gabrion J et al. Polarized multicellular structures designed for the in vitro study of thyroid cell function and polarization. Methods Enzymol 1983; 98:477-486.
67. Koefoed-Johnsen V, Ussing HH. The nature of the frog skin potential. Acta Physiol Scand 1958; 42:298-308.
68. Cereijido M, Ehrenfeld J, Meza I et al. Structural and functional membrane polarity in cultured monolayers of MDCK cells. J Membr Biol 1980; 52:147-159.
69. Cereijido M, Shoshani L, Contreras RG. The polarized distribution of Na+, K+-ATPase and active transport across epithelia. J Membr Biol 2001; 184:299-304.
70. Louvard D. Apical membrane aminopeptidase appears at site of cell-cell contact in cultured kidney epithelial cells. Proc Natl Acad Sci USA 1980; 77:4132-4136.
71. Shoshani L Contreras RG, Roldan ML et al. The polarized expression of Na$^+$,K$^+$-ATPase in epithelia depends on the association between β-subunits located in neighboring cells. Mol Biol Cell 2005; 16:1071-1081.
72. Blanco G, Mercer RW. Isozymes of the Na-K-ATPase: Heterogeneity in structure, diversity in function. Am J Physiol 1998; 275:F633-F650.
73. Pedersen PL, Carafoli E. Ion Motive ATPases I. Ubiquity, properties and significance to cell function. TIBS1987; 12:146-150.
74. Pedersen PL, Carafoli E. Ion motive ATPases: Energy coupling and work output. TIBS 1987; 12:186-189.
75. Kirley TL. Determination of three disulfide bonds and one free sulfhydryl in the beta subunit of (Na,K)-ATPase. J Biol Chem 1989; 264:7185-7192.

76. Schmalzing G, Ruhl K, Gloor SM. Isoform-specific interactions of Na,K-ATPase subunits are mediated via extracellular domains and carbohydrates. Proc Natl Acad Sci USA 1997; 94:1136-1141.
77. Gatto C, McLoud SM, Kaplan JH. Heterologous expression of Na(+)-K(+)-ATPase in insect cells: intracellular distribution of pump subunits. Am J Physiol Cell Physiol 2001; 281:C982-C992.
78. Laughery MD, Todd ML, Kaplan JH. Mutational analysis of alpha-beta subunit interactions in the delivery of Na,K-ATPase heterodimers to the plasma membrane. J Biol Chem 2003; 278:34794-34803.
79. Swann AC, Daniel A, Albers RW et al. Interactions of lectins with (Na+ + K+)-ATPase of eel electric organ. Biochim Biophys Acta. 1975; 401:299-306.
80. Ochieng J, Warfield P, Green-Jarvis B et al. Galectin-3 regulates the adhesive interaction between breast carcinoma cells and elastin. J Cell Biochem 1999; 75:505-514.
81. Malik N, Canfield VA, Beckers MC et al. Identification of the mammalian Na,K-ATPase 3 subunit. J Biol Chem 1996; 271:22754-22758.
82. Martin-Vasallo P, Dackowski W, Emanuel JR et al. Identification of a putative isoform of the Na,K-ATPase beta subunit. Primary structure and tissue-specific expression. J Biol Chem 1989; 264:4613-4618.
83. Schneider JW, Mercer RW, Gilmore-Hebert M et al. Tissue specificity, localization in brain, and cell-free translation of mRNA encoding the A3 isoform of Na+,K+-ATPase. Proc Natl Acad Sci USA 1988; 85:284-288.
84. Caplan MJ, Anderson HC, Palade GE et al. Intracellular sorting and polarized cell surface delivery of (Na+,K+)ATPase, an endogenous component of MDCK cell basolateral plasma membranes. Cell 1986; 46:623-631.
85. Caplan MJ, Palade GE, Jamieson JD. Newly synthesized Na,K-ATPase alpha-subunit has no cytosolic intermediate in MDCK cells. J Biol Chem 1986; 261:2860-2865.
86. Gottardi CJ, Caplan MJ. Molecular requirements for the cell-surface expression of multisubunit ion-transporting ATPases. Identification of protein domains that participate in Na,K-ATPase and H,K-ATPase subunit assembly. J Biol Chem 1993; 268:14342-14347.
87. Gottardi CJ, Caplan MJ. Delivery of Na+,K(+)-ATPase in polarized epithelial cells. Science 1993; 260:552-554.
88. Dunbar LA, Caplan MJ. The cell biology of ion pumps: sorting and regulation. Eur J Cell Biol 2000; 79:557-563.
89. Dunbar LA, Aronson P, Caplan MJ. A transmembrane segment determines the steady-state localization of an ion-transporting adenosine triphosphatase. J Cell Biol 2000; 148:769-778.
90. Kraemer D, Koob R, Friedrichs B et al. Two novel peripheral membrane proteins, pasin 1 and pasin 2, associated with Na+,K(+)-ATPase in various cells and tissues. J Cell Biol 1990; 111:2375-2383.
91. Kraemer DM, Strizek B, Meyer HE et al. Kidney Na+,K(+)-ATPase is associated with moesin. Eur J Cell Biol 2003; 82:87-92.
92. Devarajan P, Scaramuzzino DA, Morrow JS. Ankyrin binds to two distinct cytoplasmic domains of Na,K-ATPase alpha subunit. Proc Natl Acad Sci USA 1994; 91:2965-2969.
93. Zhang Z, Devarajan P, Dorfman AL et al. Structure of the ankyrin-binding domain of alpha-Na,K-ATPase. J Biol Chem 1998; 273:18681-18684.
94. Baumann O, Lautenschlager B, Takeyasu K. Immunolocalization of Na,K-ATPase in blowfly photoreceptor cells. Cell Tissue Res 1994; 275:225-234.
95. Baumann O. Biogenesis of surface domains in fly photoreceptor cells: fine-structural analysis of the plasma membrane and immunolocalization of Na+, K+ ATPase and alpha-spectrin during cell differentiation. J Comp Neurol 1997; 382:429-442.
96. Dubreuil RR, Maddux PB, Grushko TA et al. Segregation of two spectrin isoforms: polarized membrane-binding sites direct polarized membrane skeleton assembly. Mol Biol Cell 1997; 8:1933-1942.
97. Dubreuil RR, Wang P, Dahl S et al. Drosophila beta spectrin functions independently of alpha spectrin to polarize the Na,K ATPase in epithelial cells. J Cell Biol 2000; 149:647-656.
98. Lee JK, Coyne RS, Dubreuil RR et al. Cell shape and interaction defects in alpha-spectrin mutants of Drosophila melanogaster. J Cell Biol 1993; 123:1797-1809.
99. Lee JK, Brandin E, Branton D et al. alpha-Spectrin is required for ovarian follicle monolayer integrity in Drosophila melanogaster. Development 1997; 124:353-362.
100. Bok D, O'Day W, Rodriguez-Boulan E. Polarized budding of vesicular stomatitis and influenza virus from cultured human and bovine retinal pigment epithelium. Exp Eye Res. 1992; 55:853-860.
101. Caldwell RB, McLaughlin BJ. Redistribution of Na-K-ATPase in the dystrophic rat retinal pigment epithelium. J Neurocytol 1984; 13:895-910.
102. Gundersen D, Orlowski J, Rodriguez-Boulan E. Apical polarity of Na,K-ATPase in retinal pigment epithelium is linked to a reversal of the ankyrin-fodrin submembrane cytoskeleton. J Cell Biol 1991; 112:863-872.

103. Steinberg RH, Miller SS. Transport and membrane properties of the retinal pigment epithelium. In: Zinn KM, Marmor MF, eds. The Retinal Pigment Epithelium. Cambridge: Harvard University Press, 1979:205-225.
104. Bok D. Autoradiographic studies on the polarity of plasma membrane receptors in retinal pigment epithelial cells. In: J.G.Hollyfield, ed. The Structure of the Eye. New York: Elsevier Biomedical, 1982:245-256.
105. Rizzolo LJ, Zhou S. The distribution of Na+,K(+)-ATPase and 5A11 antigen in apical microvilli of the retinal pigment epithelium is unrelated to alpha-spectrin. J Cell Sci 1995; 108 (Pt 11):3623-3633.
106. Axelsen KB, Palmgren MG. Evolution of substrate specificities in the P-type ATPase superfamily. J Mol Evol 1998; 46:84-101.
107. Vilsen B, Andersen JP, Petersen J et al. Occlusion of 22Na+ and 86Rb+ in membrane-bound and soluble protomeric alpha beta-units of Na,K-ATPase. J Biol Chem 1987; 262:10511-10517.
108. Okamura H, Denawa M, Ohniwa R et al. P-type ATPase superfamily: evidence for critical roles for kingdom evolution. Ann NY Acad Sci 2003; 986:219-223.
109. Okamura H, Yasuhara JC, Fambrough DM et al. P-type ATPases in Caenorhabditis and Drosophila: implications for evolution of the P-type ATPase subunit families with special reference to the Na,K-ATPase and H,K-ATPase subgroup. J Membr Biol 2003; 191:13-24.
110. Krogh A. Osmotic Regulation in Aquatic Animals. Cambridge: University Cambridge Press; 1939.
111. Diamond JM. Transport mechanisms in the gallbladder. American Physiological Society Handbook 5. The Alimentary Canal 1968:2451.
112. Ahearn GA, Duerr JM, Zhuang Z et al. Ion transport processes of crustacean epithelial cells. Physiol Biochem Zool 1999; 72:1-18.
113. Cornell JC. Sodium and chloride transport in the isolated intestine of the earthworm, Lumbricus terrestris (L.). J Exp Biol 1982; 97:197-216.
114. Hudson RL. Ion transport by the isolated mantle epithelium of the freshwater clam, Unio complanatus. Am J Physiol 1992; 263:R76-R83.
115. Jungreis AM, Hoges TK, Kleinzeller A et al. Water Relations in Membrane Transport in Plants and Animals. New York, N.Y.: Academic Press; 2004.
116. Prusch R, Otter T. Annelid transepithelial ion transport. Comp Biochem Physiol 1977; 57A:87-92.
117. Filshie BK, Flower NE. Junctional structures in hydra. J Cell Sci 1977; 23:151-172.
118. McNutt NS. A thin-section and freeze-fracture study of microfilament-membrane attachments in choroid plexus and intestinal microvilli. J Cell Biol 1978; 79:774-787.
119. Staehelin LA. Structure and function of intercellular junctions. Int Rev Cytol 1974; 39:191-283.
120. Asano A, Asano K, Sasaki H et al. Claudins in Caenorhabditis elegans: their distribution and barrier function in the epithelium. Curr Biol 2003; 13:1042-1046.
121. Wu VM, Paul S, Schutle J et al. A cl-audin and specific isoforms of Na,K-ATPase are required for Drosophila epithelial tube-size control and septate junction organization. 43rd Annual Meeting of the American Society for Cell Biology, L308. 2003.
122. Matter K, Balda MS. Signalling to and from tight junctions. Nat Rev Mol Cell Biol 2003; 4:225-236.
123. Tepass U, Tanentzapf G, Ward R et al. Epithelial cell polarity and cell junctions in Drosophila. Annu Rev Genet 2001; 35:747-784.
124. Gierer A, Meinhardt H. A theory of biological pattern formation. Kybernetik 1972; 12:30-39.
125. Fei K, Yan L, Zhang J et al. Molecular and biological characterization of a zonula occludens-1 homologue in Hydra vulgaris, named HZO-1. Dev Genes Evol 2000; 210:611-616.
126. Wikramanayake AH, Hong M, Lee PN et al. An ancient role for nuclear beta-catenin in the evolution of axial polarity and germ layer segregation. Nature 2003; 426:446-450.
127. Adell T, Nefkens I, Muller WE. Polarity factor 'Frizzled' in the demosponge Suberites domuncula: identification, expression and localization of the receptor in the epithelium/pinacoderm(1). FEBS Lett 2003; 554:363-368.
128. Takeyasu K, Lemas V, Fambrough DM. Stability of Na(+)-K(+)-ATPase alpha-subunit isoforms in evolution. Am J Physiol 1990; 259:C619-C630.
129. Takeyasu K, Okamura H, Yasuhara JC et al. P-type ATPase diversity and evolution: the origins of ouabain sensitivity and subunit assembly. Cell Mol Biol (Noisy -le-grand) 2001; 47:325-333.
130. Fambrough DM, Lemas MV, Hamrick M et al. Analysis of subunit assembly of the Na-K-ATPase. Am J Physiol 1994; 266:C579-C589.
131. Lemas MV, Hamrick M, Takeyasu K et al. 26 amino acids of an extracellular domain of the Na,K-ATPase alpha-subunit are sufficient for assembly with the Na,K-ATPase beta-subunit. J Biol Chem 1994; 269:8255-8259.
132. Wang SG, Farley RA. Valine 904, tyrosine 898, and cysteine 908 in Na,K-ATPase alpha subunits are important for assembly with beta subunits. J Biol Chem 1998; 273:29400-29405.

133. Kriebel PW, Barr VA, Parent CA. Adenylyl cyclase localization regulates streaming during chemotaxis. Cell 2003; 112:549-560.
134. Grimson MJ, Coates JC, Reynolds JP et al. Adherens junctions and beta-catenin-mediated cell signalling in a non-metazoan organism. Nature 2000; 408:727-731.
135. Perez-Moreno M, Jamora C, Fuchs E. Sticky business: orchestrating cellular signals at adherens junctions. Cell 2003; 112:535-548.
136. Korswagen HC, Herman MA, Clevers HC. Distinct beta-catenins mediate adhesion and signalling functions in C. elegans. Nature 2000; 406:527-532.
137. Kaplan DD, Meigs TE, Kelly P et al. Identification of a role for beta-catenin in the establishment of a bipolar mitotic spindle. J Biol Chem 2004; 279:10829-10832.
138. Simske JS, Koppen M, Sims P et al. The cell junction protein VAB-9 regulates adhesion and epidermal morphology in C. elegans. Nat Cell Biol 2003; 5:619-625.
139. Hua VB, Chang AB, Tchieu JH et al. Sequence and phylogenetic analyses of 4 TMS junctional proteins of animals: connexins, innexins, claudins and occludins. J Membr Biol 2003; 194:59-76.
140. Kollmar R, Nakamura SK, Kappler JA et al. Expression and phylogeny of claudins in vertebrate primordia. Proc Natl Acad Sci USA 2001; 98:10196-10201.
141. Uemura T. The cadherin superfamily at the synapse: more members, more missions. Cell. 1998; 93:1095-1098.
142. Nollet F, Kools P, van Roy F. Phylogenetic analysis of the cadherin superfamily allows identification of six major subfamilies besides several solitary members. J Mol Biol 2000; 299:551-572.
143. Wu Q, Maniatis T. Large exons encoding multiple ectodomains are a characteristic feature of protocadherin genes. Proc Natl Acad Sci USA. 2000; 97:3124-3129.
144. Gallin WJ. Evolution of the "classical" cadherin family of cell adhesion molecules in vertebrates. J Mol Biol Evol 1998; 15:1099-1107.
145. Pouliot Y. Phylogenetic analysis of the cadherin superfamily. Bioessays. 1992; 14:743-748.
146. Oda H, Wada H, Tagawa K et al. A novel amphioxus cadherin that localizes to epithelial adherens junctions has an unusual domain organization with implications for chordate phylogeny. Evol Dev 2002; 4:426-434.
147. King N, Hittinger CT, Carroll SB. Evolution of key cell signaling and adhesion protein families predates animal origins. Science 2003; 301:361-363.
148. Lane NJ, Dallai R, Burighel P et al. Tight and gap junctions in the intestinal tract of tunicates (Urochordata): a freeze-fracture study. J Cell Sci 1986; 84:1-17.
149. Sasakura Y, Shoguchi E, Takatori N et al. A genomewide survey of developmentally relevant genes in Ciona intestinalis. X. Genes for cell junctions and extracellular matrix. Dev Genes Evol 2003; 213:303-313.
150. Moroi S, Saitou M, Fujimoto K et al. Occludin is concentrated at tight junctions of mouse/rat but not human/guinea pig Sertoli cells in testes. Am J Physiol 1998; 274:C1708-C1717.
151. Saitou M, Fujimoto K, Doi Y et al. Occludin-deficient embryonic stem cells can differentiate into polarized epithelial cells bearing tight junctions. J Cell Biol 1998; 141:397-408.
152. Conway Morris S. The fossil record and the early evolution of metazoa. Nature 1993; 361:219-225.
153. Runnegar B. Collagen gene construction and evolution. J Mol Evol. 1985; 22:141-149.
154. Ax P. The Hierarchy of life, molecules and morphology in Phyligenetic Analisis. In: B.B.K.a J.H.Fernholm, ed. Amsterdam Excepta Medica, 1989:229-45.
155. Muller WE. Review: How was metazoan threshold crossed? The hypothetical Urmetazoa. Comp Biochem Physiol A Mol Integr Physiol 2001; 129:433-460.
156. Muller WE, Schroder HC, Skorokhod A et al. Contribution of sponge genes to unravel the genome of the hypothetical ancestor of Metazoa (Urmetazoa). Gene 2001; 276:161-173.
157. Schram FR. The Early Evolution of Metazoa and the Significance of Problematic Taxa. Cambridge University Press; 1991.
158. Wilimer P. Invertebrate Relationships Patterns in Animal Evolution XIII. Cambridge, MA.: Cambridge University Press; 1990.
159. Green RC. A clarification of the two types of invertebrate pleated septate junction. Tissue Cell 1981; 13:173-188.
160. Green RC, Bergquist PR. Cell membrane specializations in the Porifera. Biologie des Spongiaires. Paris: Colloques Internationaux du Centre National de la Recherche Scientific.; 1978:153-8.
161. Green RC, Bergquist PR. Phylogenetic relationships within the invertebrata in relation to the structure of septate junctions and the development of /324 occluding' junctional types. J Cell Sci 1982; 53:270-305.
162. Hammerton RW, Krzeminski KA, Mays RW et al. Mechanism for regulating cell surface distribution of Na+,K(+)-ATPase in polarized epithelial cells. Science. 1991; 254:847-850.
163. Gloor S, Antonicek H, Sweadner KJ et al. The adhesion molecule on glia (AMOG) is a homologue of the beta subunit of the Na,K-ATPase. J Cell Biol 1990; 110:165-174.

CHAPTER 2

Occludin, a Constituent of Tight Junctions

Yan-Hua Chen, Daniel A. Goodenough and Qun Lu

Abstract

Occludin was the first tight junction (TJ) integral membrane protein identified. This ~65 kDa protein specifically localizes at TJs of epithelial and endothelial cells and is incorporated into the network of TJ strands. Occludin has a predicted tetraspan membrane topology with two extracellular loops and a large COOH-terminal cytoplasmic domain. Both extracellular domains are enriched with tyrosine residues, and in the first domain, more than half of the residues are tyrosines and glycines. Multiple domains of occludin are responsible for its localization and functions. Occludin interacts with many structural as well as signaling molecules and participates in the regulation of TJ functions. Although occludin-deficient mice show various phenotypes such as significant postnatal growth retardation, chronic inflammation, hyperplasia of the gastric epithelium, and calcification in the brain, the roles of occludin in vivo remain unclear since many epithelia in occludin null mice seem unaffected.

Identification of Occludin from Different Species

The first TJ integral membrane protein was identified in 1993 by Tsukita's laboratory. Furuse et al[1] used the membrane fraction of adherens junctions isolated from the chick liver as antigen and obtained monoclonal antibodies (mAb) against a ~65 kDa protein named occludin. Immunofluorescence and immunoelectron microscopy revealed that occludin was exclusively localized at TJs between both epithelial and endothelial cells. Subsequently, Tsukita's group cloned occludin cDNAs from human, mouse, dog, and rat-kangaroo.[2] By sequence analysis, the three mammalian (human, murine, and canine) occludins are highly homologous to each other, but diverge considerably from those of chicken and rat-kangaroo. Three years later, Cordenonsi et al[3] obtained *Xenopus laevis* occludin cDNA, which showed ~58% sequence identity to mammalian occludins and ~41% identity to chicken and rat-kangaroo. Table 1 compares the occludin sequence homology from different species to human occludin at the amino acid level.

Several occludin splice variants have been reported. Muresan et al[4] isolated a transcript encoding an alternatively spliced form of occludin (occludin 1B) from MDCK cells. Occludin 1B transcript contains a 193 base pair (bp) insertion encoding a unique amino terminal sequence of 56 amino acids. Occludin 1B is co-expressed with occludin in cultured MDCK and T84 cells as well as in various mouse tissues by indirect immunofluorescence method.[4] Another alternative spliced isoform of occludin has been reported.[5] This occludin variant (occludin TM4⁻) lacks the fourth transmembrane domain containing 162 bp, which coincides precisely with occludin exon 4. Therefore, occludin TM4⁻ could be generated by the exon 4 deletion with the reading in frame with the downstream exon 5. Occludin TM4⁻ is present at low levels in human Caco-2 cells, monkey BSC1 cells, canine MDCK cells, murine CMT64/61 cells and lung tissue by immunoblot analysis.[5] Mankertz et al[6] found by PCR using human

Tight Junctions, edited by Lorenza Gonzalez-Mariscal. ©2006 Landes Bioscience and Springer Science+Business Media.

Table 1. Comparison of occludin sequence identity from different species to human occludin

Species	Amino Acids in ORF (Open Reading Frame)	Sequence Identity (%)	Accession Number
Human	522	100	U49184
Dog	521	91.9	U49221
Mouse	521	91.5	U49185
Rat	523	88	NM_031329
Xenopus	493	58	AF170275
Chicken	504	45.6	D21837
Rat-kangaroo	489	45	U49183

colon cDNA that there are three additional forms of occludin mRNAs with different in frame deletions in addition to the original full-length occludin mRNA. These products show altered subcellular distribution when expressed in human intestinal HT-29/B6 cells.[6] The functions of these occludin variants are unknown.

The identification of occludin presents the first experimental evidence supporting the model that TJ strands contain integral membrane proteins. This has opened the door for studying TJ structure and function at the molecular level and led to the discovery of the claudin family (see Chapter 3).

Localization of Occludin at TJs

Immunoelectron microscopy, freeze-fracture immunoreplica electron microscopy and indirect immunofluorescence microscopy have been applied to visualize the location of occludin at TJs. Using a monoclonal antibody specific for occludin, Furuse et al[1] showed that occludin was precisely localized at the TJ region between chick intestinal epithelial cells as revealed by immunogold labeling (Fig. 1A). On the freeze-fracture image, the immunogold particles were clustered along the TJ fibrils in chick hepatocytes (Fig. 1B)[7] and in MDCK cells (Fig. 1C), indicating that occludin is a structural component of the TJ network. Figure 2 shows the immunofluorescence staining of occludin in cultured cells as well as in tissues. Occludin is localized at the cell-cell junction of LLC-RK1 cells (Fig. 2A), at the apical surface of blastomeres of early *Xenopus* embryo (Fig. 2B), in the epithelial cells lining the kidney tubules (Fig. 2C), and in the blood vessels of adult mouse brain (Fig. 2D). Besides its expression in brain endothelial cells, occludin is also present at cell-cell contacts of cultured endothelial cells[8-10] and localized at TJs of retinal pigment epithelial cells (RPE) in chick and mouse.[11-13]

The staining pattern of occludin in tissues and the mechanism of occludin incorporation into TJs can be quite different in different tissues and cell types. Gonzalez-Mariscal et al[14] observed very different immunostaining patterns of occludin in proximal and collecting tubules of rabbit kidney. In proximal tubules, occludin displayed a punctuated and discontinuous pattern along the cellular boundaries, while in the collecting tubules, occludin continuously labeled cellular borders. The epithelium of collecting tubules has much higher TER value than the proximal tubule, but the relationship of occludin distribution to this physiology is unknown. Expression of occludin was also found in primary and secondary cultures of astrocytes.[15] Following treatment with 1% Triton X-100, occludin was completely extracted from astrocytic membranes, but not from MDCK cell membranes, suggesting a difference in the cytoplasmic and/or plasma membrane anchoring of occludin between these two cell types.

Developmental regulation of occludin expression and localization have been observed in several tissue types.[16-18] For example, in the gastrointestinal tract of 3- to 21-day-old chick embryos, the immunoreactivity for occludin is not detected until day 4, and it gradually increases with

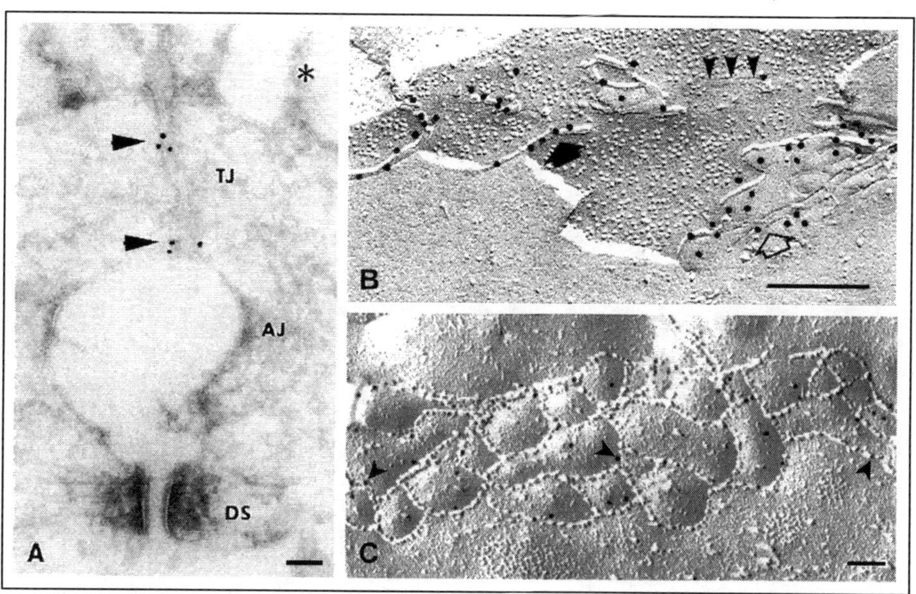

Figure 1. Localization of occludin at TJs by immunoelectron microscopy and freeze-fracture immunoreplica electron microscopy. A) Chick intestinal epithelial cells were fixed, ultrathin cryosetioned, and labeled with an anti-occludin mAb. The immunogold particles indicated by the arrowheads were only present in the TJ region, not in the adherens junction (AJ) and desmosome (DS) regions. *microvilli. (Reproduced from Furuse et al, J Cell Biol 1993; 123:1777-1788,[1] by copyright permission of the Rockefeller University Press.) B) On the freeze-fracture replica of chick hepatocytes, the immunogold labeling for occludin was observed along the network of TJ strands. Filled arrow: strand on the P-face; Open arrow: groove on the E-face; Arrowheads: groove on the P-face. (From Fujimoto J Cell Sci 1995; 108:3443-3449,[7] with the permission of The Company of Biologists Ltd.) C) Freeze-fracture replica of MDCK cells labeled by immunogold for occludin. Arrowheads indicate the gold particles associated with TJ strands. (Courtesy of Dr. Eveline Schneeberger, Massachusetts General Hospital). Bars: 100 nm in A; 200 nm in B; 80 nm in C.

development.[17] By day 11, occludin immunoreactivity is observed only at the apical surfaces of the epithelial cells. Interestingly, the immunoreactivity for occludin is clearly detected at the apical end of the lateral membrane of neuroepithelial cells throughout the chick neural plate, but this activity is lost during neural tube closure.[16] Moreover, by injecting horseradish peroxidase into the amniotic cavity of mouse embryos, functional TJs are present in the neural plate, but not in the neural tube. The loss of occludin staining during the neural tube formation indicates the change of these neuroepithelial cells from the epithelial type into the neural precursor cell type.

Expression of Occludin in Vitro

Occludin has the ability to form TJ-like structures when expressed in vitro. For example, when chicken occludin is overexpressed in Sf9 insect cells by recombinant baculovirus infection, multilamellar structures are induced in the cytoplasm.[19] Thin section electron microscopy reveals that these multi-lamellar structures formed TJ-like structures. Moreover, on the freeze-fracture replica of these multi-lamellar structures, short TJ-like strands are specifically labeled by anti-occludin mAb.

When mouse occludin is co-transfected with claudin -1 or -2 into mouse L fibroblast cells lacking TJs, occludin is recruited to freeze-fracture strands at cell-cell contact sites.[20] McCarthy et al[21] used an inducible system to transfect chick occludin into MDCK cells. By freeze-fracture analysis, they found that the number of parallel TJ fibrils shifts from three strands in control

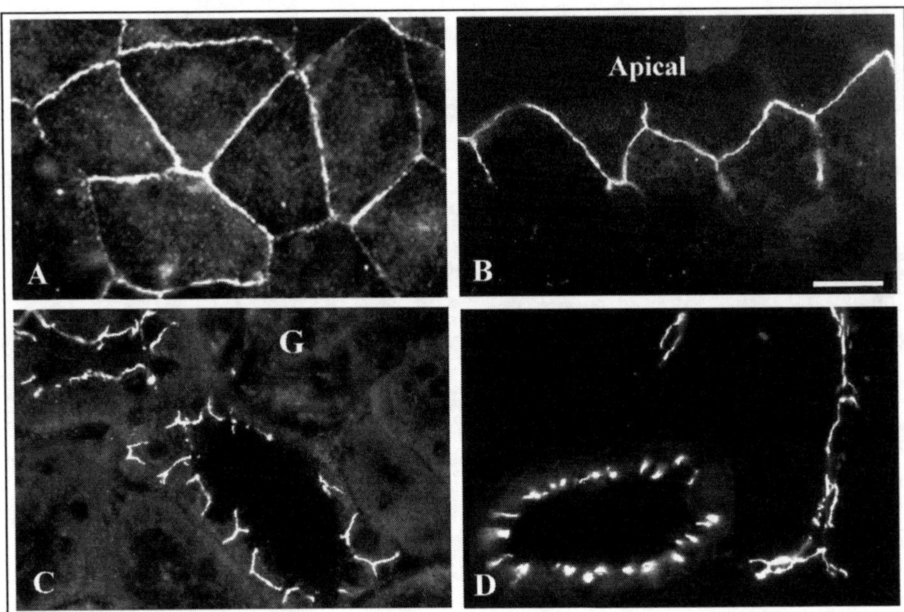

Figure 2. Immunofluorescent light microscopy of occludin localization in cultured cells as well as in tissues. A) LLC-RK1 cells (rabbit kidney proximal epithelial cells) were fixed in 100% methanol and immunostained with an anti-occludin antibody. B) Frozen sections of unfixed gastrulae of *Xenopus* embryos were labeled with an anti-occludin antibody. The fluorescent signal surrounded the apices of surface blastomeres in the partial side-view of the embryonic surface. (Reprinted from Merzdorf et al, Dev Biol 1998; 195:187-203,[18] with permission from Elsevier.) C) Frozen sections of mouse kidney were fixed in 100% methanol and stained with an anti-occludin antibody. Anti-occludin signals were shown in the epithelial cells lining the kidney tubules. G, glomerulus. D) Frozen sections of mouse brain were fixed in 4 % paraformaldehyde and stained with an anti-occludin antibody. Occludin was observed in the endothelial cells of blood vessels and capillaries. Bar, 15 μm in A, 20 μm in B, C, and D.

cells to four strands in cells expressing chick occludin. The mean width of the TJ network increases from 175 ± 11 nm to 248 ± 16 nm. Overexpression of full length occludin in MDCK cells significantly increases the transepithelial electrical resistance (TER), suggesting that occludin is a functional component of TJs.[21,22]

Occludin also confers adhesiveness when expressed in fibroblasts.[23] Stable expression of human occludin in NRK and Rat-1 fibroblast cells lacking endogenous occludin and TJ structure is able to induce cell adhesion in the absence of calcium. This indicates that the cell adhesion induced by the occludin expression is independent of cadherin-cadherin contacts. Synthetic peptides containing the amino acid sequences of the first extracellular loop inhibit the cell adhesion suggesting the direct involvement of the first extracellular loop of occludin in cell-cell adhesion.

Functional Analysis of Occludin Domains

Hydrophilicity plots predict that occludin contains four transmembrane segments, two extracellular domains, and a short intracellular loop. Both the NH_2- and COOH-termini are located in the cytoplasm.[1-3] Interestingly, the first extracellular domain of occludin contains high content of tyrosine and glycine residues (~ 60% in human, mouse, dog, rat-kangaroo, and rat occludins, 46% in *Xenopus* occludin). The two extracellular loops have no net charge at neutral pH consistent with their participation in an extracellular hydrophobic barrier.

Many studies indicate that different structural domains of occludin are important for different functions. It has been shown that the extracellular domains are critical for occludin localization and tight junction sealing.[24-26] The TER of *Xenopus* kidney epithelial cells (A6) is greatly reduced following incubation with a synthetic peptide corresponding to the second extracellular loop of occludin.[24] This decrease in TER is associated with an increase in paracellular flux of membrane-impermeable tracers, indicating the disruption of TJ barrier function. However, there are no changes in cell morphology as examined by scanning electron microscopy. Lacaz-Vieira, et al[25] applied small synthetic peptides homologous to segments of the first extracellular domain of occludin to A6 cell monolayers, and found that these peptides impaired the TJ resealing. Occludin proteins lacking the second, or both, extracellular domains are absent from TJs and found only on the basolateral surface of MDCK cells.[26] This experiment suggests that the presence of the second extracellular domain is required for integration of occludin into TJs. Tavelin et al[27] reported that peptides corresponding to the NH_2-terminus of the first extracellular domain of occludin increase the permeability of TJs as judged by the increased paracellular flux of [^{14}C] mannitol. Interestingly, these peptides have to be added to the basolateral side of the monolayer in order to have an effect. This could be partially due to the degradation of peptides by apical peptidases and aggregate formation since a lipopeptide, which protects the peptide from degradation and aggregation, is effective when added to the apical side of the monolayer.[27] Blaschuk et al[28] demonstrated that a LYHY sequence located in the second extracellular domain of occludin is an occludin cell adhesion recognition sequence. This short peptide inhibits the establishment of endothelial cell barriers in vitro and in vivo, and also prevents the aggregation of fibroblasts stably transfected with occludin cDNA.

The NH_2-terminal half of occludin also plays an important role in TJ assembly and barrier function. For example, Bamforth et al[29] showed that an occludin construct lacking NH_2-terminus and extracellular domains exerts a dramatic effect on TJ integrity. Cell monolayers transfected with this deletion construct have a lower TER and increased paracellular flux to small molecular tracers although the mutant protein is correctly targeted to the TJ. Furthermore, gaps are found to have been induced in the P-face associated TJ strands, as visualized by freeze-fracture electron microscopy. The study reported by Huber et al[30] indicated that the NH_2-terminal domain of occludin is important for the transmigration of neutrophils across epithelial sheets, but does not affect the paracellular permeability. Introduction of N-linked glycosylation sites into the two extracellular domains causes the glycosylated occludin unable to integrate into TJs.[31] It was found that glycosylated occludin accumulate in the basolateral membrane, which suggests that occludin is first inserted into basolateral membrane.

Studies from different laboratories indicate that the COOH-terminus of occludin is required for TJ function.[22,30-32] When COOH-terminally truncated chicken occludin is stably expressed in MDCK II cells, mutant occludin is incorporated into TJs, but exhibits a discontinuous junctional staining pattern as revealed by confocal immunofluorescence light microscopy.[22] However, no abnormal TJ morphology is observed on thin section electron micrographs or freeze-fracture electron microscopic images. More interestingly, overexpression of this mutant occludin in MDCK II cells causes an increase of both TER and paracellular permeability. This paradox has not been resolved. One explanation could be that the overexpression of mutant occludin may lead to the increased numbers of paracellular pores formed by occludin that are permeable only to neutral molecules and the decreased numbers of paracellular channels formed by claudins that are permeable only to ions. In this scenario, both increased TER and paracellular permeability could be observed. Another explanation could be that recent transepithelial flux measurements have suggested that mannitol may not be a reliable reporter of tight junction small molecule flux since its flux across the epithelial cell lines does not correlate well with the TER.[33] Introduction of COOH-terminally truncated occludin into *Xenopus* embryos results in an increased paracellular leakage of low molecular tracers as shown in Figure 3.[32] The leakage induced by the mutant occludin can be rescued by coinjection with full-length occludin mRNA. Immunoprecipitation analysis of detergent-solubilized embryo membranes reveals that the exogenous occludin is bound

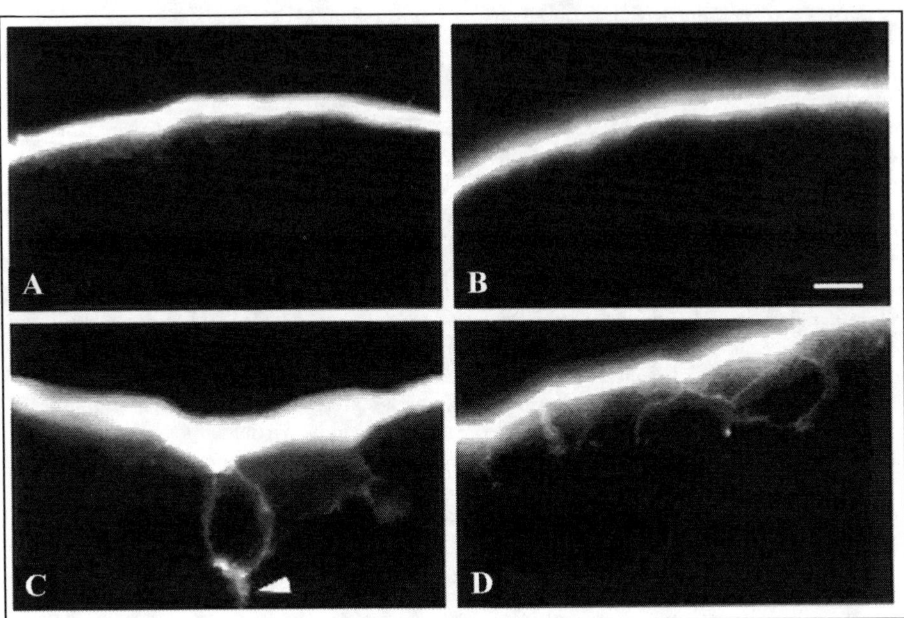

Figure 3. Expression of mutant occludin disrupted the barrier function of TJ in *Xenopus* embryos. mRNAs transcribed from full-length or mutant occludins were microinjected into the antero-dorsal blastomere of eight-cell embryos. 6 hours after injection (2,000 cell blastula), the embryos were labeled by incubation in 1 mg/ml NHS-LC-biotin for 12 minutes at 10°C, then washed and fixed. Frozen sections were stained with RITC-avidin. The staining of NHS-LC-biotin appeared as a thick continuous line on the surface of blastomeres. The TJs in the embryos injected with full-length (A, 504 amino acids), or the least COOH-terminally truncated (B, 486 amino acids) occludin mRNAs were impermeable to the biotin tracer. In contrast, in the embryos injected with two COOH-terminally truncated occludins (C, 386 amino acids, and D, 336 amino acids), the TJs were leaky and therefore, the biotin tracer penetrated into the intercellular spaces. Bar, 10 μm. (Reproduced from Chen et al, J Cell Biol 1997; 138(4):891-899,[32] by copyright permission of the Rockefeller University Press.)

to endogenous *Xenopus* occludin in vivo, indicating that occludin forms oligomers during the normal process of TJ assembly. These data demonstrate that the COOH-terminus of occludin is required for the correct assembly of TJ barrier function. COOH-terminal domain of occludin is also sufficient to mediate direct basolateral targeting of a reporter protein, indicating that it contains a basolateral-targeting determinant.[31] Further support for the function of COOH-terminus of occludin in its TJ targeting comes from a connexin-occludin chimera study. Connexins are the integral components of gap junctions that are between the lateral membranes of epithelial cells. However, connexin-occludin chimeras are localized within TJ fibrils when expressed in MDCK cells as assessed by immunofluorescence and immunogold freeze-fracture imaging.[34] This chimera contains transmembrane and extracellular domains of connexin 32 and the COOH-terminal half of occludin. Therefore, it seems likely that the localization of chimeras at TJs depends on the COOH-terminal domain of occludin.

Occludin Interacting Proteins

Occludin is able to interact with many proteins. Table 2 is a list of known occludin interacting proteins.

Besides the above molecules, occludin also interacts directly with F-actin and does not require other scaffolding proteins for this interaction.[40] All proteins listed in Table 2 interact

Table 2. Occludin interacting proteins

	M.W. kDa	Binding Site/ Method Used	Possible Roles	References
TJ associated proteins				
ZO-1	220	GUK domain/GSTLocalization of pull-down	occludin at TJ	Furuse et al[35] Fanning et al[3]
ZO-2	160	N-terminal dlg-like domain/Co-IP	Form a complex to Establish TJ	Itoh et al[37] Peng et al[38]
ZO-3	130	Unknown/Affinity resin	Linkage between occludin and F-actin	Haskins et al[39] Wittchen et al[40]
Cingulin	140-160	Unknown/GST pull-down	Unknown	Cordenonsi et al[41]
Gap junction (GJ) proteins				
Connexin 32	32	Unknown/Co-IP	Regulation of occludin expression	Kojima et al[42,43]
Connexin 26	26	Unknown/Occludin bait peptide	Proximity of TJ and GJ	Nusrat et al,[44]
Signaling molecules				
PKCξ(Protein kinase C)	80	Unknown/Occludin bait peptide	Regulation of actin function	Nusrat et al[44]
PKC	80	Unknown/Kinase assay; Peptide mass fingerprint	Regulation of occludin function at TJ	Andreeva et al[45]
c-Yes	62	Unknown/Occludin bait peptide; Co-IP	TJ formation and regulation	Nusrat et al[44] Chen et al, 2002[46,47]
c-Src	60	Unknown/GST pull-down	Disruption of TJ	Kale et al[48]
Subunit of PI3-kinase	85	Unknown/Occludin bait peptide; GST pull-down	Affecting actin polymeration; Oxidative stress-induced disruption of TJ	Nusrat et al[44] Sheth et al[49]
CK2 (Casein kinase 2)	39	Unknown/Kinase assay; Peptide mass fingerprint	Regulation of occludin phosphorylation	Cordenonsi et al[3] Smales et al[50]
CPE (Clostridium perfringens enterotoxin)	90	Unknown/IP	CPE-induced cytotoxicity	Singh et al[51]
E3 ubiquitin-protein ligase Itch	120	WW motifs/Yeast two-hybrid screen; Co-IP	Ubiquitination of occludin	Traweger et al[52]
VAP-33	33	Unknown/Yeast two-hybrid screen; Subcellular fractionation	Vesicle targeting; Regulate occludin localization	Lapierre et al[53]

with the COOH-terminal cytoplasmic domain of occludin except the E3 ubiquitin-protein ligase Itch, which binds to NH_2-terminal portion of occludin.[52]

Roles of Occludin Phosphorylation

Occludin migrates as a cluster of multiple bands (62-82 kDa) on SDS gels resulting from multiple phosphorylation of serine and threonine residues.[3,54,55] Using a monoclonal antibody specific for phosphorylated occludin, Sakakibara et al[54] found that highly phosphorylated occludin selectively concentrates at the TJ region of chick intestinal epithelial cells. Occludin can be directly phosphorylated by several protein kinases including PKC and CK2 in vitro.[45,50] Smales et al[50] used a recombinant COOH-terminal fragment of occludin as a substrate, and incubated the fragment with a kinase obtained from crude extracts of brain. This kinase was later identified as CK2 by peptide mass fingerprinting, immunoblotting, and mutation of CK2 sites within the occludin sequence.

Phosphorylation of occludin may be important for tight junction formation since occludin phosphorylation and dephosphorylation are closely associated with TJ assembly and disassembly.[46,56] In a calcium-switch model, when monolayers of MDCK cells are exposed to prolonged Ca^{2+} starvation, the TJ complex is disassembled.[56] In this case, occludin is no longer localized at cell surface, and the phosphorylated 71 kDa band of occludin disappears on western blots. For endothelial cells, Kevil et al[57] reported that occludin is heavily phosphorylated on serine residues upon H_2O_2 administration. The H_2O_2-mediated elevation in endothelial solute permeability and alterations in occludin localization and phosphorylation, are all blocked by MEK-1 inhibitor, PD 98059. This suggests the involvement of ERK1/ERK2 signaling pathway in the regulation of endothelial TJ integrity.

Occludin is phosphorylated upon the addition of the PKC activators PMA or diC8 to MDCK cells incubated in low calcium medium.[45] The redistribution of occludin from the cytoplasm to the lateral plasma membrane is observed after the treatment of MDCK cells with PMA or diC8. The PKC inhibitor GF-109203X markedly inhibits both the phosphorylation of occludin and its incorporation into TJs induced by Ca^{2+}- switch. In vitro experiments show that the recombinant COOH-terminal domain of murine occludin can be phosphorylated by purified PKC. Further experiments identify Ser (338) of occludin as an in vitro PKC phosphorylation site by using peptide mass fingerprint analysis and electrospray ionization tandem mass spectroscopy. These findings indicate that PKC may regulate the phosphorylation and cellular localization of occludin at TJs. However, other studies show that PKC activation by TPA or PMA leads to dephosphorylation of occludin, and an increase in TJ permeability in LLC-PK1 and human corneal epithelial cells.[58,59] Since there are so many PKC isoforms, different isoforms could have opposite effects on occludin phosphorylation and TJ function in different types of cells.

Studies of tyrosine phosphorylation have also revealed contradictory data on TJ physiology. Inhibition of tyrosine phosphatase activity has been reported to result in decreases in TER and increases in paracellular permeability in both epithelial and endothelial cells.[60-63] For example, the tyrosine phosphatase inhibitor, pervanadate, causes a concentration- and time-dependent decrease in TER in both MDCK and brain endothelial cells.[60] Tyrosine phosphorylation of occludin decreases its interaction with ZO-1, ZO-2, and ZO-3 in Caco-2 cells.[48] In other studies, PI3-kinase plays a role in oxidative stress-induced disruption of TJs.[49] PI3-kinase inhibitor, LY294002, prevents oxidative stress-induced tyrosine phosphorylation of occludin and dissociation of occludin from the actin cytoskeleton.

A number of studies have shown that tyrosine phosphorylation can be positively and temporally correlated with tight junction assembly and function.[46,64,65] Tsukamoto and Nigam[65] provided the first evidence that occludin is tyrosine phosphorylated and that tyrosine kinase activity is necessary for TJ reassembly during ATP repletion. Chen et al[46] demonstrated that occludin is tyrosine phosphorylated when localized at TJs. When Ca^{2+} is depleted from the culture medium, occludin tyrosine phosphorylation is diminished in minutes from MDCK cells, which correlates with a significant reduction in TER. Reconstitution of Ca^{2+} restores occludin tyrosine phosphorylation that is temporally associated with an increase in TER. Moreover, occludin forms an immunoprecipitable complex with nonreceptor tyrosine kinase c-Yes, and this complex dissociates when the cells are incubated in Ca^{2+}-free medium or treated with a c-Yes inhibitor, CGP77675. In the presence of c-Yes inhibitor after Ca^{2+} repletion, occludin tyrosine phosphorylation is abolished and the increase of TER inhibited. These results provide strong evidence that tyrosine phosphorylation of occludin is tightly linked to TJ formation in epithelial cells and that tyrosine kinase c-Yes is involved in the regulation of this process.

In epithelia of Xenopus embryos, occludin dephosphorylation is correlated with the de novo assembly of TJs.[66] Occludin migrates as a 61 kDa protein on the immunoblot in unfertilized eggs, but presents as multiple bands with 57-60 kDa in fertilized eggs and in early cleavages up to blastula stage 8. In gastrulae, neurulae and tailbud stage embryos, occludin migrates as a 57 kDa protein. This mobility downshift was specifically reproduced by treatment of egg extracts with a phosphatase, indicating that it is due to dephosphorylation. These studies, together with the findings discussed above, suggest that occludin phosphorylation may play different roles in different biological systems. In addition, phosphorylation of specific residues by different kinases may have different functional consequences.

Regulation of Occludin by Signaling Molecules

Occludin localization and function can be regulated by many signaling molecules and pathways including the Rho family small GTPases,[67-71] Par proteins and aPKC,[71,72] G proteins,[73] [74,75] proinflammatory cytokines and cytotoxic necrotizing factor-1,[76-79] hormones and growth factors,[80-82] ERK1/2 and p38 MAP kinase pathway,[59,83-86] PKA- and PKC-dependent pathway[87-89] and protein phosphatases and PLCγ.[90,91] Not surprisingly, occludin in different cells responds differently to the activation and inactivation of different signaling pathways.

In MDCK cells, Rho signaling is required for TJ assembly.[68] Constitutive Rho activation causes an accumulation of occludin at the cell junction, while Rho inhibition results in decreased localization of occludin at the cell junction. Expression of an activating Rho mutant protein increases levels of occludin phosphorylation, indicating that occludin is a target for Rho signaling. In a different study, however, Jou et al[92] reported that occludin distribution in MDCK cells is disrupted by constitutively active mutants RhoAV14 and Rac1V12, but not by dominant negative mutants RhoAN19 and Rac1N17. All RhoA and Rac1 mutants result in decreased TER and increased paracellular flux. Differences in occludin localization induced by Rho inhibition might result from different methods used in the studies: Rho inhibitor (C3 transferase) versus dominant negative construct (RhoAN19).

Activation of the MAP kinase pathway disrupts TJ structure and function in several cell lines, such as rat salivary gland epithelial cell line Pa-4, Ras-transformed MDCK cells, in addition to human corneal epithelial cells.[59,83,84] Li and Mrsny[83] demonstrated that transfection of a constitutively active construct of Raf-1, a MEK-1 kinase, into Pa-4 cells results in a complete loss of TJ function. The cells transfected with Raf-1 display a stratified phenotype that lacked cell-cell contact growth control. The expression of occludin is downregulated in these cells. More importantly, introduction of occludin into Raf-1 transfected cells results in reacquisition of normal epithelial phenotype and functionally intact TJs. Similarly, in *ras*-transformed MDCK cells, TJ structure is absent and occludin is present only in the cytoplasm.[84] These cells show fibroblastic morphology and grow on top of each other. When Ras signaling is attenuated by the inhibition of MAP kinase pathway using MEK1 inhibitor, TJ structure and function are restored, occludin is concomitantly recruited to the cell-cell contact areas and cells resume their normal epithelial phenotype. Therefore, constitutive activation of MAP kinase pathway negatively regulates TJ assembly and function. However, Macek et al[93] recently reported that in claudin-1 expressing T47-D cells (low Ras activity) and claudin-1 negative MCF-7 cells (elevated Ras activity), no quantitative changes of mRNA or protein levels of occludin can be detected after inhibition of MAP kinase pathway by MEK1 inhibitor. Also, there is no difference in TER and paracellular flux between these two cell lines. It seems likely that the regulatory machinery of TJ is complex and different in different cell lines depending on the origin of the cell, the physiological and pathological state of the cell, and the genetic background of the cell.

Phenotypes of Occludin-Deficient Mice

Occludin-deficient mice display complex phenotypes.[94] These mice show significant postnatal growth retardation with ~75% of normal weight at 8 weeks of age. Interestingly, occludin$^{-/-}$ females do not suckle their litters, resulting in neonatal death. Histological examination reveals a chronic inflammation and hyperplasia in the gastric epithelium of occludin null mice. Gastritis develops around 10 weeks of age and becomes very severe around 28 weeks of age. A progressive accumulation of calcium and phosphorus deposits is observed in the cerebellum and basal ganglia of occludin null mice using electron microscopy with energy dispersive X-ray microanalysis. Abnormalities are also found in the bone. In occludin null mice, the compact bone is significantly thinner than that of wild-type controls. However, despite these complex phenotypes, occludin-deficient mice form morphologically normal TJ strands. The TJ barrier function in intestinal epithelium is also normal as measured by TER. These results are consistent with the previous report that occludin-deficient ES cells can differentiate into polarized epithelial cells bearing a well-developed network of TJ strands.[95] It is clear from these studies that occludin is not required for the formation of TJ strands and its role in TJ barrier function can partially be compensated by other TJ membrane proteins in vivo.

Occludin in Cancer Cells

TJs are crucial for maintaining the cell polarity and junctional integrity. In many cancer tissues, cells lose their polarity and cell junctions are disrupted. These cells then become undifferentiated and begin to migrate. Occludin expression is often downregulated in human cancers.[96-100] For example, in human prostate cancer, occludin expression is completely lost in unpolarized cells of high Gleason grade tumors (Gleason 4 and above), but remains expressed in cells facing a lumen in all grades of cancer.[99] Therefore, the loss of occludin expression is closely associated with the loss of cell polarity. Malignant brain tumors cause cerebral edema because they have leaky endothelial TJs, which allow plasma fluid to enter the brain from the microvessel lumen. Using immunohistochemistry and immunoblotting methods, Papadopoulos et al[96] found that occludin expression is decreased in high grade (III or IV) brain tumors compared to that of non-neoplastic brain tissue samples. Similarly, in human gastric cancer, the expression of occludin mRNA in moderately and poorly differentiated groups is gradually reduced when compared with well-differentiated groups, suggesting a significant correlation between tumor differentiation and the expression of occludin mRNA.[97] Most recently, Tobioka et al[100] also reported that occludin expression decreases progressively in parallel with the increase in carcinoma grade, and the decreased occludin expression correlates with myometrial invasion and lymph node metastasis. All these studies suggest that the loss of occludin expression is closely associated with the loss of cell polarity, the cell de-differentiated state and the progression of human tumors.

Summary

Occludin was the first integral membrane protein identified in TJ strands, but it is not required for the formation of TJ strands. Studies of occludin assembly, domain function, and regulation by phosphorylation and signaling molecules yield rich, although sometimes contradictory information that suggests its regulations in different biological systems are complex.

References

1. Furuse M, Hirase T, Itoh M et al. Occludin-a novel integral membrane protein localizing at tight junctions. J Cell Biol 1993; 123:1777-1788.
2. Ando-Akatsuka Y, Saitou M, Hirase T et al. Interspecies diversity of the occludin sequence-cDNA cloning of human, mouse, dog, and rat-kangaroo homologues. J Cell Biol 1996; 133:43-47.
3. Cordenonsi M, Turco F, D'Atri F et al. Xenopus laevis occludin. Identification of in vitro phosphorylation sites by protein kinase CK2 and association with cingulin. Eur J Biochem 1999; 264(2):374-384.

4. Muresan Z, Paul DL, Goodenough DA. Occludin 1B, a variant of the tight junction protein occludin. Mol Biol Cell 2000; 11(2):627-634.
5. Ghassemifar MR, Sheth B, Papenbrock T et al. Occludin TM4(-): an isoform of the tight junction protein present in primates lacking the fourth transmembrane domain. J Cell Sci 2002; 115(Pt 15):3171-3180.
6. Mankertz J, Stefan WJ, Hillenbrand B et al. Gene expression of the tight junction protein occludin includes differential splicing and alternative promoter usage. Biochem Biophys Res Commun 2002; 298(5):657-666.
7. Fujimoto K. Freeze-fracture replica electron microscopy combined with SDS digestion for cytochemical labeling of integral membrane proteins. Application to the immunogold labeling if intercellular junctional complexes. J Cell Sci 1995; 108:3443-3449.
8. Hirase T, Staddon JM, Saitou M et al. Occludin as a possible determinant of tight junction permeability in endothelial cells. J Cell Sci 1997; 110(Part 14):1603-1613.
9. Kevil CG, Okayama N, Trocha SD et al. Expression of zonula occludens and adherens junctional proteins in human venous and arterial endothelial cells: role of occludin in endothelial solute barriers. Microcirculation 1998; 5(2-3):197-210.
10. Antonetti DA, Barber AJ, Hollinger LA et al. Vascular endothelial growth factor induces rapid phosphorylation of tight junction proteins occludin and zonula occluden 1. A potential mechanism for vascular permeability in diabetic retinopathy and tumors. J Biol Chem 1999; 274(33):23463-23467.
11. Konari K, Sawada N, Zhong Y et al. Development of the blood-retinal barrier in vitro-formation of tight junctions as revealed by occludin and ZO-1 correlates with the barrier function of chick retinal pigment epithelial cell. Exp Eye Res 1995; 61:99-108.
12. Williams CD, Rizzolo LJ. Remodeling of junctional complexes during the development of the outer blood-retinal barrier. Anat Rec 1997; 249(3):380-388.
13. Tserentsoodol N, Shin BC, Suzuki T et al. Colocalization of tight junction proteins, occludin and ZO-1, and glucose transporter GLUT1 in cells of the blood-ocular barrier in the mouse eye [In Process Citation]. Histochem Cell Biol 1998; 110(6):543-551.
14. Gonzalez-Mariscal L, Namorado MC, Martin D et al. Tight junction proteins ZO-1, ZO-2, and occludin along isolated renal tubules. Kidney Int 2000; 57(6):2386-2402.
15. Bauer H, Stelzhammer W, Fuchs R et al. Astrocytes and neurons express the tight junction-specific protein occludin in vitro. Exp Cell Res 1999; 250(2):434-438.
16. Aaku-Saraste E, Hellwig A, Huttner WB. Loss of occludin and functional tight junctions, but not ZO-1, during neural tube closure-remodeling of the neuroepithelium prior to neurogenesis. Dev Biol 1996; 180(2):664-679.
17. Kawasaki K, Hayashi Y, Nishida Y et al. Developmental expression of the tight junction protein, occludin, in the gastrointestinal tract of the chick embryo. Histochem Cell Biol 1998; 109(1):19-24.
18. Merzdorf CS, Chen Y-H, Goodenough DA. Formation of functional tight junctions in Xenopus embryos. Dev Biol 1998; 195:187-203.
19. Furuse M, Fujimoto K, Sato N et al. Overexpression of occludin, a tight junction-associated integral membrane protein, induces the formation of intracellular multilamellar bodies bearing tight junction-like structure. J Cell Sci 1996; 109:429-435.
20. Furuse M, Sasaki H, Fujimoto K et al. A single gene product, claudin-1 or -2, reconstitutes tight junction strands and recruits occludin in fibroblasts. J Cell Biol 1998; 143(2):391-401.
21. McCarthy KM, Skare IB, Stankewich MC et al. Occludin is a functional component of the tight junction. J Cell Sci 1996; 109:2287-2298.
22. Balda MS, Whitney JA, Flores C et al. Functional dissociation of paracellular permeability and transepithelial electrical resistance and disruption of the apical-basolateral intramembrane diffusion barrier by expression of a mutant tight junction membrane protein. J Cell Biol 1996; 134(4):1031-1049.
23. Van Itallie CM, Anderson JM. Occludin confers adhesiveness when expressed in fibroblasts. J Cell Sci 1997; 110:1113-1121.
24. Wong V, Gumbiner BM. A synthetic peptide corresponding to the extracellular domain of occludin perturbs the tight junction permeability barrier. J Cell Biol 1997; 136:399-409.
25. Lacaz-Vieira F, Jaeger MM, Farshori P et al. Small synthetic peptides homologous to segments of the first external loop of occludin impair tight junction resealing. J Membr Biol 1999; 168(3):289-297.
26. Medina R, Rahner C, Mitic LL et al. Occludin localization at the tight junction requires the second extracellular loop. J Membr Biol 2000; 178(3):235-247.
27. Tavelin S, Hashimoto K, Malkinson J et al. A new principle for tight junction modulation based on occludin peptides. Mol Pharmacol 2003; 64(6):1530-1540.

28. Blaschuk OW, Oshima T, Gour BJ et al. Identification of an occludin cell adhesion recognition sequence. Inflammation 2002; 26(4):193-198.
29. Bamforth SD, Kniesel U, Wolburg H et al. A dominant mutant of occludin disrupts tight junction structure and function. J Cell Sci 1999; 112(Pt 12):1879-1888.
30. Huber D, Balda MS, Matter K. Occludin modulates transepithelial migration of neutrophils. J Biol Chem 2000; 275(8):5773-5778.
31. Matter K, Balda MS. Biogenesis of tight junctions-the C-terminal domain of occludin mediates basolateral targeting. J Cell Sci 1998; 111(Part 4):511-519.
32. Chen Y-H, Merzdorf C, Paul DL et al. COOH terminus of occludin is required for tight junction barrier function in early Xenopus embryos. J Cell Biol 1997; 138(4):891-899.
33. Tang VW, Goodenough DA. Paracellular ion channel at the tight junction. Biophys J 2003; 84(3):1660-1673.
34. Mitic LL, Schneeberger EE, Fanning AS et al. Connexin-occludin chimeras containing the ZO-binding domain of occludin localize at MDCK tight junctions and NRK Cell Contacts. J Cell Biol 1999; 146(3):683-693.
35. Furuse M, Itoh M, Hirase T et al. Direct association of occludin with ZO-1 and its possible involvement in the localization of occludin at tight junctions. J Cell Biol 1994; 127:1617-1626.
36. Fanning AS, Jameson BJ, Jesaitis LA et al. The tight junction protein ZO-1 establishes a link between the transmembrane protein occludin and the actin cytoskeleton. J Biol Chem 1998; 273(45):29745-29753.
37. Itoh M, Furuse M, Morita K et al. Direct binding of three tight junction-associated MAGUKs, ZO-1, ZO-2, and ZO-3, with the COOH termini of claudins. J Cell Biol 1999; 147(6):1351-1363.
38. Peng BH, Lee JC, Campbell GA. In vitro protein complex formation with cytoskeleton anchoring domain of occludin identified by limited proteolysis. J Biol Chem 2003.
39. Haskins J, Gu L, Wittchen ES et al. ZO-3, a novel member of the MAGUK protein family found at the tight junction, interacts with ZO-1 and occludin. J Cell Biol 1998; 141(1):199-208.
40. Wittchen ES, Haskins J, Stevenson BR. Protein interactions at the tight Junction. Actin has multiple binding partners, and ZO-1 forms independent complexes with ZO-2 and ZO-3. J Biol Chem 1999; 274(49):35179-35185.
41. Cordenonsi M, D'Atri F, Hammar E et al. Cingulin contains globular and coiled-coil domains and interacts with ZO-1, ZO-2, ZO-3, and Myosin. J Cell Biol 1999; 147(7):1569-1582.
42. Kojima T, Kokai Y, Chiba H et al. Cx32 but Not Cx26 Is Associated with Tight Junctions in Primary Cultures of Rat Hepatocytes. Exp Cell Res 2001; 263(2):193-201.
43. Kojima T, Sawada N, Chiba H et al. Induction of tight junctions in human connexin 32 (hCx32)-transfected mouse hepatocytes: connexin 32 interacts with occludin [In Process Citation]. Biochem Biophys Res Commun 1999; 266(1):222-229.
44. Nusrat A, Chen JA, Foley CS et al. The coiled-coil domain of occludin can act to organize structural and functional elements of the epithelial tight junction. J Biol Chem 2000; 275:29816-29822.
45. Andreeva AY, Krause E, Muller EC et al. Protein kinase C regulates the phosphorylation and cellular localization of occludin. J Biol Chem 2001; 276:38480-38486.
46. Chen Y-H, Lu Q, Goodenough DA et al. Nonreceptor tyrosine kinase c-Yes interacts with occludin during tight junction formation in canine kidney epithelial cells. Mol Biol Cell 2002; 13(4):1227-1237.
47. Chen YH, Lu Q. Association of nonreceptor tyrosine kinase c-yes with tight junction protein occludin by coimmunoprecipitation assay. Methods Mol Biol 2002; 218:127-132.
48. Kale G, Naren AP, Sheth P et al. Tyrosine phosphorylation of occludin attenuates its interactions with ZO-1, ZO-2, and ZO-3. Biochem Biophys Res Commun 2003; 302(2):324-329.
49. Sheth P, Basuroy S, Li C et al. Role of phosphatidylinositol in oxidative stress-induced disruption of tight junction. J Biol Chem 2003.
50. Smales C, Ellis M, Baumber R et al. Occludin phosphorylation: identification of an occludin kinase in brain and cell extracts as CK2. FEBS Lett 2003; 545(2-3):161-166.
51. Singh U, Van Itallie CM, Mitic LL et al. CaCo-2 Cells treated with Clostridium perfringens enterotoxin form multiple large complex species, one of which contains the tight junction protein occludin. J Biol Chem 2000; 275:18407-18417.
52. Traweger A, Fang D, Liu YC et al. The tight junction-specific protein occludin is a functional target of the E3 ubiquitin protein ligase itch. J Biol Chem 2002.
53. Lapierre LA, Tuma PL, Navarre J et al. VAP-33 localizes to both an intracellular vesicle population and with occludin at the tight junction. J Cell Sci 1999; 112(Pt 21):3723-3732.
54. Sakakibara A, Furuse M, Saitou M et al. Possible involvement of phosphorylation of occludin in tight junction formation. J Cell Biol 1997; 137(6):1393-1401.

55. Wong V. Phosphorylation of occludin correlates with occludin localization and function at the tight junction. Am J Physiol 1997; 273(6):C1859-C1867.
56. Farshori P, Kachar B. Redistribution and phosphorylation of occludin during opening and resealing of tight junctions in cultured epithelial cells. J Membr Biol 1999; 170(2):147-156.
57. Kevil CG, Oshima T, Alexander B et al. H_2O_2-mediated permeability: role of MAPK and occludin. Am J Physiol Cell Physiol 2000; 279(1):C21-C30.
58. Clarke H, Soler AP, Mullin JM. Protein kinase C activation leads to dephosphorylation of occludin and tight junction permeability increase in LLC-PK1 epithelial cell sheets. J Cell Sci 2000; 113(Pt 18):3187-3196.
59. Wang Y, Zhang J, Yi X et al. Activation of ERK1/2 MAP kinase pathway induces tight junction disruption in human corneal epithelial cells. Exp Eye Res 2004; 78(1):125-136.
60. Staddon JM, Herrenknecht K, Smales C et al. Evidence that tyrosine phosphorylation may increase tight junction permeability. J Cell Sci 1995; 108:609-619.
61. Collares-Buzato CB, Jepson MA, Simmons NL et al. Increased tyrosine phosphorylation causes redistribution of adherens junction and tight junction proteins and perturbs paracellular barrier function in MDCK epithelia. Eur J Cell Biol 1998; 76(2):85-92.
62. Wachtel M, Frei K, Ehler E et al. Occludin proteolysis and increased permeability in endothelial cells through tyrosine phosphatase inhibition. J Cell Sci 1999; 112(Pt 23):4347-4356.
63. Atkinson KJ, Rao RK. Role of protein tyrosine phosphorylation in acetaldehyde-induced disruption of epithelial tight junctions. Am J Physiol Gastrointest Liver Physiol 2001; 280(6):G1280-G1288.
64. Kurihara H, Anderson JM, Farquhar MG. Increased tyr phosphorylation of ZO-1 during modification of tight junctions between glomerular foot processes. Am J Physiol 1995; 37:F 514-F 524.
65. Tsukamoto T, Nigam SK. Role of tyrosine phosphorylation in the reassembly of occludin and other tight junction proteins. Am J Physiol 1999; 276(5 Pt 2):F737-F750.
66. Cordenonsi M, Mazzon E, De Rigo L et al. Occludin dephosphorylation in early development of Xenopus laevis. J Cell Sci 1997; 110:3131-3139.
67. Jou TS, Schneeberger EE, Nelson WJ. Structural and functional regulation of tight junctions by RhoA and Rac1 small GTPases. J Cell Biol 1998; 142(1):101-115.
68. Gopalakrishnan S, Raman N, Atkinson SJ et al. Rho GTPase signaling regulates tight junction assembly and protects tight junctions during ATP depletion. Am J Physiol 1998; 275(3 Pt 1):C798-C809.
69. Rojas R, Ruiz WG, Leung SM et al. Cdc42-dependent modulation of tight junctions and membrane protein traffic in polarized madin-darby canine kidney cells. Mol Biol Cell 2001; 12(8):2257-2274.
70. Stamatovic SM, Keep RF, Kunkel SL et al. Potential role of MCP-1 in endothelial cell tight junction 'opening': signaling via Rho and Rho kinase. J Cell Sci 2003; 116(Pt 22):4615-4628.
71. Hirase T, Kawashima S, Wong EY et al. Regulation of Tight Junction Permeability and Occludin Phosphorylation by RhoA-p160ROCK-dependent and -independent Mechanisms. J Biol Chem 2001; 276:10423-10431.
72. Gao L, Joberty G, Macara IG. Assembly of epithelial tight junctions is negatively regulated by Par6. Curr Biol 2002; 12(3):221-225.
73. Balda MS, González-Mariscal L, Contreras RG et al. Assembly and sealing tight junctions: possible participation of G-proteins, phospholipase C, protein kinase C and calmodulin. J Membr Biol 1991; 122:193-202.
74. Denker BM, Saha C, Khawaja S et al. Involvement of a heterotrimeric G protein alpha subunit in tight junction biogenesis. J Biol Chem 1996; 271(42):25750-25753.
75. Saha C, Nigam SK, Denker BM. Involvement of Gα2 in the maintenance and biogenesis of epithelial cell tight junctions. J Biol Chem 1998; 273(34):21629-21633.
76. Hopkins AM, Walsh SV, Verkade P et al. Constitutive activation of Rho proteins by CNF-1 influences tight junction structure and epithelial barrier function. J Cell Sci 2003; 116(Pt 4):725-742.
77. Abe T, Sugano E, Saigo Y et al. Interleukin-1beta and barrier function of retinal pigment epithelial cells (ARPE-19): Aberrant expression of junctional complex molecules. Invest Ophthalmol Vis Sci 2003; 44(9):4097-4104.
78. Bruewer M, Luegering A, Kucharzik T et al. Proinflammatory cytokines disrupt epithelial barrier function by apoptosis-independent mechanisms. J Immunol 2003; 171(11):6164-6172.
79. Coyne CB, Vanhook MK, Gambling TM et al. Regulation of airway tight junctions by proinflammatory cytokines. Mol Biol Cell 2002; 13(9):3218-3234.
80. Wong V, Ching D, McCrea PD et al. Glucocorticoid Down-regulation of Fascin Protein Expression Is Required for the Steroid-induced Formation of Tight Junctions and Cell-Cell Interactions in Rat Mammary Epithelial Tumor Cells. J Biol Chem 1999; 274(9):5443-5453.

81. Grande M, Franzen A, Karlsson JO et al. Transforming growth factor-beta and epidermal growth factor synergistically stimulate epithelial to mesenchymal transition (EMT) through a MEK-dependent mechanism in primary cultured pig thyrocytes. J Cell Sci 2002; 115(Pt 22):4227-4236.

82. Lui WY, Wong CH, Mruk DD et al. TGF-beta3 regulates the blood-testis barrier dynamics via the p38 mitogen activated protein (MAP) kinase pathway: An in vivo study. Endocrinology 2003; 144(4):1139-1142.

83. Li D, Mrsny RJ. Oncogenic Raf-1 disrupts epithelial tight junctions via downregulation of occludin. J Cell Biol 2000; 148(4):791-800.

84. Chen Y-H, Lu Q, Schneeberger EE et al. Restoration of tight junction structure and barrier function by down-regulation of the mitogen-activated protein kinase pathway in Ras-transformed Madin-Darby Canine Kidney cells. Mol Biol Cell 2000; 11(3):849-862.

85. Kevil CG, Oshima T, Alexander JS. The role of p38 MAP kinase in hydrogen peroxide mediated endothelial solute permeability. Endothelium 2001; 8(2):107-116.

86. Zhang L, Deng M, Kao CW et al. MEK kinase 1 regulates c-Jun phosphorylation in the control of corneal morphogenesis. Mol Vis 2003; 9:584-593.

87. Klingler C, Kniesel U, Bamforth SD et al. Disruption of epithelial tight junctions is prevented by cyclic nucleotide-dependent protein kinase inhibitors. Histochem Cell Biol 2000; 113(5):349-361.

88. Marano CW, Garulacan LA, Ginanni N et al. Phorbol ester treatment increases paracellular permeability across IEC- 18 gastrointestinal epithelium in vitro. Dig Dis Sci 2001; 46(7):1490-1499.

89. Yoo J, Nichols A, Mammen J et al. Bryostatin-1 enhances barrier function in T84 epithelia through PKC-dependent regulation of tight junction proteins. Am J Physiol Cell Physiol 2003.

90. Nunbhakdi-Craig V, Machleidt T, Ogris E et al. Protein phosphatase 2A associates with and regulates atypical PKC and the epithelial tight junction complex. J Cell Biol 2002; 158(5):967-978.

91. Ward PD, Kline RR, Troutman MD et al. Phospholipase C-gamma modulates tight junction permeability through hyperphosphorylation of tight junction proteins. J Biol Chem 2002.

92. Jou TS, Nelson WJ. Effects of regulated expression of mutant RhoA and Rac1 small GTPases on the development of epithelial (MDCK) cell polarity. J Cell Biol 1998; 142(1):85-100.

93. Macek R, Swisshelm K, Kubbies M. Expression and function of tight junction associated molecules in human breast tumor cells is not affected by the Ras-MEK1 pathway. Cell Mol Biol (Noisy -le-grand) 2003; 49(1):1-11.

94. Saitou M, Furuse M, Sasaki H et al. Complex phenotype of mice lacking occludin, a component of tight junction strands. Mol Biol Cell 2000; 11(12):4131-4142.

95. Saitou M, Fujimoto K, Doi Y et al. Occludin-deficient embryonic stem cells can differentiate into polarized epithelial cells bearing tight junctions. J Cell Biol 1998; 141(2):397-408.

96. Papadopoulos MC, Saadoun S, Woodrow CJ et al. Occludin expression in microvessels of neoplastic and non-neoplastic human brain. Neuropathol Appl Neurobiol 2001; 27(5):384-395.

97. Yin F, Qiao T, Shi Y et al. In situ hybridization of tight junction molecule occludin mRNA in gastric cancer. Zhonghua Zhong Liu Za Zhi 2002; 24(6):557-560.

98. Davies DC. Blood-brain barrier breakdown in septic encephalopathy and brain tumours. J Anat 2002; 200(6):639-646.

99. Busch C, Hanssen TA, Wagener C et al. Down-regulation of CEACAM1 in human prostate cancer: Correlation with loss of cell polarity, increased proliferation rate, and Gleason grade 3 to 4 transition. Hum Pathol 2002; 33(3):290-298.

100. Tobioka H, Isomura H, Kokai Y et al. Occludin expression decreases with the progression of human endometrial carcinoma. Hum Pathol 2004; 35(2):159-164.

CHAPTER 3

Tight Junction Channels

James Melvin Anderson and Christina M. Van Itallie

Abstract

Transport across epithelia occurs through both transcellular and paracellular pathways. Transcellular movement of solutes and ions through transporters and channels is energy-dependent, directional, highly selective and regulated. Paracellular transport occurs through tight junctions, which behave like charge- and size-selective channels, but with a lower degree of selectivity than transcellular channels and lack their directional rectification. Until recently the molecular nature of tight junction channels was unknown. Now it appears that a large family of transmembrane proteins called claudins creates the tight junctions' selective barriers and channels. Manipulation of claudins in cell culture models and mice support this role, as do the phenotypes of disease-causing mutations in several human claudin genes. Claudins are related to other tetraspan proteins, including PMP-22, epithelial membrane proteins (EMPs) and the lens fiber protein MP20, whose diverse functions suggests tight junctions may also have a role in cell growth, apoptosis and cancer.

Tight junctions are continuous adhesive cell-cell contacts, which form selective barriers in the paracellular space between epithelial cells. They are not complete barriers but operate like channels, selectively limiting transepithelial movements of ions, solutes and water. Barrier properties vary among epithelia in terms of electrical conductance and size and charge selectivity and are regulated by both physiologic and pathophysiologic stimuli. Although the physiologic properties of paracellular transport were well described by the late 1960 to early 1970s,[1-4] the first major insight into the molecular basis for the barrier was made in 1998 with discovery of the claudins.[5] In keeping with the goals of this book, we will present the evidence that claudins create tight junction channels, including some of our own work, and speculate on the role of the extended claudin family. Other contributions in this book describe other transmembrane, cytosolic signaling and scaffolding proteins and the possible role of tight junctions in cell signaling, disease and cancer. Several excellent reviews are available describing molecular components of the junction,[6-8] its regulation[9-11] and pathophysiology.[12]

Channel-Like Properties of Tight Junctions

Paracellular transport through the tight junction is passive, driven by electro-osmotic gradients and shows identical selectivity and conductance in both the mucosal and serosal directions.[3] Solute and ion movements represent either "back leak" from gradients established by active transcellular transport or external gradients created for example in the gut by ingestion of a meal. The three major variable characteristics of tight junctions are electrical conductance, size and charge selectivity. Conductance is the most variable and ranges over five orders of magnitude between so-called "leaky" and "tight" epithelia. Leaky epithelia are defined as having half or more of their total conductance through the tight junction. They are found in epithelia like the small intestine, which move large volumes of isosmotic fluid.[3] "Tight" tight junctions are found where high electro-osmotic gradients are required, as in the distal nephron.[3] Although all tight

Tight Junctions, edited by Lorenza Gonzalez-Mariscal. ©2006 Landes Bioscience and Springer Science+Business Media.

junctions show charge selectivity, tissue specific differences are only meaningful in leaky epithelia, where there exists sufficient ion movement to make paracellular discrimination relevant. Experimentally, selectivity is most often measured as the permeability ratio for sodium to chloride (P_{Na+} / P_{Cl-}) and varies among tissues and cultured epithelial cell lines by about 30-fold;[13] most epithelia in the body are cation-selective.[3]

The tight junction behaves as a barrier perforated by relatively large aqueous pores capable of discriminating charge and size.[2,4,9] This is supported by the observation that the permeability for hydrophilic nonelectrolytes is inversely related to size, up to a cut-off characteristic for each tissue,[3] suggesting passage through restricted aqueous spaces. The range for size cut-off is relatively narrow among epithelia ($\approx$7-15 Å), while some endothelial tight junctions permit larger solutes (40-60 Å).[14] The permeability differences among the monovalent alkali-metal cations are small and such a low discrimination for similarly charged but differently sized cations is also consistent with a relatively large pore size.[3] Though larger and less discriminating, paracellular pores share with transmembrane ion channels an influence of ion concentration on permeability and competition between different transported molecules.[4]

The physical tight junction barrier occurs where continuous rows of transmembrane proteins from adjacent cells contact in the intercellular space (Fig. 1A,B). These proteins can be visualized by freeze-fracture electron microscopy as interconnected rows of particles; different numbers of rows and varying degrees of cross-bridging among rows are seen in different epithelia.[15] It was once thought the number and complexity of strand networks correlate with electrical resistance[15] but it now appears the protein composition of the strands is the most important determinant. The physiologic implications of strand complexity remain obscure.

The Claudin Family and Its Extended Relatives

A fundamental breakthrough in understanding the barrier structure came in 1998 when the group led by S. Tsukita isolated the first claudins, and showed that they both reconstitute strands when expressed in claudin-null fibroblasts[5] and confer cell-to-cell adhesion.[16]

Claudins are tetraspan proteins, ranging from 20-25kDa and are recognized by the WGLWCC motif of conserved amino acid residues in their first extracellular loops (Figs. 1, 2). There is no direct evidence the cysteines form a disulfide bond but it seems possible given the oxidizing extracellular environment and their total conservation among claudins. The first extracellular loops range from 41-55 aa residues, the second 10-21 aa residues and the cytoplasmic tails 21-44 aa residues. They all end in PDZ binding motifs, which bind PDZ domains in the cytoplasmic scaffolding proteins ZO-1 (2 and 3)[17] and MUPP1[18,19] and possibly other proteins. ZO-1 has three PDZ domains and MUPP1 thirteen, suggesting they can link several claudins together along the strand. There are no published data on the significance of the conserved GLW motif. Conceivably they participate in claudin-to-claudin adhesion since several claudins, which lack other common sequences are capable of heterotypic interactions.[16] Some mammalian claudins, for example claudin-1, appear rather ubiquitously expressed while others are restricted to specific cell types[20,21] or periods of development.[22,23] Two other distinct transmembrane proteins are present in mammalian tight junctions but neither is convincingly involved in forming the channel. Occludin,[24] was discovered before claudins. Occludin null mice have tight junctions, are viable and develop a complex set of phenotypes with age that are not easy to rationalize as barrier defects.[25] The second are the junction adhesion proteins (JAM 1,2,3). They are single-pass membrane proteins of the Ig-superfamily and may be involved in adhesion.[26]

Beyond the conventional claudins are a large number of structurally related proteins but whose potential functional similarities are largely unexplored. They can be found at the NCBI (National Center for Biotechnology Information) website designated as the pfam00822 protein family, http://www.ncbi.nlm.nih.gov/Structure/cdd/cddsrv.cgi?uid=1378. They too are about 20-25kD and share the tetraspan topology and WGLWCC signature motif residues (Figs. 1, 2) and a majority of pfam00822 members end in a PDZ binding motif (Fig. 1). Although originally described by nonbarrier properties and having low sequence identity, some the pfam00822 proteins appear to share barrier-forming functions with the conventional claudins.

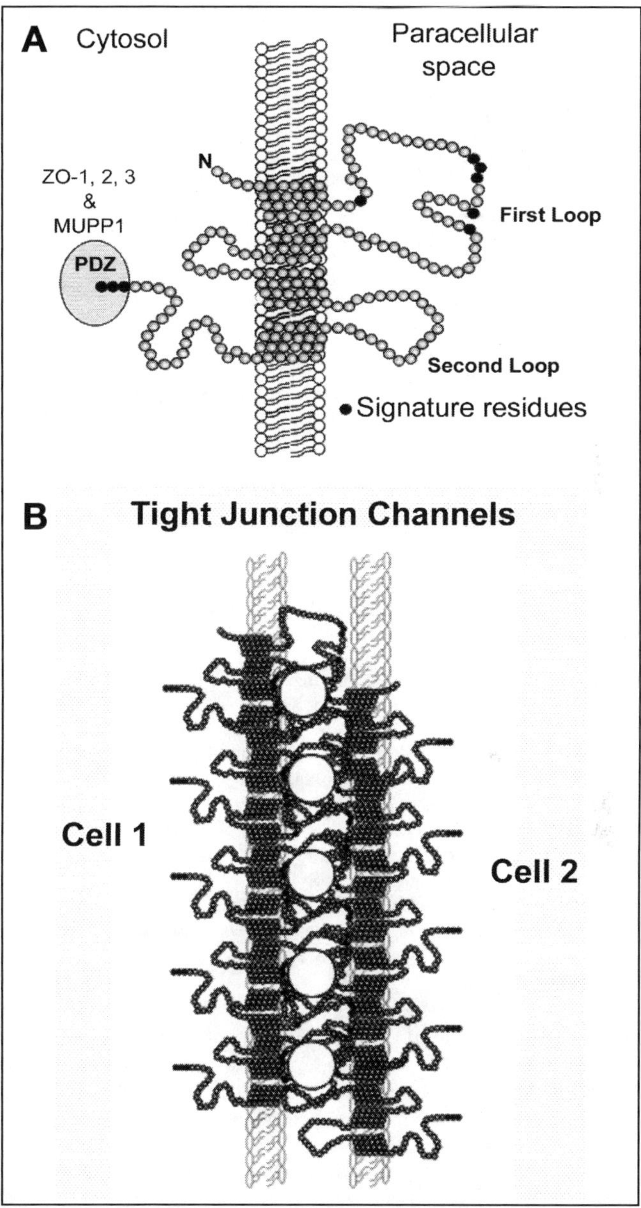

Figure 1. A) Molecular model of claudin. Claudins cross the member four times and have intracellular N- and C-termini. The first extracellular loops range from 41-55 aa residues and contain the signature WGLWCC motif (black circles). The second loops are smaller, 10-21 aa residues. The cytoplasmic tails are 21-44 aa residues and all end in a PDZ binding motif, which bind PDZ domains in the cytoplasmic scaffolding proteins ZO-1 (2 and 3) and MUPP1. This topology and the signature residues are shared by all members of the PMP-22/EMP/MP20/Claudin family, although not all have PDZ binding motifs. B) Hypothetical model depicting how claudins might form continuous barrier rows and adhere in the paracellular space between adjacent cells. The extracellular loops form aqueous channels, possibly of different sizes. The charged amino acid side chains lining the channels can influence conductance and charge selectivity.

```
A

Cln-1    PQWRIYS-----YAGDNIVTAQAMYEGLWMSCVSQS-TGQIQCKVF(14)
EMP22    NVWLVSN-----TVDAS--------VGLWKNCTNI------SCSDS(16)
PMP22    -QWIVGN----GHAT----------DLWQNCSTSSSGNVHHCFSS(10)
GAMMA5   DYWLYLEEGVIVPQNQSTEIKMSLHSGLWRVCFLAGEERGR-CFTI(24)
MP20     DHWMQYR---LSG-S----FA---HQGLWRYCLGN------KCFLQ(15)
CLP24    SNWVCQTL-----EDG-----RRRSVGLWRSCWLVD(14)HDCEAL(23)


B
                                  1         2         3
              +                   -         -         -
Cln-15 wt SYWRVSTVHGNVITTNTIF-ENLWFSCATDSLGVYNCWEF(14)
Cln-1     PQWRIYSYAGDNIVTAQAMYEGLWMSCVSQSTGQIQCKVF(14)


C
          -       -- -   + +       -       - + --
Cln-16  DCWMVN--DDSLEVSTKCR--GLWWECVTNAFDGIRTCDEY(15)
Cln-1   PQWRIYSYAGDNIVTAQAMYEGLWMSCVSQSTGQIQ-CKVF(14)
```

Figure 2. A) Sequence alignment of the first extracellular loop of members of the human PMP22/EMP22/ MP20/Claudin family. Signature residues are shaded, WGLWCC. Claudin-1 is shown as a representative of the orthodox claudins. EMP22 (epithelia membrane protein 22), PMP22 (peripheral myelin protein), MP20 (lens intrinsic membrane protein), the gamma-5 subunit of the neuronal voltage-gated Ca^{++} channel and CLP24 (claudin-like protein-24). The C-terminal most sequences are truncated by the number of residues shown in parentheses. B) First extracellular loop of claudin-15 showing signature residues (shaded) and charged residues (+ & -). Positions 1, 2 and 3 represent acidic residues mutated to basic residues and tested for effects on paracellular ionic charge selectivity of the channels. Positions 2 and 3, but not 1, were found to enhance cation permeability in the WT protein and reverse to anion selectivity when mutated to cationic residues. See text. C) Charged residues in the first extracellular loop of claudin-16. This loop has a high negative charge relative to other claudins. Mutations in human claudin-16 suggest it acts as a cation channel, perhaps consistent with the negative charges lining an aqueous channel and enhancing cation relative to anion permeability.

The most homologous to claudins are eye lens specific membrane proteins (MP20);[27] epithelial membrane proteins (EMP-1, -2, -3)[28] and peripheral myelin protein 22 (PMP22)[29] (Fig. 2A). MP20 is one of the most prevalent proteins of the lens fiber cells; human mutations in MP20 cause cataracts.[27] Although lens junctions lack the strands of tight junctions, the location of MP20 in the eye corresponds to the location of the paracellular diffusion barrier for dyes.[30] PMP22 is highly expressed in Schwann cell junctions, required for myelinization and studied for many years for its role in apoptosis and cell proliferation.[31] Mutations in PMP22 account for two types of peripheral polyneuropathies[32] (Table 1). While only 19% identical to human claudin-1 it was recently shown to be a component of tight junctions in liver, intestine[29] and the blood–brain-barrier[33] and could arguably be designated a claudin. CLP24 was recently cloned as a transcript upregulated by hypoxia. Although only 8% identical to claudin-1, when expressed in cultured MDCK cells it increases paracellular flux of tracer molecules and localizes to the apical junction complex.[34] Interestingly, it colocalizes at the adherens junction but not tight junction. The relationship of EMPs to claudins remains unresolved. Their subcellular location have not been reported and the focus of research has been on their role in apoptosis.[35] They were recently described as binding partners for the purineurgic

Table 1. *Human hereditary diseases*

Gene	Disease	Pathology/Mechanism	Reference
Tight junctions			
Cln-14	Deafness, *DFNB29*	Cochlear hair cell degeneration	Wilcox[51]
Cln-16	Hypomagnesemia hypercalciuria with nephrocalcinosis	Defective renal Mg++ reabsorption	Simon[20]
	Idiopathic hypercalciuria	Decrease ZO-1 binding	Muller[52]
ZO-2 (TJ2)	Familial hypercholanemia	Reduced claudin binding	Carlton[53]
Pfam00822 family members			
PMP22	Peripheral polyneuropathies Charcot-Maire-Tooth Type 1A Dejerine-Sottas syndrome	Defective myelinization	Brancolini[32]
MP20	Cataracts	?	Pras[27]

$P2X_7$ ATP-gated family of ion channels and required for ATP to induce apoptosis in a culture cell model.[36] Very distant members of the pfam00822 are gamma subunits of voltage-dependent calcium channels. They are larger but share the topology and signature residues (Fig. 2). Gamma subunits bind pore-forming alpha subunits and are required for proper membrane delivery of alpha.[37] A quick database search reveals even more claudin-like proteins and it will be interesting to see how far their functional similarities extend.

Claudin gene families are found across a wide range of multicellular animals. The human and mouse genomes contain at least twenty four.[38] The puffer fish *Takifugu* contains fifty six claudin genes, although this animal has other examples of gene expansions that lack obvious biologic significance.[39] The barrier-forming junctions of invertebrates, septate junctions, are structurally quite different from tight junctions, thus it is surprising to find claudin-like proteins in both flies and worms and learn they are required to form epithelial barriers. *Drosophila* has six claudin sequences.[40,41,42] Two of them, Megatrachea (*Mega*)[40] and Sinuous (*Sinu*)[42] are expressed at the barrier-forming septate junctions; mutations in *Mega* disrupt the intercellular barrier. Mutations of either result in developmental defects in the size and shape of the tracheal epithelium. While only 16% identical to human cln-1 the fly proteins contain the conserved WGLWCC motif. Five claudin-like sequences have been identified in *C. elegans*[43] and mutation of claudin-like protein 3 results in disruption of the barrier between epithelial cells of the intestine.

Lessons from Cell Culture Models

Many lines of evidence point to claudins as the basis for the selective size, charge and conductance properties of the paracellular pathway. This has been most directly tested by introducing selected claudins in cultured epithelial cell models, typically the renal tubule lines MDCK and LLC-PK$_1$ cells. A drawback of this approach is that cultured epithelial cells already express a background of multiple claudins. Thus, one cannot define a unit of activity but can only document changes from the background, for example, does monolayer conductance go up or down, or selectivity change for cations or anions. Without the equivalent of patch clamping, this will continue to be a limitation in the field of tight junction channels.

The first such experiments were conducted in very low resistance MDCK type II cells. Expression of claudin-1 resulted in an increase in transepithelial electrical resistance,[44,45] which was interpreted as due to high cell-cell adhesion. In contrast, expression of claudin-2 in high resistance MDCK type I monolayers made them leakier.[46] Rather than propose that claudin-2 forms leaky channel the authors offered the alternative explanation that claudin-2 could not pair with existing claudins and created breaks in the seal.

Transepithelial resistance is experimentally measured as the impediment to charged solutes moving through the junction under an imposed electrical field. Thus, the changes following expression of claudin-1 and-2 might be due to their contributing different ion permeabilities from the background, not that they close or open the paracellular space, respectively. Our group began to test this model by measuring the separate permeabilities for Na^+ and Cl^- following expression of claudin-4 in MDCK cells. At baseline these cell are very low resistance and have a higher Na^+ than Cl^- permeability. Expression under a regulated promoter induced a dose-dependent increase in resistance that was completely explained by discrimination against cations.[47] These results suggested that claudin-4 creates channels that discriminate against cations. Subsequently, similar discrimination against cations relative to anions was observed for claudin-8[19,48] and the opposite charge selectivity was demonstrated for claudin-2.[49] These results support a model where claudins create charge-selective channels.

We reasoned that charges on the extracellular loops of claudins might line the aqueous pores and electrostatically influence passage of soluble ions. To test this directly, we created a series of charge reversing mutations in the first extracellular loop of claudin-15 and expressed them in MDCK cells.[50] Positions are shown in Figure 2B. Claudin-15 has three negative residues in the first loop, in addition to one positive charge that is relatively conserved among claudins. Consistent with the model, expression of claudin-15 increases the permeability for cations and decreases that for anions. Reversal of the charge at position 1 had no effect on selectivity. Mutation of either position 2 or 3 decreased cation permeability and a double mutation gave an additive discrimination, turning a cation selective into an anion selective epithelium.

Collectively, these results support the charged channel model. Further indirect support is provided by the phenotype resulting from human mutations in claudin-16 (paracellin-1). Homozygous null individuals have reduced cation permeability through tight junctions of the thick ascending loop of Henle,[20] suggesting claudin-16 is a cation channel. Coincidently, claudin-16 has many negative residues in the first loop (Fig. 2C). Models of how pores might be organized within strands are well reviewed by Yu.[9]

Lessons from Heritable Human Diseases

Presently, human diseases are known to result from mutations in genes encoding claudin-14[51] and 16[20,52] and the cytoplasmic claudin-binding scaffold ZO-2,[53] but it is likely that more will be described in the future.

The claudin-16 mutations are the easiest to rationalize as channel defects. Claudin-16 is largely restricted to the kidney, where it is found in tight junctions in the thick ascending loop of Henle.[20] This segment of the nephron develops an intraluminal positive electrical potential, which drives cations, such as Mg^{++} back through the tight junction to the blood. Total body Mg^{++} homeostasis is largely controlled by passive electro-diffusion at this site. In 1998, positional cloning identified claudin-16 (paracellin-1) as the gene responsible for a rare Mg^{++} wasting disease called familial hypomagnesaemia with hypercalciuria and nephrocalcinosis (FHHN).[20] It presents in childhood with weakness and seizures due to low blood Mg^{++} levels and later progresses to renal failure because of interstitial calcium deposition. Urinary Mg^{++} and Ca^{++} losses appear to be due to inability of tight junctions in the TALH to allow paracellular reabsorption of cations from the tubule. This is consistent with the charge-selective channel model. Coincidently claudin-16 has a large number of negative residues in its extracellular loops, compatible with forming a cation-selective channel (Fig. 2C).

A distinct form of abnormal renal calcium loss, grouped under idiopathic hypercalcuria, was recently shown to result from novel mutations in the claudin-16 gene, which interrupts binding to the PDZ domain of ZO-1.[52] In contrast to the symptoms of FHHN describe above, the patients display self-limited childhood hypercalciuria. When the mutant protein was tested in cultured cells it no longer localized to tight junctions but accumulated in lysosomes. The reason for distinct clinical syndromes resulting from null versus targeting-defective mutants is not clear.

Claudin-14 mutations are more difficult to rationalize as channel defects. They were first identified as the cause of nonsyndromic recessive deafness DFNB29 in two large consanguineous Pakistani families.[51] Immunolocalization demonstrated claudin-14 in tight junctions of the sensory epithelium of the organ of Corti. It was initially assumed claudin-14 serves as a paracellular barrier to maintain the very high extracellular K^+ concentration found in the endolymph compartment, and which is required for hair cell depolarization following mechanical stimulation. Loss of the ion gradient due to potassium channel mutations is known to cause deafness.[54] Consistent with this proposed function, when expressed in MDCK cells claudin-14 forms very tight junctions of low cation permeability.[55] However, claudin-14-null mice have a normal endocochlear potential and become deaf because of degeneration of cochlear outer hair cells.[55] It remains unresolved whether cell degeneration is secondary to a barrier defect.

Human tight junction diseases have now extended beyond the claudins to the cytoplasmic scaffolding protein ZO-2. A point mutation in the PDZ domain of ZO-2 is the basis of Amish familial hypercholanemia;[52] a bile secretory defect characterized by elevated serum bile acid concentrations, itching, and fat malabsorption. This mutation disables claudin binding to the PDZ domain. Conceivably interruption of this connection disrupts the barrier and allows back-leak of bile through the tight junction with accumulation of serum bile acids. Morphologic changes are observed in hepatocyte tight junctions, although these are very nonspecific and seen in all forms of liver injury. Mutations in a bile acid-conjugating enzyme enhance the penetrance of disease in patients homozygous recessive for the ZO-2 mutations. This would be the first example of a polygenic human disease in which tight junctions are involved. It seems likely we will see other tight junction diseases influencing transport and drug pharmacokinetics.

Lessons from Mutant Mice

Several informative mouse mutants have been generated to investigate the physiologic role of specific claudins or to provide mouse models for human disease (Tables 1, 2). Claudin-1 is prevalent in junctions of the skin. When deleted by homologous recombination, homozygous mice die shortly after birth, apparently from rapid evaporative water loss from the skin.[56] Additional evidence that claudins participate in skin barriers comes from experiments where claudin-6 was overexpressed under a skin-specific promoter. Presumably through a dominant negative mechanism this too results in a similar barrier defect, dehydration and death.[57] Claudin-14 null mice phenocopy the human deafness syndrome DFNB29 and should provide a useful model to study the pathophysiologic mechanism.[55]

Tight junction-like structures were documented long ago between glial cells, where they wrap around axons. They were assumed to represent ion barriers required to insulate the axon and allow rapid saltatory conduction of the action potential. This function is served by oligodendrocytes in the central nervous system and Schwann cells in peripheral neurons. Claudin-11 (originally called oligodendrocyte-specific protein, OSP) is concentrated along these sealing junctions in the central nervous system. Consistent with a barrier forming role, claudin-11 null mice show delayed conduction velocity and absence of the junctional membrane particle rows characteristically seen in freeze-fracture electron micrographic images.[21] This was the first evidence that a claudin creates a barrier in a nonepithelial cell type. The claudin-like protein PMP22 (Fig. 2) appears to provide an analogous barrier in the peripheral nerves. Human mutations in PMP22 lead to demyelinating polyneuropathies[32] and PMP22 null mice phenocopy the human disease.[58]

The most provocative knock out so far is that of claudin-5, which provides the first evidence that claudins influence size selectivity.[59] Endothelial tight junctions in the brain express at least claudins 5 and 12. They are among the tightest endothelia in the body and the structural basis of the blood-brain-barrier. To test the role of claudins in the blood-brain-barrier Nitta et al[59] created claudin-5 null mice. They are born with normal appearing tight junctions in brain endothelia, but they die within several hours. When the vascular space was perfused with a range of size markers, the junctions in vessels from wild-type animals retained markers

Table 2. Mouse claudin knock-outs and transgenics

Gene		Phenotype	Reference
Cln-1	ko	skin barrier defect	Furuse et al[56]
Cln-5	ko	size-selective BBB defect	Nitta et al[59]
Cln-6	tg	skin barrier defect	Turksen[57]
Cln-11	ko	CNS myelin defect	Gow[21]
Cln-14	ko	phenocopy of human	Ben-Yosef[55]
Pfam00822 transgenic			
PMP22	tg	phenocopies Charcot-Maire-Tooth	Robaglia-Schlupp[58]

ranging from albumin (68 kDa) down to the Hoechst dye 33258 (562 Da). Intriguingly, the endothelia of -/- mice become leaky to the 562Da marker yet still restrict the next largest (1,862 Da) marker. A role in size-selectivity is supported by recent work of McLaughlin et al who were investigating the mechanism by which the mycotoxin ochratoxin A alters the intestinal barrier.[60] Application of toxin to a cultured intestinal cell line induced selective removal of claudins 3 and 4, but not 1, and increased permeability for small but not larger fluorescently-labeled dextrans. While these studies reveal a role for claudins in size discrimination the molecular mechanism remains unclear.

Conclusions

A growing body of evidence supports a role for claudins in forming tight junction channels. A tentative model proposes that claudins form rows of adhesive contacts sealing the paracellular and creating channels through their extracellular loops (Fig. 1B). Among the unanswered questions, do other proteins participate in forming channels and are the channels acutely regulated? This information can be used to better understand normal paracellular transport, changes in disease and manipulation of the channels to improve therapeutic drug delivery.

Acknowledgements

The authors thank the State of North Carolina, the Broad Medical Research Program of the Eli and Edythe L. Broad Foundation and the National Institutes of Health (DK61397, DK45134) for support of their research.

References

1. Berry CA, Boulpaep EL. Nonelectrolyte permeability of the paracellular pathway in necturus proximal tubule. Am J Physiol 1975; 228(2):581-595.
2. Diamond JM. Channels in epithelial cell membranes and junctions. Fed Proc 1978; 37:2639-2644.
3. Powell DW. Barrier function of epithelia. Am J Physiol 1981; 241(4):G275-G288.
4. Tang VW, Goodenough DA. Paracellular ion channel at the tight junction. Biophys J 2003; 84(3):1660-1673.
5. Furuse M, Fujita K, Hiiragi T et al. Claudin-1 and -2: Novel integral membrane proteins localizing at tight junctions with no sequence similarity to occludin. J Cell Biol 1998; 141(7):1539-1550.
6. Gonzalez-Mariscal L, Betanzos A, Nava P et al. Tight junction proteins. Prog Biophys Mol Biol 2003; 81(1):1-44.
7. D'Atri F, Citi S. Molecular complexity of vertebrate tight junctions (Review). Mol Membr Biol 2002; 19(2):103-112.
8. Tsukita S, Furuse M. Claudin-based barrier in simple and stratified cellular sheets. Curr Opin Cell Biol 2002; 14(5):531-536.
9. Yu AS. Claudins and epithelial paracellular transport: The end of the beginning. Curr Opin Nephrol Hypertens 2003; 12(5):503-509.
10. Balda MS, Matter K. Epithelial cell adhesion and the regulation of gene expression. Trends Cell Biol 2003; 13(6):310-318.

11. Nusrat A, Chen JA, Foley CS et al. The coiled-coil domain of occludin can act to organize structural and functional elements of the epithelial tight junction. J Biol Chem 2000; 275(38):29816-29822.
12. Nusrat A, Turner JR, Madara JL. Molecular physiology and pathophysiology of tight junctions. IV. Regulation of tight junctions by extracellular stimuli: Nutrients, cytokines, and immune cells. Am J Physiol Gastrointest Liver Physiol 2000; 279(5):G851-G857.
13. Van Itallie CM, Anderson JM. The role of claudins in determining paracellular charge selectivity. Proc Amer Thoracic Soc 2004; 1:38-41.
14. Firth JA. Endothelial barriers: From hypothetical pores to membrane proteins. J Anatomy 2002; 200(6):541-548.
15. Claude P. Morphological factors influencing transepithelial permeability: A model for the resistance of the zonula occludens. J Membr Biol 1978; 39(2-3):219-232.
16. Furuse M, Sasaki H, Tsukita S. Manner of interaction of heterogeneous claudin species within and between tight junction strands. J Cell Biol 1999; 147(4):891-903.
17. Itoh M, Furuse M, Morita K et al. Direct binding of three tight junction-associated MAGUKs, ZO-1, ZO-2 and ZO-3, with the COOH termini of claudins. J Cell Biol 1999; 147(6):1351-1363.
18. Hamazaki Y, Itoh M, Sasaki H et al. Multi-PDZ domain protein 1 (MUPP1) is concentrated at tight junctions through its possible interaction with claudin-1 and junctional adhesion molecule. J Biol Chem 2002; 277(1):455-461.
19. Jeansonne B, Lu Q, Goodenough DA et al. Claudin-8 interacts with multi-PDZ domain protein 1 (MUPP1) and reduces paracellular conductance in epithelial cells. Cell Mol Biol (Noisy.-le-grand) 2003; 49(1):13-21.
20. Simon DB, Lu Y, Choate KA et al. Paracellin-1, a renal tight junction protein required for paracellular Mg2+ resorption. Science 1999; 285(5424):103-106.
21. Gow A, Southwood CM, Li JS et al. CNS myelin and sertoli cell tight junction strands are absent in osp/claudin 11-null mice. J Neurochem 2000; 74:S35-S35.
22. Fukuhara A, Shimizu K, Kawakatsu T et al. Involvement of nectin-activated Cdc42 small G protein in organization of adherens and tight junctions in Madin-Darby canine kidney cells. J Biol Chem 2003; 278(51):51885-51893.
23. Turksen K, Troy TC. Claudin-6: A novel tight junction molecule is developmentally regulated in mouse embryonic epithelium. Devel Dynamics 2001; 222(2):292-300.
24. Furuse M, Hirase T, Itoh M et al. Occludin - A novel integral membrane-protein localizing at tight junctions. J Cell Biol 1993; 123(6):1777-1788.
25. Saitou M, Furuse M, Sasaki H et al. Complex phenotype of mice lacking occludin, a component of tight junction strands. Mol Biol Cell 2000; 11(12):4131-4142.
26. Bazzoni G. The JAM family of junctional adhesion molecules. Curr Opin Cell Biol 2003; 15(5):525-530.
27. Steele EC, Lyon MF, Favor J et al. A mutation in the connexin 50 (Cx50) gene is a candidate for the No 2 mouse cataract. Curr Eye Res 1998; 17(9):883-889.
28. Jetten AM, Suter U. The peripheral myelin protein 22 and epithelial membrane protein family. Prog Nucleic Acid Res Mol Biol 2000; 64:6497-129.
29. Notterpek L, Roux KJ, Amici SA et al. Peripheral myelin protein 22 is a constituent of intercellular junctions in epithelia. Proc Natl Acad Sci USA 2001; 98(25):14404-14409.
30. Grey AC, Gonen T, Jacobs MD et al. Restricted extracellular fluxes in the lens: Roles for MP20 and MIP. Invest Ophthalmol Vis Sci 2003; 44:U501-U501.
31. Schneider C, King RM, Philipson L. Genes specifically expressed at growth arrest of mammalian-cells. Cell 1988; 54(6):787-793.
32. Brancolini C, Edomi P, Marzinotto S et al. Exposure at the cell surface is required for Gas3/PMP22 to regulate both cell death and cell spreading: Implication for the Charcot-Marie-Tooth type 1A and Dejerine-Sottas diseases. Mol Biol Cell 2000; 11(9):2901-2914.
33. Roux KJ, Amici SA, Notterpek L. The temporospatial expression of peripheral myelin protein 22 at the developing blood-nerve and blood-brain barriers. J Comp Neurol 2004; 474(4):578-588.
34. Kearsey J, Petit S, De Oliveira C et al. A novel four transmembrane spanning protein, CLP24. Eur J Biochem 2004; 271(13):2584-2592.
35. Taylor V, Suter U. Epithelial membrane protein-2 and epithelial membrane protein-3: Two novel members of the peripheral myelin protein 22 gene family. Gene 1996; 175(1-2):115-120.
36. Wilson HL, Wilson SA, Surprenant A et al. Epithelial membrane proteins induce membrane blebbing and interact with the P2X(7) receptor C terminus. J Biol Chem 2002; 277(37):34017-34023.
37. Tomita S, Fukata M, Nicoll RA et al. Dynamic interaction of stargazin-like TARPs with cycling AMPA receptors at synapses. Science 2004; 303(5663):1508-1511.
38. Tepass U. Claudin complexities at the apical junctional complex. Nat Cell Biol 2003; 5(7):595-597.

39. Loh YH, Christoffels A, Brenner S et al. Extensive expansion of the claudin gene family in the teleost fish, fugu rubripes. Gen Res 2004; 14(7):1248-1257.
40. Behr M, Riedel D, Schuh R. The claudin-like megatrachea is essential in septate junctions for the epithelial barrier function in Drosophila. Dev Cell 2003; 5(4):611-620.
41. Tepass U, Tanentzapf G, Ward R et al. Epithelial cell polarity and cell junctions in Drosophila. Annu Rev Genet 2001; 35:747-784.
42. Wu VM, Schulte J, Hirschi A et al. Sinuous is a Drosophila claudin required for septate junction organization and epithelial tube size control. J Cell Biol 2004; in press.
43. Asano A, Asano K, Sasaki H et al. Claudins in Caenorhabditis elegans: Their distribution and barrier function in the epithelium. Curr Biol 2003; 13(12):1042-1046.
44. McCarthy KM, Francis SA, McCormack JM et al. Inducible expression of claudin-1-myc but not occludin-VSV-G results in aberrant tight junction strand formation in MDCK cells. J Cell Sci 2000; 113(19):3387-3398.
45. Inai T, Kobayashi J, Shibata Y. Claudin-1 contributes to the epithelial barrier function in MDCK cells. Eur J Cell Biol 1999; 78(12):849-855.
46. Furuse M, Furuse K, Sasaki H et al. Conversion of Zonulae Occludentes from tight to leaky strand type by introducing claudin-2 into Madin-Darby canine kidney I cells. J Cell Biol 2001; 153(2):263-272.
47. Van Itallie C, Rahner C, Anderson JM. Regulated expression of claudin-4 decreases paracellular conductance through a selective decrease in sodium permeability. J Clin Invest 2001; 107(10):1319-1327.
48. Yu AS, Enck AH, Lencer WI et al. Claudin-8 expression in Madin-Darby canine kidney cells augments the paracellular barrier to cation permeation. J Biol Chem 2003; 278(19):17350-17359.
49. Amasheh S, Meiri N, Gitter AH et al. Claudin-2 forms cation-selective pores in tight junctions of epithelial cells. Gastroenterology 2002; 122(4):A18.
50. Colegio OR, Van Itallie CM, Mccrea HJ et al. Claudins create charge-selective channels in the paracellular pathway between epithelial cells. Am J Physiol-Cell Physiol 2002; 283(1):C142-C147.
51. Wilcox ER, Burton QL, Naz S et al. Mutations in the gene encoding tight junction claudin-14 cause autosomal recessive deafness DFNB29. Cell 2001; 104(1):165-172.
52. Muller D, Kausalya PJ, Claverie-Martin F et al. A novel claudin 16 mutation associated with childhood hypercalciuria abolishes binding to ZO-1 and results in lysosomal mistargeting. Am J Hum Genet 2003; 73(6):1293-1301.
53. Carlton VE, Harris BZ, Puffenberger EG et al. Complex inheritance of familial hypercholanemia with associated mutations in TJP2 and BAAT. Nat Genet 2003; 34(1):91-96.
54. Coucke PJ, Van Hauwe P, Kelley PM et al. Mutations in the KCNQ4 gene are responsible for autosomal dominant deafness in four DFNA2 families. Hum Mol Genet 1999; 8(7):1321-1328.
55. Ben Yosef T, Belyantseva IA, Saunders TL et al. Claudin 14 knockout mice, a model for autosomal recessive deafness DFNB29, are deaf due to cochlear hair cell degeneration. Hum Mol Genet 2003; 12(16):2049-2061.
56. Furuse M, Hata M, Tsukita S. Mice lacking claudin-1, a constituent of tight junction strands. Mol Biol Cell 2001; 12:133A-134A.
57. Turksen K, Troy T. Permeability barrier dysfunction in transgenic mice overexpressing Claudin-6. J Invest Dermatol 2002; 119(1):259-259.
58. Robaglia-Schlupp A, Pizant J, Norreel JC et al. PMP22 overexpression causes dysmyelination in mice. Brain 2002; 125:2213-2221.
59. Nitta T, Hata M, Gotoh S et al. Size-selective loosening of the blood-brain barrier in claudin-5-deficient mice. J Cell Biol 2003; 161(3):653-660.
60. McLaughlin J, Padfield PJ, Burt JP et al. Ochratoxin a increases permeability through tight junctions by removal of specific claudin isoforms. Am J Physiol Cell 2004; [Epub ahead of print].

CHAPTER 4

JAM Family Proteins:
Tight Junction Proteins That Belong to the Immunoglobulin Superfamily

Susumu Hirabayashi and Yutaka Hata

Abstract

Junctional adhesion molecules (JAMs) are found at the tight junctions (TJs) of polarized epithelial and endothelial cells. JAMs are characterized by the molecular structure composed of two extracellular immunoglobulin loops, a single transmembrane domain, and a cytoplasmic domain with a PDZ binding motif. JAMs interact with several ligands in both homophilic and heterophilic manners by the extracellular domain, and bind to various PDZ domain-containing proteins at the C-terminus. Through these protein-protein interactions, JAMs are implicated in diverse biological functions at TJs, including cell-cell adhesion, junctional assembly, junctional stabilization, regulation of paracellular permeability and leucocyte transmigration.

Introduction

Recent intensive studies have revealed the molecular organization of tight junctions (TJs) of epithelial and endothelial cells. Like other cell junctions, TJs comprise integral membrane proteins, membrane-associated proteins that interact with membrane proteins and link them to the cytoskeleton, and cell signaling molecules. As integral membrane proteins, three classes of proteins have been identified. All of them are a kind of cell adhesion proteins, and so far, no receptors have been reported for TJs. The first and second integral membrane proteins are occludin and claudin. Both proteins have four transmembrane domains and two extracellular loops, and are involved directly in the formation of TJ strands.[1] The characters and roles of these proteins will be discussed in other chapters of this book. The third group of membrane proteins are members of immunoglobulin superfamily (IgSF) and subjects of this chapter.

IgSF proteins are one of the largest families of proteins. Members of this family are defined by the presence of one or more copies of the immunoglobulin (Ig)-domain with a compact structure with two cysteine residues separated by 55-75 amino acids.[2] Ig-domains function as modules for protein-protein interactions and are detected in various proteins. According to the recent estimation based on the human genome project, 765 human genes encode Ig-domains.[3] IgSF proteins are conserved through evolution and can be found in flies and worms. Typical IgSF cell adhesion molecules have a large N-terminal extracellular region containing Ig-domains, a single transmembrane region, and a cytoplasmic tail.[4,5] In 1998, the first IgSF protein was identified at TJs of both epithelial and endothelial cells and was named junctional adhesion molecule (JAM).[6] Since then, several JAM-related molecules have been reported and now form a subfamily in the IgSF. In contrast to occludin and claudin, JAM-related proteins do not constitute TJ strands.[7] This raises the possibility that JAM-related proteins have distinct biological roles from occludin and claudin. In this review, we will describe JAM-related proteins as JAM family. As we will discuss below, JAM family is composed of two subclasses of proteins based on the structures. The first group contains three original members and the second group is composed of coxsackie and adenovirus receptor (CAR) and its related proteins. Although our knowledge about the functions of JAM family proteins is still fragmentary, the accumulating information suggests that JAM family proteins play roles in diverse range of biological phenomenon at TJs.

General Features of JAM Family Proteins

Nomenclature

JAM family proteins were reported from independent groups and furthermore, human and mouse homologues were cloned for each JAM family protein and named, resulting in the confused nomenclature. Original members of JAM family proteins were renamed in 2003.[8] Table 1 summarizes the names of JAM-related proteins.

Common Structures

JAM family proteins are characterized by two extracellular Ig-domains, a single transmembrane region, and a cytoplasmic region (Fig. 1). These characters are also recognized in various IgSF proteins expressed in lymphocytes and leukocytes such as CD2, CD48, LFA-3, CD80, and CD86.[9] JAM family proteins are distinct from these blood cell markers in that they have a canonical PDZ-binding motif. Thereby, we may be able to designate IgSF proteins with two extracellular Ig-domains and the C-terminal PDZ-binding motif as JAM family proteins. According to this criteria, JAM-A, JAM-B, JAM-C, CAR, endothelial cell-selective adhesion

Table 1. Nomenclature of published JAM family proteins

Name	Year	Original Designation	Species	Reference
JAM-A	1998	JAM	mouse	6
	1998	106 antigen	mouse	68
	1999	JAM	human	58
	2000	F11 R	human	62
	2001	JAM-1	human	69
JAM-B	2000	VE-JAM	mouse	19
	2000	VE-JAM	human	19
	2000	JAM2	human	28
	2000	JAM-3	mouse	70
JAM-C	2001	JAM-2	mouse	21
	2001	JAM3	human	33
	2002	JAM-2	human	57
CAR	1997	CAR	human	71
	1997	HCAR	human	72
	1997	MCAR	mouse	72
ESAM	2001	ESAM	mouse	29
	2001	ESAM	human	29
JAM4	2003	JAM4	mouse	17
BT-IgSF	2002	BT-IgSF	human	13
	2002	BT-IgSF	mouse	13
CLMP	2003	CLMP	human	14
	2003	CLMP	mouse	14

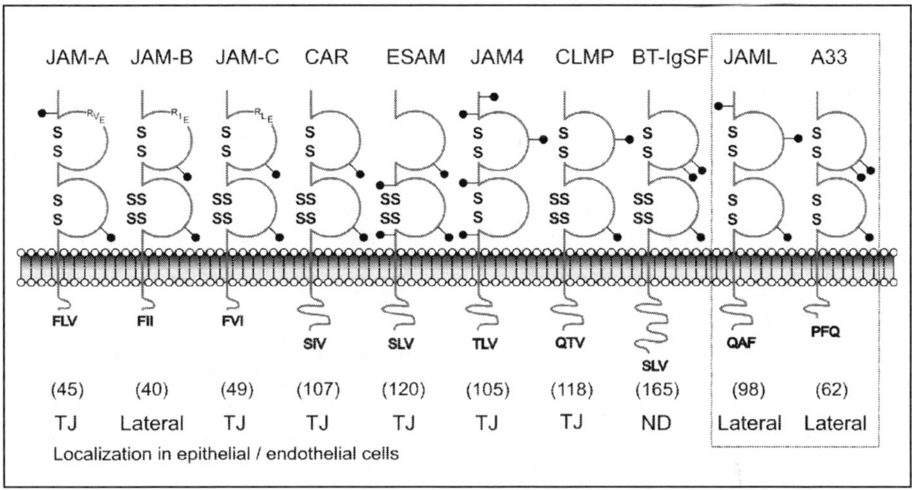

Figure 1. Structural model of JAM family proteins. The extracellular region contains two immunoglobulin domains with intramolecular disulfide bonds. R (V, I, L) E dimerization motif at the membrane-distal Ig-domain is indicated. Putative N-linked glycosylation sites are depicted by dots. The C-terminal three amino-acid residues are shown by single-letter codes. The numbers of amino-acid residues of the cytoplasmic regions are indicated in brackets. Localization of each molecule in epithelial or endotheial cells is shown at the bottom. TJ, tight junction; ND, not determined; and Lateral, lateral cell borders.

molecule (ESAM), and JAM4 are counted as JAM family proteins. Junctional adhesion molecule-like protein (JAML) is expressed in granulocytes and lacks the PDZ-binding motif.[10] We will omit this protein from the family, although it was reported as JAM-like protein. A33 is related in the sequence to JAM family proteins but is devoid of the PDZ-binding motif.[11,12] Because A33 is detected in epithelial cells and is accumulated at cell-cell contacts, it may be a member of JAM family, but we do not discuss A33 in this review. Brain and testis-specific immunoglobulin superfamily (BT-IgSF) fits to the criteria and should be considered as a member.[13] Coxsackie and adenovirus receptor-like membrane protein (CLMP), which has been very recently reported, has QTV (single letters indicate amino acid residues) at the C-terminus.[14] This sequence does not fully match the consensus sequence of the PDZ-binding motif. However, the last valine residue that is important to bind a PDZ domain is conserved and CLMP may interact with PDZ-containing proteins. Furthermore, CLMP has been clearly shown to localize at TJs.[14] For these reasons, we tentatively add CLMP to the JAM family proteins.

As described later, crystal structures have been solved for the Ig-domains of JAM-A and CAR.[15,16] JAM-A was first reported to harbor two tandem V-type Ig-domains.[6] The crystal structure, however, revealed that the membrane-proximal Ig-domain of JAM-A is not a V-type Ig-domain.[15] The membrane-distal Ig-domain of CAR, termed CAR-D1, was confirmed to be V-type.[16] The membrane-proximal Ig-domain of CAR is regarded as C2-type based on the sequence. Similarly, membrane-distal and membrane-proximal loops of JAM-B, JAM-C, ESAM, BT-IgSF, and CLMP are assigned to be V-type and C2-type, respectively. The membrane-distal Ig-domain of JAM4 is V-type, whereas the membrane-proximal Ig-domain cannot be specified as C2-type. Another common feature for JAM family proteins is N-glycosylation. The extracellular domain has two or more potential N-glycosylation sites, and the direct evidence of N-glycosylation is reported for BT-IgSF and JAM4.[13,17] Because glycosylation depends on cell types as shown for JAM4, the physiological significance of glycosylation for the function and localization of JAM family proteins is an intriguing question.

Subclasses

JAM family proteins are classified into two subgroups based on the type of PDZ-binding motif and the sequence similarity. The C-terminal amino acids of JAM-A, JAM-B, and JAM-C represent type II PDZ-binding motifs, while those of CAR, ESAM, JAM4, and BT-IgSF represent type I. The length of the cytoplasmic tail also differs between these two subgroups. JAM-A, JAM-B, and JAM-C have a short cytoplasmic tail of 45-50 residues, but CAR and its related proteins have a cytoplasmic domain with 105-165 residues. Therefore the two subgroups probably associate with different sets of molecules through the cytoplasmic domain and the C-terminal PDZ-binding motif.

Tissue and Subcellular Distribution

Table 2 summarizes the tissue distribution of JAM family proteins. JAM-A is found in endothelial cells of different origins and in a variety of polarized epithelial cells.[6] Immunoelectron microscopy reveals that JAM-A is concentrated at TJs of mouse small intestinal epithelium, while no specific staining is observed at adherens junction or desmosomes.[6] Furthermore, JAM-A is expressed at hematopoietic cells including monocytes, lymphocytes and red blood cells.[18] Immunohistochemistry indicates the expression of JAM-B at high endothelial venules of tonsil, lymph nodes, and in endothelium of arterioles.[19,20] JAM-C is also shown to be expressed at endothelial cells such as mesenteric lymph nodes.[21] There is no evidence that JAM-B and JAM-C are expressed in epithelial cells. Currently the ultrastructural analysis on JAM-B and JAM-C has not been addressed. However, since ectopic expression of JAM-C in epithelial cells results in colocalization with ZO-1, JAM-C is also suggested to localize at TJs in endothelial cells.[21] The subcellular localization of JAM-B is less clear, because JAM-B is not concentrated at TJs when expressed in epithelial cells.[22]

CAR is expressed at both epithelial and endothelial cells.[23,24] It has been reported to localize at TJs of well differentiated airway epithelial cells, human colonic cell line, and human umbilical vein endothelial cells. The expression of CAR in neurons is also reported.[25] ESAM is proposed to be selectively expressed on endothelial cells and associated with TJs in blood capillaries of mouse brain and skeletal muscle as judged by immunogold labeling.[26] When ESAM is expressed ectopically in epithelial cells, it localizes to TJs.[26] No evidence for the expression of JAM4 in endothelial cells has been obtained. The ectopic expression in Madine Darby canine kidney (MDCK) cells and the detection of the endogenous protein in mammary gland epithelial cells support the localization of JAM4 at TJs of epithelial cells.[17] Furthermore, the electron microscopical studies on rat kidney and mouse intestine indicate the localization of JAM4 at slit diaphragm and TJs.[17,27] However, JAM4 is also on apical membranes in several epithelial cells. The newest member, CLMP, is also localized at TJs in CACO2 and MDCK cells.[14] BT-IgSF is so far detected only in neurons and glias, and there is no evidence that BT-IgSF is localized at TJs.[13]

These different expression patterns of JAM family proteins may result from cell type specific expression of individual proteins and contribute to functional differences. The fact that all JAM family proteins except for JAM-B and BT-IgSF are recruited to TJs when ectopically expressed in polarized epithelial cells leads us to speculate that some molecule in epithelial cells recognizes the common structure of JAM family proteins and tethers them to TJs.

Extracellular Ligands

Homophilic Interactions

JAM family proteins mediate homophilic *trans*-interactions between adjacent opposing cells. JAM family proteins are accumulated at cell-cell contacts in various cells, supporting the homophilic binding.[6,14,17,21,24,26,28,29] However, it is not clear whether homophilic interactions of JAM-A, JAM-B, and JAM-C contribute to cell adhesion. In contrast, CAR, ESAM, JAM4, and CLMP have been shown to induce aggregation in fibroblasts and Chinese hamster ovary (CHO) cells, which implies the role in cell adhesion.[14,17,24,25,29]

Table 2. Expression of JAM family genes in mouse and human tissues

Name	Species	Brain	Heart	Lung	Liver	Spleen	Kidney	Intestine	Testis	Others	Reference
JAM-A	mouse	+	+	+	+	+	+	ND	+	Peyer patches, lymph nodes	6, 21
	human	−	+	+	+	ND	+	ND	ND	pancreas, placenta	69
JAM-B	mouse	ND	+	+	+	−	+	+	+	Peyer patches, lymph nodes	22
	human	−	+	−	−	−	−	−	ND	placenta	19, 28
JAM-C	mouse	ND	+	+	+	+	+	ND	+	Peyer patches, lymph nodes	21
	human	+	+	+	−	+	+	+	ND	placenta, thymus	33, 57
CAR	mouse	+	+	+	+	−	+	ND	−		72
	human	+	+	+	+	−	−	+	+	pancreas, prostate	72
ESAM	mouse	+	+	+	−	ND	+	ND	ND	skin	29
	human	ND	ND	ND	ND	ND	ND	ND	ND		
JAM4	mouse	−	−	+	+	−	+	+	−	skeletal muscle, stomach	17
	human	ND	ND	ND	ND	ND	ND	ND	ND		
BT-IgSF	mouse	ND	ND	ND	ND	ND	ND	ND	ND		
	human	+	−	−	−	−	+	−	+	adrenal gland	13
CLMP	mouse	+	+	+	+	+	+	ND	+		14
	human	−	+	+	+	+	+	+	ND	placenta	14

ND: not determined

Heterophilic Interactions

JAM family proteins interact with other ligands in a heterophilic manner. JAM-A binds to β2 integrin leukocyte function-associated antigen-1 (LFA-1, integrin αLβ2) through its membrane-proximal Ig-domain.[30] JAM-B and JAM-C bind T cell-associated integrin very large antigen-4 (VLA-4, integrin α4β1) and β2 integrin Mac-1 (integrin αMβ2), respectively.[31,32] These heterophilic interactions probably regulate leukocyte-endothelial cell and leukocyte-platelet interactions. It remains to be defined whether CAR, ESAM, JAM4, and CLMP interact with integrin family proteins. Heterophilic interaction in *trans* between JAM family proteins has been demonstrated for JAM-B and JAM-C.[33]

Crystal Structures of Extracellular Domains

Based on the crystal structure of the extracellular region of mouse JAM-A, a model for homophilic adhesion has been proposed.[34] In this model, two JAM-A molecules are connected by membrane-distal V-type Ig-domain and form U-shaped homodimers in *cis*. At the interface between V-type Ig-domains, two salt bridges are formed in a complementary manner by a dimerization motif R (V, I, L) E. This dimerization motif is essential for the formation of a JAM-A dimer in solution and is conserved in JAM-B and JAM-C, suggesting that JAM-B and JAM-C also form dimers with a similar mechanism. The analysis of the crystal structure of human JAM-A demonstrated several differences in the dimeric structure from that of mouse JAM-A.[15] Interestingly, the structure of human JAM-A dimer is close to that of CAR dimer. Because reovirus and adenovirus attachment proteins are similar, it is reasonable that the structures of human JAM-A (reovirus receptor) and CAR (adenovirus receptor) resemble to each other.

Intracellular Ligands

Since all members of the JAM family proteins have a PDZ-binding motif at the C-terminus, they interact with several PDZ proteins. To date, JAM-A has been reported to associate with five PDZ proteins, ZO-1, AF-6/afadin, CASK, PAR-3/ASIP, and MUPP1.[7,35-39] JAM-A binds to a protein without a PDZ domain named cingulin.[35] JAM-B and JAM-C also interacts with ZO-1 and PAR-3/ASIP.[40]

JAM family proteins that have type I PDZ-binding motif do not bind to ZO-1. ESAM has been shown not to interact with ZO-1, ZO-2, ZO-3, AF-6/afadin, or PAR-3/ASIP.[26] CAR interacts with neither ZO-1 nor PAR-3/ASIP.[40] Recently CAR has been reported to bind to Ligand-of-Numb protein-X (LNX) with four PDZ domains.[41] JAM4 interacts with MAGI-1, a membrane-associated guanylate kinase that specifically localizes to TJs.[17] JAM4 does not bind to ZO-1 or PAR-3/ASIP (unpublished observation).[17] Intracellular partners of ESAM and CLMP are currently unidentified.

Functions of JAM Family Proteins

Regulation of TJ Assembly

The most straightforward evidence of physiological functions of JAM family proteins may come from gene targeting mice. However, as we write this review, such a line of study is reported only for ESAM. Thereby, we have to speculate functions of JAM molecules based on the biochemical properties of the proteins and the experiments done at cellular level. Because JAM-A interacts with various proteins that may function as scaffolds at TJs, it is tempting to hypothesize that JAM-A is involved in TJ assembly. JAM-A binds to the third PDZ domain of ZO-1, whereas claudin and ZO-2 interact with the other PDZ domains and occludin binds the guanylate kinase domain.[42-45] Thereby, these proteins can interact with ZO-1 simultaneously and form a complex. Similarly, JAM-A and claudin bind to distinct PDZ domains of MUPP1.[39] Recent studies have revealed two macromolecular complexes that play essential roles in the polarity of epithelial cells. One is Crumbs-Pals-PATJ complex and the other is PAR complex.[46-49] JAM-A indirectly interacts with PATJ complex, because PATJ binds to ZO-3 and ZO-3 binds to ZO-1.[43,50] Moreover, MUPP1 is similar to PATJ and may form a complex like PATJ.[39] If this is the case, JAM-A will be more directly involved in the complex. PAR complex is composed of PAR-6, aPKC, and PAR-3/ASIP. As described above, JAM-A binds to the PDZ domain of PAR-3/ASIP.[7,40] These findings support the assumption that JAM-A together with ZO-1, ZO-2, and ZO-3 are well positioned to link two macromolecular complexes and are important to form TJs. The importance of JAM-A in the establishment of cell polarity is reinforced by the interaction with AF-6/afadin.[36] AF-6/afadin interacts with ZO-1 directly as well as indirectly via α-actinin.[51] Thus, JAM-A, ZO-1, and AF-6/afadin have multiple interactions with each other. AF-6/afadin binds to the C-termini of nectins, which are also members of IgSF with three Ig-domains.[52] The series of studies using MDCK and L cells imply the important roles of AF-6/afadin and nectins in the formation of adherens junctions and TJs.[53] Because AF-6/afadin is not concentrated at TJs in polarized epithelial cells, this interaction may take place only during the assembly of TJs. Likewise, although JAM-A binds to CASK in vitro, CASK is not an element of TJs. CASK has a similar molecular organization to Pals, a component of PATJ-complex, but the subcellular localization is quite different. In polarized epithelial cells, CASK is distributed along the lateral membranes. JAM-A may interact with CASK before TJs develop. Other JAM family proteins are not studied so much intensively. The interaction of JAM-B and JAM-C with PAR-3/ASIP suggests that they play roles in endothelial cells as JAM-A does in epithelial cells.[40] JAM4 is shown to recruit ZO-1 and occludin via MAGI-1 to newly formed cell-cell contacts in L cells.[17] This finding in the model system insinuates that JAM4 functions in the assembly of TJs. However, the expression of JAM4 and MAGI-1 is detected in rather limited kinds of epithelial cells. It remains to be determined whether the regulation of TJ assembly is a major role of JAM4.

Regulation of Paracellular Permeability

Growing evidence suggests that JAM family proteins functionally contribute to the regulation of paracellular permeability in endothelial and epithelial cells. The paracellular permeability is evaluated by two methods, the measurements of transepithelial electrical resistance (TER) and the flux of tracer solutes. JAM-A-specific monoclonal antibody and a recombinant extracellular fragment of JAM-A fused to the Fc domain of IgG markedly inhibited TER recovery of epithelial monolayers after intracellular junctions were disrupted by the transient calcium depletion or the trypsin-EDTA treatment.[18,54] A recent study reveals that the antibody against membrane-distal V-type Ig-domain inhibits the barrier recovery after the calcium switch.[55] This result is consistent with that membrane-distal V-type Ig-domain is involved in formation of *cis*-dimers, which occurs prior to any adhesive *trans*-interaction. JAM-A, JAM-C, CAR, and JAM4 reduce the paracellular permeability of tracers such as FITC-dextran in the monolayer of CHO cells.[6,17,21,24] Furthermore, the coexpression of MAGI-1 with JAM4 augments the sealing effect of JAM4 in CHO cells.[17] Because JAM4 and MAGI-1 are coexpressed at podocytes, they may cooperatively regulate the permeability in kidney glomeruli.[17]

Leukocyte Trans-Endothelial Migration

JAM family proteins are the first integral proteins of TJs that were demonstrated to play a direct role in the control of immune cell transport across the endothelial cells. Monoclonal antibody against JAM-A inhibited the migration of human monocytes through mouse endothelial cell layers in vitro.[6] In addition, the intravenous injection of the same anti-JAM-A antibody significantly blocked leukocyte accumulation in the cerebrospinal fluid and infiltration in the brain parenchyma.[56] JAM-C also regulates leukocyte transendothelial migration. When JAM-C is transfected in endothelioma cells, the rate of lymphocyte transmigration increases.[57] Conversely, anti-JAM-C antibody and the recombinant protein containing the extracellular domain of JAM-C decrease the transmigration of leukocytes across endothelial cells.[57] Furthermore, as described above, JAM-A binds to LFA-1, a crucial molecule for leukocyte transmigration expressed on T cells and neutrophils.[30] Leukocytes bind JAM-A expressed on the apical membrane, tightly adhere to endothelial cells, and then transmigrate depending on the interaction with JAM-A localized at junctional sites.[30] The expression of JAM-A on the apical surface of endothelial cells is enhanced after stimulation with inflammatory cytokines.[58] Taken together, under inflammatory conditions, JAM-A traps leucocytes and promote their transendothelial migration. The interaction of endothelial JAM-B with hematopoietic JAM-C and VLA-4 may be similarly involved in leukocyte migration.[20,31,33]

However, several lines of evidence challenge the hypothesis that JAM-A on endothelial cells actively controls leukocyte transmigration. Different antibodies and methods give contentious results. Several anti-JAM-A antibodies do not inhibit leukocyte transendothelial migration.[18,59,60] Cytokine activation induces the redistribution of JAM-A in endothelial cells, but the inhibition of neutrophil or monocyte transmigration is not observed in the in vitro flow assay unlike in the static assay.[60] Therefore, whether JAM-A regulates transendothelial migration is still controversial.

Other Putative Functions

In view of physiological significance of JAM family proteins in cell migration, sertoli cell TJs is an interesting model to investigate. During spermatogenesis, developing spermatocytes translocate from the basal to the apical adluminal compartment of the seminiferous epithelium so that fully developed spermatids are released to the tubular lumen at spermiation.[61] JAM-A, JAM-B, JAM-C, CAR and CLMP are detected at testis, and may be involved in spermatocytes migration, although the precise localization of these proteins in testis remains to be examined. JAM-A and JAM-C are found at platelets, and may be relevant to platelet aggregation and binding to endothelial cells and leukocytes.[32,62,63] Various stimuli to activate platelets cause JAM-A phosphorylation via protein kinase C.[64] The expression of CAR in brain changes depending on the developmental stages, suggesting its role in the formation of neural network.[25]

JAM family proteins may be implicated in several pathological process. JAM-A is a receptor for reovirus, while CAR is a receptor for coxsackie and adenovirus.[65] CagA protein from *Helicobacter pylori* recruits JAM-A to sites of bacteria attachment, which may help to target and retain *H. pylori* at epithelial TJs.[66] Junctional regions are sites of enhanced membrane cycling, endocytic uptake, and intracellular signaling. Therefore, virus and bacteria may have selected integral proteins of TJs as their receptors during the evolution.

Conclusion

JAM family proteins are still a growing family of proteins. Although much work has been done, our knowledge about this family is still far from complete. In polarized epithelial and endothelial cells JAM family proteins may control the paracellular permeability. In the immune system they may help to orchestrate the recruitment of leukocytes to sites of inflammation. Some members may provide entry for virus and bacteria. These properties are apparently related to the regulation of TJ assembly. However, it remains unclear whether all TJs have JAM family proteins as essential components. JAM family proteins may exert modifying roles only at some population of TJs. Vascular endothelial cells from ESAM-deficient mice have less migratory and angiogenic activity, but the overall development and reproduction of the mice are normal.[67] This may indicate that JAM family proteins have functionally overlapping roles. Moreover, it is disputing whether the subgroups with type I and type II PDZ-binding motifs should be discussed together. Both groups are similar in the overall structures but different in protein-protein interactions, suggesting that they have biological functions in distinct context. Accumulation of data using a series of gene targeting mice will be enlightening for these questions.

49. Straight SW, Shin K, Fogg VC et al. Loss of PALS1 expression leads to tight junction and polarity defects. Mol Biol Cell 2004.
50. Roh MH, Liu CJ, Laurinec S et al. The carboxyl terminus of zona occludens-3 binds and recruits a mammalian homologue of discs lost to tight junctions. J Biol Chem 2002; 277(30):27501-27509.
51. Yamamoto T, Harada N, Kano K et al. The ras target AF-6 interacts with ZO-1 and serves as a peripheral component of tight junctions in epithelial cells. J Cell Biol 1997; 139(3):785-795.
52. Takahashi K, Nakanishi H, Miyahara M et al. Nectin/PRR: An immunoglobulin-like cell adhesion molecule recruited to cadherin-based adherens junctions through interaction with Afadin, a PDZ domain-containing protein. J Cell Biol 1999; 145(3):539-549.
53. Takai Y, Nakanishi H. Nectin and afadin: Novel organizers of intercellular junctions. J Cell Sci 2003; 116(Pt 1):17-27.
54. Liang TW, DeMarco RA, Mrsny RJ et al. Characterization of huJAM: Evidence for involvement in cell-cell contact and tight junction regulation. Am J Physiol Cell Physiol 2000; 279(6):C1733-1743.
55. Mandell KJ, McCall IC, Parkos CA. Involvement of the junctional adhesion molecule-1 (JAM1) homodimer interface in regulation of epithelial barrier function. J Biol Chem 2004.
56. Del Maschio A, De Luigi A, Martin-Padura I et al. Leukocyte recruitment in the cerebrospinal fluid of mice with experimental meningitis is inhibited by an antibody to junctional adhesion molecule (JAM). J Exp Med 1999; 190(9):1351-1356.
57. Johnson-Leger CA, Aurrand-Lions M, Beltraminelli N et al. Junctional adhesion molecule-2 (JAM-2) promotes lymphocyte transendothelial migration. Blood 2002; 100(7):2479-2486.
58. Ozaki H, Ishii K, Horiuchi H et al. Cutting edge: Combined treatment of TNF-alpha and IFN-gamma causes redistribution of junctional adhesion molecule in human endothelial cells. J Immunol 1999; 163(2):553-557.
59. Lechner F, Sahrbacher U, Suter T et al. Antibodies to the junctional adhesion molecule cause disruption of endothelial cells and do not prevent leukocyte influx into the meninges after viral or bacterial infection. J Infect Dis 2000; 182(3):978-982.
60. Shaw SK, Perkins BN, Lim YC et al. Reduced expression of junctional adhesion molecule and platelet/endothelial cell adhesion molecule-1 (CD31) at human vascular endothelial junctions by cytokines tumor necrosis factor-alpha plus interferon-gamma does not reduce leukocyte transmigration under flow. Am J Pathol 2001; 159(6):2281-2291.
61. Lui WY, Mruk D, Lee WM et al. Sertoli cell tight junction dynamics: Their regulation during spermatogenesis. Biol Reprod 2003; 68(4):1087-1097.
62. Sobocka MB, Sobocki T, Banerjee P et al. Cloning of the human platelet F11 receptor: A cell adhesion molecule member of the immunoglobulin superfamily involved in platelet aggregation. Blood 2000; 95(8):2600-2609.
63. Babinska A, Kedees MH, Athar H et al. F11-receptor (F11R/JAM) mediates platelet adhesion to endothelial cells: Role in inflammatory thrombosis. Thromb Haemost 2002; 88(5):843-850.
64. Ozaki H, Ishii K, Arai H et al. Junctional adhesion molecule (JAM) is phosphorylated by protein kinase C upon platelet activation. Biochem Biophys Res Commun 2000; 276(3):873-878.
65. Barton ES, Forrest JC, Connolly JL et al. Junction adhesion molecule is a receptor for reovirus. Cell 2001; 104(3):441-451.
66. Amieva MR, Vogelmann R, Covacci A et al. Disruption of the epithelial apical-junctional complex by Helicobacter pylori CagA. Science 2003; 300(5624):1430-1434.
67. Ishida T, Kundu RK, Yang E et al. Targeted disruption of endothelial cell-selective adhesion molecule inhibits angiogenic processes in vitro and in vivo. J Biol Chem 2003; 278(36):34598-34604.
68. Malergue F, Galland F, Martin F et al. A novel immunoglobulin superfamily junctional molecule expressed by antigen presenting cells, endothelial cells and platelets. Mol Immunol 1998; 35(17):1111-1119.
69. Naik UP, Naik MU, Eckfeld K et al. Characterization and chromosomal localization of JAM-1, a platelet receptor for a stimulatory monoclonal antibody. J Cell Sci 2001; 114(Pt 3):539-547.
70. Aurrand-Lions MA, Duncan L, Du Pasquier L et al. Cloning of JAM-2 and JAM-3: An emerging junctional adhesion molecular family? Curr Top Microbiol Immunol 2000; 251:91-98.
71. Bergelson JM, Cunningham JA, Droguett G et al. Isolation of a common receptor for coxsackie B viruses and adenoviruses 2 and 5. Science 1997; 275(5304):1320-1323.
72. Tomko RP, Xu R, Philipson L. HCAR and MCAR: The human and mouse cellular receptors for subgroup C adenoviruses and group B coxsackieviruses. Proc Natl Acad Sci USA 1997; 94(7):3352-3356.

CHAPTER 5

Cingulin, a Cytoskeleton-Associated Protein of the Tight Junction

Laurent Guillemot and Sandra Citi

Abstract

Cingulin is a M_r ~140 kDa phosphoprotein component of the cytoplasmic plaque of vertebrate tight junctions (TJ), and was identified as an actomyosin-associated protein. The cingulin sequence predicts that the molecule is a parallel homodimer, each subunit consisting of a coiled-coil "rod" domain separating a large globular N-terminal "head" domain from a small globular C-terminal "tail". The head domain mediates direct in vitro interaction with ZO-1, ZO-2, ZO-3 and actin, and is critical for the junctional recruitment of cingulin in transfected cells, whereas the rod domain mediates dimerization. Cingulin forms complexes with ZO-1, ZO-2 and the immunoglobulin-like adhesion molecule JAM-1. These complexes are only detected in Triton-insoluble cell fractions, suggesting that one function of cingulin is to link TJ-associated proteins to the actomyosin cytoskeleton during TJ biogenesis and regulation. In fact, purified cingulin acts in vitro as an actin-binding and -bundling protein. Cingulin shows a cortical, apical localization in early vertebrate embryos, and in *Xenopus* embryos is recruited into nascent TJ starting at the 2-cell stage, suggesting that cingulin may play an important role in the development of vertebrate epithelia. Cingulin may represent the first member of a family of proteins, since a novel protein, paracingulin, with similar domain organization and significant sequence homology to cingulin has recently been identified.

Identification of Cingulin as a Tight Junction-Associated Protein

Cingulin was identified by monoclonal antibodies raised against a preparation of subfragment-1 of chicken brush border myosin, which contained cingulin as a minor contaminant.[1] Cingulin copurified with the actomyosin fraction from brush border cell lysates, and was eluted by gel filtration in a peak that partially overlapped with that of nonmuscle myosin II.[2] Immunofluorescence microscopy showed that the antigen was localized in a circumferential belt around the apical area of chicken intestinal brush border cells.[1] Because of its belt-like localization, the antigen was named cingulin (from latin "cingere"=to form a belt around). Immunoelectron microscopy demonstrated that cingulin is localized on the cytoplasmic face of TJ,[1,2] and double-labeling with anti-ZO-1 antibodies showed that cingulin is localized at a further distance from the plasma membrane than ZO-1.[3] Additional evidence that cingulin is a TJ-specific protein is provided by analysis of its tissue distribution (see below).

In lysates of chicken brush border cells analyzed by SDS-PAGE and immunoblotting, cingulin was detected as two polypeptides of an apparent molecular weight of 140 kDa and 108 kDa, the latter form representing a degradation product of the full-length protein.[1] The 108 kDa chicken cingulin fragment was purifed from intestinal cells, and shown to be stable to heat and ethanol treatment, indicating a coiled-coil structure.[1] Consistent with this prediction, gel filtration elution profiles and migration in nonreducing SDS-PAGE indicated that cingulin

---

Tight Junctions, edited by Lorenza Gonzalez-Mariscal. ©2006 Landes Bioscience and Springer Science+Business Media.

is a dimer.[2] Rotary-shadowing electron microscopy of purified cingulin fragment (108 kDa) revealed elongated molecules, with a mean contour length of 130 nm ± 32 nm and a width of 2-3 nm, very similar to the coiled-coil region ("rod") of conventional myosin.[1]

Molecular Structure and Evolution of Cingulin

The complete amino acid sequences of cingulin from human, mouse and *Xenopus laevis* are shown in Figure 1. Sequences coding for cingulin from other vertebrate species (including rat and zebrafish) are available in public databases. Analysis of the predicted secondary structure of cingulin reveals that the protein is characterized by three distinct domains: a N-terminal globular "head" region, a central α-helical coiled-coil "rod" domain and a small C-terminal globular "tail" region (Fig. 2).[4,5] Although the head-rod-tail organization appears common to cingulin from all species sequenced so far, the size of the head and rod domains is slightly larger in *Xenopus laevis* cingulin than in human or mouse cingulin (Fig. 2), accounting for the larger size of *Xenopus* cingulin (~160 kDa, versus ~140 kDa in mammalian). The cingulin head sequence is not homologous to any known protein, whereas the rod is most homologous to the coiled-coil region of conventional nonmuscle myosins. Human and *Xenopus laevis* cingulin show 43% sequence identity, and the overall degree of homology is greater in the rod and tail domains

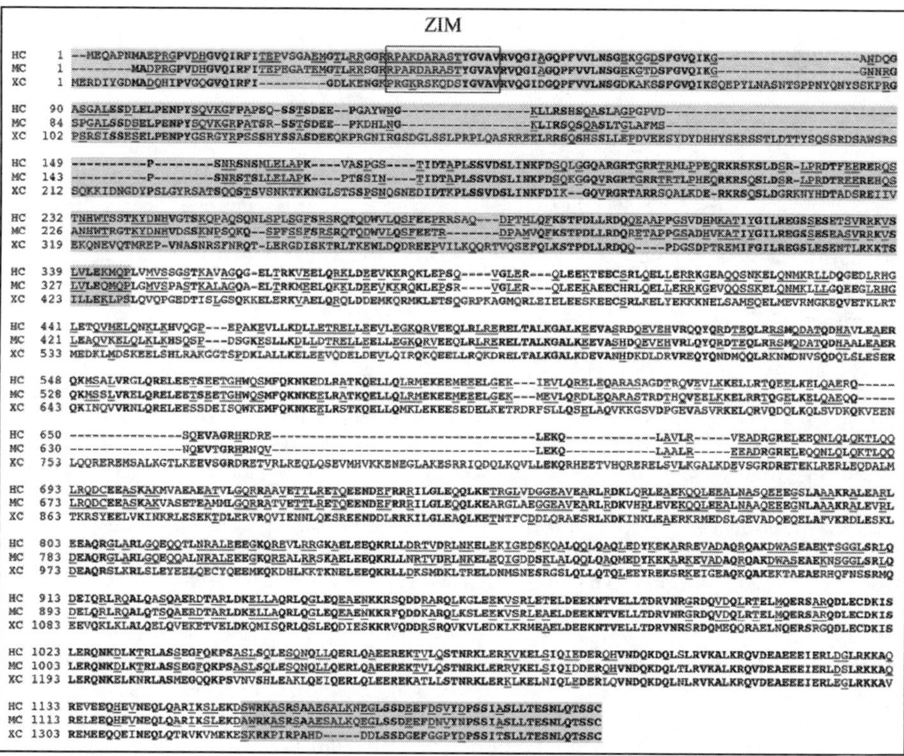

Figure 1. Alignment of cingulin amino acid sequences from human (HC), mouse (MC) and *Xenopus laevis* (XC) species. The N-terminal head and the C-terminal tail domains are shown in grey boxes. Bold letters indicate identity in all three sequences. Underlined letters indicate identity in only two sequences. Note that stretches of highly conserved sequences are detectable in the N-terminal part of the head (including the ZIM=ZO-1 Interaction Motif) and in the C-terminal region. The alignment was made using the ClustalW algorithm. Genbank accession numbers for the cingulin sequences are AF263462 (human), XM_131052 (mouse), and AF207901 (*Xenopus*).

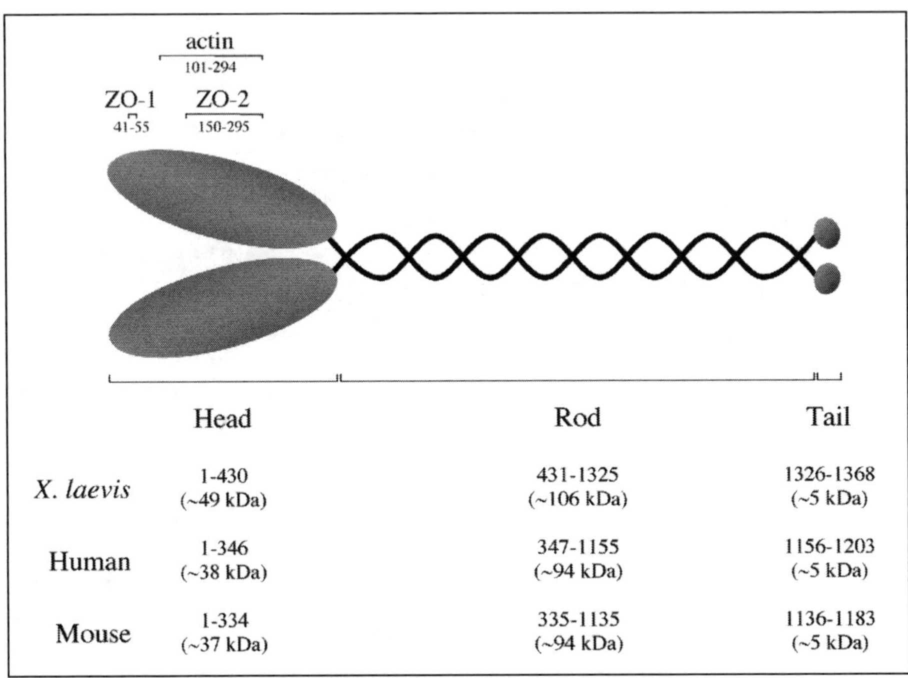

	Head	Rod	Tail
X. laevis	1-430 (~49 kDa)	431-1325 (~106 kDa)	1326-1368 (~5 kDa)
Human	1-346 (~38 kDa)	347-1155 (~94 kDa)	1156-1203 (~5 kDa)
Mouse	1-334 (~37 kDa)	335-1135 (~94 kDa)	1136-1183 (~5 kDa)

Figure 2. Schematic representation of the cingulin molecule as a parallel dimer with globular "head" and "tail", and coiled-coil "rod" domains. Amino acid boundaries and predicted molecular weights for each domain in *Xenopus laevis*, human and mouse cingulin are shown below the diagram. Amino acid positions of *Xenopus laevis* cingulin, which are required for in vitro interaction with ZO-1, ZO-2 and actin are indicated above the diagram. Additional protein interactions for cingulin have been identified (see text), but are not shown in this diagram.

than in the head region (Fig. 1). However, the head domain contains a few stretches of highly conserved sequence, one of which includes the sequence motif (ZIM = ZO-1 Interaction Motif) responsible for ZO-1 interaction (Fig. 1). Detailed analysis of the coiled-coil sequence of human and *Xenopus laevis* cingulin also confirms that the molecule is a parallel dimer, with chains in axial register (Fig. 2).[4,5] However, the sequence of the coiled-coil region suggests that, unlike conventional myosins, cingulin has a relatively low probability of forming inter-molecular, higher-order aggregates.[5] Confirming this notion, purified recombinant cingulin does not form aggregates under conditions where myosin assembles into filaments.[6] The coiled-coil region may also mediate intra-molecular interactions with the head domain, as suggested by GST pull-down experiments.[4,6]

Based on sequence homology, thus far no cingulin-related protein has been identified in invertebrate organisms, such as *D. melanogaster* and *C. elegans*. Thus, cingulin appears to be one of few vertebrate TJ proteins for which no invertebrate ortholog has been described. If cingulin is indeed a vertebrate-specific protein, its function may be related to physiological processes that are exclusively associated with vertebrate organisms. Another possibility is that cingulin function is conserved, but the proteins that carry out this function in invertebrates are different or have extremely divergent sequences. One such protein may be AJM-1, a junctional *C. elegans* protein containing coiled-coil and globular domains, mutation of which leads to late embryonic growth arrest.[7] However, AJM-1 shows greater sequence homology to JEAP,[8] another coiled-coil TJ-associated protein, than to cingulin.

Interaction of Cingulin with the Actomyosin Cytoskeleton

Purified recombinant full-length cingulin binds directly to purified F-actin.[9] However, recombinant cingulin shows no actin-activated Mg-ATPase activity, suggesting that it is not an actin-dependent motor protein. Among TJ proteins (cingulin, ZO-1, ZO-2, ZO-3 and occludin) that have been shown to bind directly to F-actin, cingulin is the only one which displays actin-bundling activity in vitro.[9] Thus, cingulin may function to help organize actin filaments during TJ formation and functional regulation. The actin-binding region of cingulin has been mapped to sequences within the central part of the head domain (Fig. 2).[9] In addition, cingulin interacts in vitro with nonmuscle myosin II, though both rod and head domains.[4] The observation that cingulin copurifies with the actomyosin fraction from lysates of chicken brush border cells further indicates a link between function of cingulin and organization of the actin cytoskeleton.[2] However, evidence for a direct role of cingulin in cytoskeletal regulation in vivo is missing.

Cingulin Associates with Several Tight Junction Proteins

Cingulin interacts directly or indirectly with several protein components of TJ. GST pull-down assays show that a head fragment of *Xenopus laevis* cingulin interacts in vitro with ZO-1, ZO-2, ZO-3, and AF-6/afadin.[4] Measurement of the dissociation constants for ZO-1/cingulin and ZO-2/cingulin interactions shows that these interactions display a high affinity.[4,6] Furthermore, immunoprecipitation experiments show that cingulin exists in a complex with ZO-1 and ZO-2,[4] indicating the physiological relevance of this interaction. Although the cingulin-binding regions of ZO-1 and ZO-2 have not been mapped, in the case of ZO-3 both the N-terminal half and the C-terminal half of the molecule contain binding sites for cingulin.[10] The ZO-1 and ZO-2 interacting regions of cingulin were mapped by analysis of mutant GST fusion proteins.[6] A conserved ~15-residues stretch in N-terminal region of cingulin (ZIM, residues 41-55 of *Xenopus laevis* cingulin and 48-61 of human cingulin) (Figs. 1, 2) was found to be required for in vitro interaction with ZO-1, whereas ZO-2 and actin interaction requires sequences in the central part of the cingulin head region (Fig. 2).[6] The functional relevance of cingulin/ZO-1 interaction is indicated by the observation that overexpression of full-length *Xenopus* cingulin in *Xenopus* A6 cells results in the redistribution of endogeneous ZO-1,[6] suggesting that cingulin overexpression can disrupt the association of ZO-1 with other TJ components. Cingulin also forms a complex with the TJ membrane protein JAM-1, as shown by immunoprecipitation experiments.[11] In addition, a direct in vitro interaction has been observed between the head region of cingulin and the C-terminal fragments of JAM-1 and occludin.[11,12] Remarkably, the ZO-1/ZO-2/cingulin and the JAM-1/cingulin interactions are almost exclusively detectable in the Triton X-100-insoluble fraction (associated with the cytoskeleton), despite the fact that these proteins are found in both Triton X-100-soluble and -insoluble fractions. Thus, the association of cingulin with the TJ protein network appears linked to its ability to bind actin filaments or other actin-binding proteins.

The Head Domain of Cingulin Is Essential for Its Junctional Recruitment

Transfection of mutant cingulin proteins into *Xenopus* A6 epithelial cells shows that the head domain can be recruited into junctions, whereas the rod displays an exclusively cytoplasmic localization.[6] Thus, the head domain of cingulin has a critical role in targeting cingulin to junctions, presumably through its interaction with specific TJ proteins. In the presence of the coiled-coil domain, targeting of the head to junctions is considerably more efficient, suggesting that rod-mediated dimerization stabilizes cingulin association with the junctional plaque.[6] Alternatively, the rod domain may facilitate optimal folding or conformation of the head. Analysis of N-terminal deletion mutants shows that removal of the ZIM motif does not abolish junctional recruitment, whereas removal of the first 294 residues does. On the other hand, the first 100 residues of the cingulin head, which contain the ZIM motif, are sufficient, when fused to the coiled-coil region, for junctional recruitment.[6] Thus, the N-terminal portion of the cingulin head is critically important for cingulin junctional recruitment, and redundant protein interactions can help to target cingulin to TJ even in the absence of the high affinity ZO-1 binding site.

The key role of the ZO-1 interaction in cingulin junctional recruitment is further highlighted by studies on Rat-1 fibroblasts, which lack TJ, but display ZO-1 containing adherens-like junctions.[6] In these cells, transfected cingulin localizes at ZO-1 containing junctions, demonstrating that cingulin can be recruited into junctions even in the absence of the molecular context of TJ, provided that ZO-1 is present. Interestingly, deletion of the ZIM domain abolishes junctional targeting of cingulin in Rat-1 cells, and the protein shows a cytoplasmic localization.[6] Thus, ZO-1 interaction is essential for cingulin recruitment into adherens-type junctions. If TJ are formed by the "maturation" of adherens-like junctions,[13] ZO-1 interaction may therefore be crucial for physiological targeting of cingulin to developing TJ.

Cingulin Is Maternally Expressed and Is an Early Component of Cell-Cell Junctions in Developing Vertebrate Embryos

Cingulin is one of the first proteins to assemble into junctions of early vertebrate embryos.[14-18] During mouse oogenesis, cingulin shows a cortical, submembrane localization both in oocytes and in adjacent cumulus cells, suggesting a role in cumulus/oocyte interactions, which are essential for oocyte growth and the regulation of meiotic maturation.[14] Cingulin is expressed in oocytes by the maternal programme, which is maintained after fertilization and starts to decline from the 2-cell stage.[15] Following fertilization of the egg, the trophectoderm forms the wall of the blastocyst, an epithelium where TJ become first detectable. In compacted 8-cell embryos, cytocortical cingulin is localized at the apical microvillous pole on the outer surface of each bastomere, but is not detectable at the nascent apicolateral TJ sites.[14] The cytocortical staining of cingulin is evident during early cleavage stages, up to the 16-cell morulae, when cingulin becomes faintly detectable at the nascent TJ sites.[14] Anti-cingulin antibodies also label punctate cytoplasmic foci, derived from apical endocytosed membrane. This cytoplasmic presence of cingulin gradually declines after 8 hours post onset of cavitation, coincident with the increase in cingulin level in TJ sites.[14] In late morulae, cingulin staining is increasingly detected at junctions, while blastocoele expansion progresses. The early localization of cingulin at the sites of newly developing junctions suggests that cingulin is important in the establishment of functional TJ.[14]

As in the mouse, cingulin shows a cytocortical distribution in *Xenopus laevis* eggs and early embryos.[16] Starting from the first cleavage, maternal cingulin accumulates in the regions of cell-cell contact between the dividing blastomeres, and appears to be recruited from the neighbouring membrane areas (Fig. 3A), since they become less intensely stained for cingulin.[16] Electron microscopy has not revealed typical TJ before the 32-cell stage in *Xenopus* embryos.[19] Thus, the presence of junctional cingulin may indicate the formation of an immature TJ.[16] Interestingly, labeling of early embryos with antibodies to ZO-1 shows that ZO-1 is recruited into the new basolateral membrane, through the fusion of membrane vesicles which are distributed in a linear array in the cytoplasm (Fig. 3B), presumably along microtubule tracks.[17] Immunofluorescent labeling for occludin, a 4-pass transmembrane component of TJ, is detected in the cytoplasm as small membrane vesicles (Fig. 3C), which are topologically distinct from ZO-1-containing vesicles.[17] Following vesicle fusion with the new basolateral membrane, ZO-1 and occludin labeling become distributed along the lateral membrane up to the TJ, and they remain excluded from the cingulin-containing apical membrane (Fig. 3).[17] The observation that occludin recruitment occurs later than cingulin and ZO-1 is consistent with the notion that assembly of the membrane domain of TJ occurs only after the cytoplasmic scaffold has been established. Thus, TJ formation can schematically be viewed as resulting from the recruitment of apical cingulin by the basolateral, ZO-1-containing complex, followed by recruitment of occludin (Fig. 3).

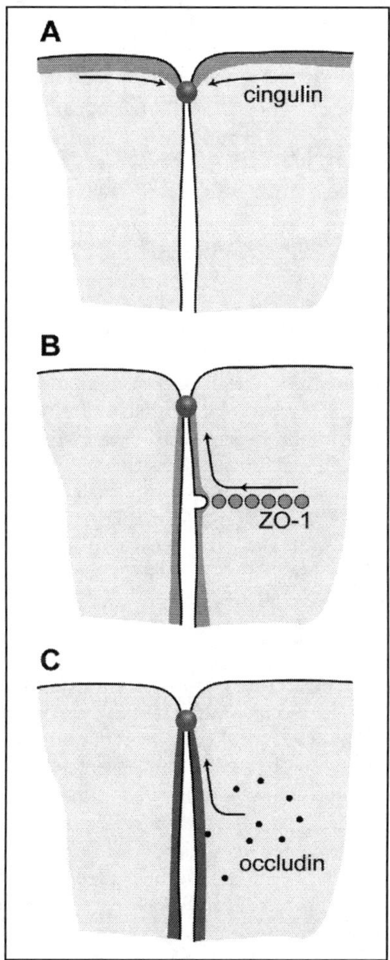

Figure 3. Schematic representation of the assembly of cingulin (A), ZO-1 (B) and occludin (C) into developing TJ of *Xenopus laevis* embryos. Cingulin shows an apical, cortical localization in eggs, and is recruited from this pool into new sites of cell-cell contacts generated following division of blastomeres (A). ZO-1 (B) and occludin (C) are not detected in the apical cortex, and targeted to the new lateral membrane through the fusion of distinct populations of membrane vesicles. ZO-1 and especially occludin are clearly detected along the lateral membrane even at later stages of development.

When *Xenopus* blastomeres are incubated in a solution lacking calcium, they divide but they fail to adhere through their lateral membranes, probably because of inactivation of the cadherin-based adhesion system. Under these conditions, cingulin still becomes recruited at the border between the old, apical membrane, and the new basolateral membrane generated between the two daughter cells.[16] Thus, establishment of the intramembrane fence and cingulin junctional recruitment are independent of lateral cell adhesion.

In summary, both in mouse and *Xenopus* embryos maternal cingulin appears to be associated with the cytocortex, which subsequently defines the apical domain of blastomeres generated after cleavage. The functional significance of cytocortical cingulin is unclear. One possibility is that cingulin participates in the structural organization of the cytocortex during early development, prior to fertilization. Cortical cingulin may also serve as a reservoir of protein to be recruited into junctions after fertilization, prior to the onset of the embryonic expression programme. Finally, apical cingulin may participate in establishing the axis of division of blastomeres, by acting in concert with polarity complexes.[20]

Cingulin as a Marker for Differentiated Epithelial Cells

Cingulin appears to be a good marker for differentiated epithelial cells. Indeed, the human cingulin gene is localized on chromosome 1q21, within a cluster of genes called the "epidermal differentiation complex" (EDC), which includes epidermal genes such as loricrin, involucrin and profilaggrin.[21] In tissue sections cingulin has been detected exclusively in differentiated epithelial tissues, including single-layered epithelia (intestine, stomach, kidney, liver, pancreas, retina, epididymis, inner ear), and stratified epithelia (tongue, skin, cornea, oesophagus) but not in tissues of mesenchymal and myogenic origin, such as cardiac and smooth muscle tissues.[1,2,22-26] The pattern of tissue distribution of the cingulin protein is reflected in the detection of cingulin mRNA as a ~5.2 kb message in human pancreas, kidney, liver and lung but not in skeletal muscle, placenta, brain and heart.[5] Moreover, cingulin is not detected in lens cells, which lack TJ but are rich in adherens and gap junctions.[2] With regard to endothelial tissues, cingulin expression appears restricted to "tight" endothelia, since cingulin protein has been detected by immunofluorescence in capillaries of the heart and brain,[2,27] but not in cultured HUVEC cells.[28] However, cingulin is also expressed in cultured embryonic cells with endothelial and myogenic potential,[29] and in some fibroblast-like cultured cell lines (our unpublished observations), suggesting that cingulin expression is associated with an early step of commitment to an epithelial phenotype.

If cingulin expression is restricted to epithelial tissues, one important question is whether such expression is lost or altered upon neoplastic transformation. In some epithelial tumors, cells lose their polarized phenotype and proliferate within connective tissue. To address this question, cingulin expression has been studied in benign and malignant tumors of the colon, and several other neoplasias.[28] Cingulin expression was detected both in villous adenomas and metastatic colon adenocarcinomas, but not in tumors of mesenchymal origin.[28] In metastatic adenocarcinomas cingulin was localized on the border of nests of cancerous cells which do not show any spatial relationship to a lumen or a body surface.[28] Thus, while cingulin expression in other epithelial tumors remains to be evaluated, the evidence so far is that cingulin expression persists in malignant epithelial cells, which have lost their differentiated polarized phenotype. Thus, cingulin expression could be used as a tool to identify malignant epithelial cells in diagnostic pathology.

Putative Functions and Regulation of Cingulin

The apparently ubiquitous expression of cingulin in vertebrate epithelial TJ suggests that it is an important component of the multiprotein complex underlying the TJ membrane. The observation that cingulin binds both to specific TJ proteins and to actin filaments suggests that cingulin might serve as a cytoskeletal adaptor protein. Linkage of TJ proteins to the cytoskeleton is crucial to transduce the mechanical force generated by the contraction of the actomyosin cytoskeleton into functional regulation of the paracellular barrier function of TJ.[30] Interestingly, overexpression of human cingulin in MDCK cells leads to a ~50% decrease in transepithelial resistance (Citi et al, unpublished observations), supporting the idea that cingulin may affect the organization and contractility of the actomyosin cytoskeleton. In early embryos, apically localized cingulin may contribute to the establishment of structural asymmetry in the plasma membrane, through interaction with the actin cytoskeleton and polarity complexes. Cingulin may control cytoskeletal organization directly, by bundling actin filaments, but also indirectly, by interacting with different actin-binding or microfilament-regulating proteins.

Another putative, yet unexplored function of cingulin is related to the observation that cingulin can be localized by immunofluorescence in the nuclei of cultured cells. Cingulin has been detected in the nucleus of sparsely grown MDCK cells,[31] and in the human cancer cell lines CaCo2 and IGROV1.[32] Studies on the ZO-1-binding protein ZONAB have established that TJ proteins can serve as regulators of epithelial cell proliferation and cell density.[33] If cingulin interacts with transcription factors and can be shuttled between nucleus and cytoplasm, it could act as a regulator of transcription and/or translation. This notion is supported by studies on gene targeted mouse embryoid bodies, showing that cingulin mutation results in changes in the mRNA levels for several TJ proteins.[37]

Nothing is known about the transcriptional regulation of the cingulin gene. However, the observation that cingulin is phosphorylated suggests the existence of post-translational regulatory mechanisms. Phosphorylation of cingulin occurs on serine residues,[34] but the phosphorylation site(s) and the identity of cingulin kinase(s) are not known. Since treatment of cells with the protein kinase C activator phorbol 12-myristate 13-acetate and the protein kinase C inhibitor H-7 do not affect the specific phosphorylation of cingulin, protein kinase C does not appear to be responsible for cingulin phosphorylation.[34] By analogy to nonmuscle myosins,[35] cingulin phosphorylation may control its conformation and protein interaction properties.

Paracingulin

Cingulin may represent the first member of a family of genes. By searching EST sequences with protein sequence homology to cingulin, we identified a cDNA whose 5' sequence was known, and coded for a protein with partial homology to the N-terminal region of human cingulin. We determined the complete nucleotide sequence of this cDNA (GenBank accession number AY610514) and denoted it as paracingulin, because of its similarities to cingulin (see below). Paracingulin is predicted to comprise 1302 amino acids, with a molecular weight of 148,596 daltons. Paracingulin shows a domain organization with a globular domain (residues 1-598), followed by a coiled-coil domain (residues 599-1262) and a small globular domain (residues 1263-1302) (Fig. 4). Thus, the N-terminal globular domain is larger than that cingulin, whereas the coiled-coil domain is shorter (Fig. 4). Although the overall amino acid identity between paracingulin and cingulin is only ~30%, several stretches of sequence around the ZIM motif and in the C-terminal part of the coiled-coil region are highly homologous. Since these regions are particularly well conserved across cingulins from different vertebrate species, and the ZIM has important roles in cingulin junctional recruitment and protein interactions, paracingulin appears significantly related to cingulin. Furthermore, when paracingulin is transfected into epithelial cells, it shows a junctional localization (Citi et al, unpublished results). Current studies are aimed at exploring the subcellular localization, tissue distribution, and functions of paracingulin.

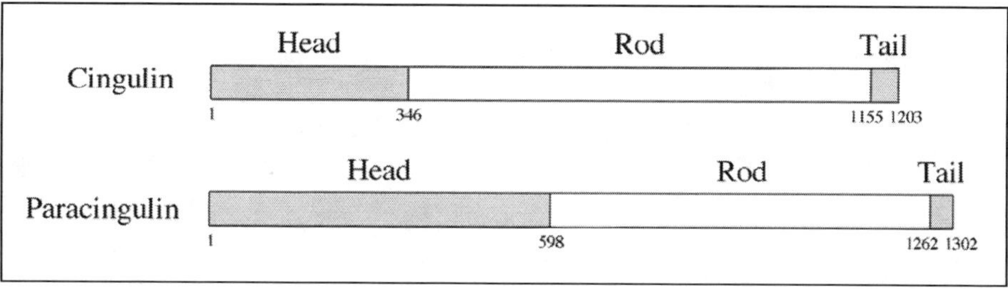

Figure 4. Comparative diagrams showing the predicted domain organization of cingulin and paracingulin. The boundaries between globular and coiled-coil domains in the predicted amino acid sequence of paracingulin (GenBank Accession number AY610514) were mapped using the Macstripe software.[36]

Conclusion

Cingulin is a specific component of vertebrate TJ. The precise physiological role of cingulin, like that of the majority of cytoplasmic TJ proteins, is unknown. However, studies on cingulin molecular structure, protein interactions, and expression suggest that it is a cytoskeletal adaptor protein and may be involved in epithelial differentiation and TJ regulation. The apparent lack of cingulin orthologs in invertebrate species suggests that cingulin may contribute to vertebrate-specific tissue functions. The possible roles of cingulin in gene transcription and translational regulation also deserve future attention. The identification of paracingulin suggests that cingulin is part of a gene family, with potentially redundant functions. A more detailed analysis of the function of cingulin family members by up- and down-regulation will undoubtedly enable us to understand better the molecular architecture and regulation of TJ and begin to approach how TJ components contribute to epithelial morphogenesis in development and disease.

Acknowledgements

We are grateful to the State of Geneva, the Swiss National Science Foundation, the Swiss Cancer League and the MIUR (Ministero dell'Istruzione, dell'Universita' e della Ricerca) for support. L. Guillemot is supported by a Fellowship from the Roche Research Foundation.

References

1. Citi S, Sabanay H, Jakes R et al. Cingulin, a new peripheral component of tight junctions. Nature 1988; 333:272-276.
2. Citi S, Sabanay H, Kendrick-Jones J et al. Cingulin: Characterization and localization. J Cell Sci 1989; 93:107-122.
3. Stevenson BR, Heintzelman MB, Anderson JM et al. ZO-1 and cingulin: Tight junction proteins with distinct identities and localizations. Am J Physiol 1989; 257(Cell Physiol 26):C621-C628.
4. Cordenonsi M, D'Atri F, Hammar E et al. Cingulin contains globular and coiled-coil domains and interacts with ZO-1, ZO-2, ZO-3, and Myosin. J Cell Biol 1999; 147(7):1569-1582.
5. Citi S, D'Atri F, Parry DAD. Human and Xenopus cingulin share a modular organization of the coiled-coil rod domain: Predictions for intra- and intermolecular assembly. J Struct Bio 2000; 131:135-145.
6. D'Atri F, Nadalutti F, Citi S. Evidence for a functional interaction between cingulin and ZO-1 in cultured cells. J Biol Chem 2002; 277:27757-27764.
7. Koppen M, Simske JS, Sims PA et al. Cooperative regulation of AJM-1 controls junctional integrity in Caenorhabditis elegans epithelia. Nat Cell Biol 2001; 3(11):983-991.
8. Nishimura M, Kakizaki M, Ono Y et al. JEAP: A novel component of tight junctions in exocrine cells. J Biol Chem 2001; 3:3.
9. D'Atri F, Citi S. Cingulin interacts with F-actin in vitro. FEBS Lett 2001; 507:21-24.
10. Wittchen ES, Haskins J, Stevenson BR. Exogenous expression of the amino-terminal half of the tight junction protein ZO-3 perturbs junctional complex assembly. J Cell Biol 2000; 151(4):825-836.
11. Bazzoni G, Martinez-Estrada OM, Orsenigo F et al. Interaction of junctional adhesion molecule with the tight junction components ZO-1, cingulin, and occludin. J Biol Chem 2000; 275(27):20520-20526.
12. Cordenonsi M, Turco F, D'Atri F et al. Xenopus laevis occludin: Identification of in vitro phosphorylation sites by protein kinase CK2 and association with cingulin. Eur J Biochem 1999; 264:374-384.
13. Yonemura S, Itoh M, Nagafuchi A et al. Cell-to-cell adherens junction formation and actin filament organization: Similarities and differences between nonpolarized fibroblasts and polarized epithelial cells. J Cell Sci 1995; 108(Pt 1):127-142.
14. Fleming TP, Hay M, Javed Q et al. Localisation of tight junction protein cingulin is temporally and spatially regulated during early mouse development. Development 1993; 117:1135-1144.
15. Javed Q, Fleming TP, Hay MJ et al. Tight junction protein cingulin is expressed by maternal and embryonic genomes during early mouse development. Development 1993; 117:1145-1151.
16. Cardellini P, Davanzo G, Citi S. Tight junctions in early amphibian development: Detection of junctional cingulin from the 2-cell stage and its localization at the boundary of distinct membrane domains in dividing blastomeres in low calcium. Dev Dyn 1996; 207:104-113.

17. Fesenko I, Kurth T, Sheth B et al. Tight junction biogenesis in the early Xenopus embryo. Mech Dev 2000; 96:51-65.
18. Shook DR, Majer C, Keller R. Urodeles remove mesoderm from the superficial layer by subduction through a bilateral primitive streak. Dev Biol 2002; 248(2):220-239.
19. Muller HA, Hausen P. Epithelial cell polarity in early Xenopus development. Dev Dyn 1995; 202:405-420.
20. Chalmers AD, Strauss B, Papalopulu N. Oriented cell divisions asymmetrically segregate aPKC and generate cell fate diversity in the early Xenopus embryo. Development 2003; 130(12):2657-2668.
21. Mischke D, Korge BP, Marenholz I et al. Genes encoding structural proteins of epidermal cornification and S100 calcium-binding proteins form a gene complex ("epidermal differentiation complex") on human chromosome 1q21. J Invest Dermatol 1996; 106(5):989-992.
22. Byers SW, Citi S, Anderson JM et al. Polarized functions and permeability properties of rat epididymal epithelial cells in vitro. J Reprod Fert 1992; 95:385-396.
23. Raphael Y, Altschuler RA. Reorganization of cytoskeletal and junctional proteins during cochlear hair cell degeneration. Cell Motil Cytosk 1991; 18:215-227.
24. Lora L, Mazzon E, Martines D et al. Hepatocyte tight-junctional permeability is increased in rat experimental colitis. Gastroenterology 1997; 113(4):1347-1354.
25. Goodyear R, Richardson G. Pattern formation in the basilar papilla: Evidence for cell rearrangement. J Neurosci 1997; 17(16):6289-6301.
26. Langbein L, Grund C, Kuhn C et al. Tight junctions and compositionally related junctional structures in mammalian stratified epithelia and cell cultures derived therefrom. Eur J Cell Biol 2002; 81(8):419-435.
27. Wolburg H, Lippoldt A. Tight junctions of the blood-brain barrier: Development, composition and regulation. Vascul Pharmacol 2002; 38(6):323-337.
28. Citi S, Amorosi A, Franconi F et al. Cingulin, a specific protein component of tight junctions, is expressed in normal and neoplastic human epithelial tissues. Am J Pathol 1991; 138:781-789.
29. Eisenberg CA, Bader D. QCE-6: A clonal cell line with cardiac myogenic and endothelial cell potentials. Dev Biol 1995; 167(2):469-481.
30. Turner JR. 'Putting the squeeze' on the tight junction: Understanding cytoskeletal regulation. Semin Cell Dev Biol 2000; 11(4):301-308.
31. Citi S, Cordenonsi M. The molecular basis for the structure, function and regulation of tight junctions. In: Garrod DR, North AJ, Chidgey MAJ, eds. Adhesive Interactions of Cells. Greenwich: JAI Press Inc., 1999:28:203-233.
32. Nakamura T, Blechman J, Tada S et al. huASH1 protein, a putative transcription factor encoded by a human homologue of the Drosophila ash1 gene, localizes to both nuclei and cell-cell tight junctions. Proc Natl Acad Sci USA 2000; 97(13):7284-7289.
33. Balda MS, Garrett MD, Matter K. The ZO-1-associated Y-box factor ZONAB regulates epithelial cell proliferation and cell density. J Cell Biol 2003; 160(3):423-432.
34. Citi S, Denisenko N. Phosphorylation of the tight junction protein cingulin and the effect of protein kinase inhibitors and activators in MDCK epithelial cells. J Cell Sci 1995; 108:2917-2926.
35. Tan JL, Ravid S, Spudich JA. Control of nonmuscle myosins by phosphorylation. Annu Rev Biochem 1992; 61:721-759.
36. Knight AE. The diversity of myosin-like proteins. [PhD Thesis] Cambridge 1994.
37. Guillemot L, Hammar E, Kaister C et al. Disruption of the cingulin gene does not prevent tight junction formation but alters gene expression. J Cell Sci 2004; 117:5245-5256.

CHAPTER 6

ZO Proteins and Tight Junction Assembly

Alan S. Fanning

Abstract

Tight junctions are highly structured plasma membrane domains. They are composed of transmembrane proteins, such as occludin and claudin (see Chapters 2 and 3), that are organized into highly cross-linked linear arrays of particles restricted to an apical circumferential band known as the apical junctional complex (AJC). These transmembrane proteins form a selective barrier in the paracellular space that is in turn intimately associated with a large array of tight junction-specific cytoplasmic, cytoskeletal, and signaling molecules (for a recent review see ref. 1). The high degree of order in this structure, and its ability to undergo rapid structural and functional changes in response to environmental stimuli, has created a great deal of interest in the underlying structural and organizational determinants of this array. Thus, much attention has been focused on three homologous polypeptides that are members of a large family of membrane-associated scaffolding proteins: ZO-1, ZO-2 and ZO-3, otherwise known as the ZO proteins. The emerging evidence suggests that these proteins organize not only the unique structure of tight junctions, but may also have a role in the functional organization of adhesion and signaling complexes in diverse cell types.

Introduction

The ZO proteins were the first tight junction-specific molecules to be identified. Although initially isolated in the mid 1980's as TJ-specific epitopes for antibodies raised against a crude tight junction fraction isolated from rat liver,[2] the first of these proteins, ZO-1, was not cloned until the early nineties.[3,4] Sequence analysis revealed that ZO-1, and subsequently ZO-2 and ZO-3, were all members of a larger family which became known as the membrane-associated guanylate kinase homologs (MAGUKs).[5] These are all essentially multi-domain proteins, characterized by a core domain of three protein-binding motifs: a PDZ domain, SH3 domain, and a region of homology to guanylate kinase (Fig. 1). However, each MAGUK may also contain protein binding motifs that are unique to its particular function. Conceptually, this multi-domain structure lends itself well to scaffolding activity, and studies in cellular and animal models have indicated that many, if not all, of these proteins regulate the subcellular distribution of transmembrane and membrane-associated proteins in structures as diverse as neuronal and immune synapses and cell-cell adhesive junctions of polarized cells (for review see 6,7). The ZO proteins, like other MAGUKs, are highly conserved throughout phylogeny, from *hydra vulgaris*[8] to *homo sapiens*, and are thus presumed to be components of conserved cellular function.

By analogy to other MAGUKs, it was predicted that ZO proteins might also be molecular scaffolds, and as such might recruit tight junction proteins and/or organize them within the apical junctional complex. Aside from the core domain of PDZ, SH3 and GUK domains, ZO proteins also contain an additional two PDZ domains within the amino terminal half of the protein, and a mutually conserved domain of unknown function, the "Acidic Domain", that is adjacent to the GUK domain (Fig. 1).[3,4] The ZO proteins are unique among MAGUKs in that with normal tight junction structure and permeability, and with the exception of cingulin all TJ proteins examined had a normal subcellular distribution.[22] Interestingly, as in the studies described above there was a significant delay in the assembly of other tight junction proteins into the AJC and a transient increase in paracellular flux during calcium-switch induced assembly of junctions. While the subtlety of the effects described above could, in part, be explained by the presence of endogenous wild type protein in all of the cell culture assays, or by a redundancy in function among the ZO proteins (as implied in the knock down experiments), they clearly suggest that ZO proteins are required for the normal kinetics of junction assembly. Thus, it is interesting to speculate what similar gene disruption studies in mice might reveal about the requirement of ZO proteins during morphogenetic changes characteristic of early development.

The results from the Eph cell knock out did make one clear point; that ZO-1 is required for the localization of at least one protein, cingulin, to the tight junction. This direct observation supports several other studies that indicate that ZO binding does have a role in recruiting proteins into the tight junction. For example, Roh et al have found that PATJ, a tight junction scaffolding protein that regulates apical-basal polarity, binds to ZO-3 via its sixth PDZ domain and requires this domain for localization to tight junctions.[15] In a separate series of studies, Mitic et al found that the C-terminal ZO-1-binding domain of occludin fused to connexin 43 was both necessary and sufficient to localize this chimeric molecule to the tight junction.[23] In fact, tight junction transmembrane proteins such as occludin and JAM require their respective ZO binding domains for localization to the cadherin-based cell-cell contacts in fibroblasts.[23-25] The same is also true of the cytoplasmic protein cingulin.[26] Although this system is somewhat artificial, especially since many of these proteins are not normally expressed in fibroblasts, it serves well as a reductionist system as a measure of the potential of ZO proteins to direct subcellular localization.

Surprisingly, however, it was found that occludin and claudin constructs lacking their respective ZO binding domains were still able to localize to tight junctions in polarized epithelial cells. Several groups have demonstrated that deletion of the PDZ-binding domains of claudin, which mediate interaction with the first PDZ domain of ZO-1, or deletion of the C-terminal ZO-binding domain of occludin[23,27-30] did not disrupt the ability of these proteins to localize to tight junctions. However, it was observed that over-expression of these mutant transgenes, as well as unaltered transgenes, often resulted in the accumulation of ectopic structures in the lateral plasma membrane.[29-31] These observations suggest ZO proteins may function more to organize the proteins within the AJC, perhaps serving as an "anchor" for the proteins that compose individual fibrils, than they do to recruit proteins to the plasma membrane. They also suggest that the amount of ZO protein available for this role is limited (i.e., saturable). Again, the important caveat to all these experiments is that transgene expression occurs in an otherwise normal background of occludin and claudin expression, and both of these proteins have an inherent ability to self-associate through homotypic and heterotypic interactions.[28,32] Thus, cooperative interactions among endogenous and exogenous claudin and occludin polypeptides (see Chapters 2 and 3) likely mask the underlying mechanisms of assembly in this system. Still, the available evidence supports the hypothesis that ZO-binding can direct subcellular localization.

ZO Proteins as Cytoskeletal Linkers

The observation that ZO proteins bind to both tight junction and cytoskeletal components suggests that one of the critical functions of ZO proteins may be to link the paracellular seal to the actin cytoskeleton. There is a considerable body of evidence that has accumulated to suggest that the cytoskeleton regulates different aspects of tight junction assembly and physiology (see Chapter 10). Thus, the early observation that the C-terminal halves of ZO-1 and ZO-2, and a region in the N-terminal half of ZO-3, could bind directly to F-actin and mediate localization to the cortical cytoskeleton were received with some enthusiasm.[33-36] In ZO-1, this interaction was localized to a 220 aa domain that was both necessary and sufficient for direct interaction with F-actin.[37] These initial findings were followed by the observation that several other cytoskeletal proteins could interact directly with ZO proteins. For example, Benz et al found that a tight junction-specific isoform of protein 4.1, a membrane-anchoring cytoskeletal protein, bound directly to the C-terminal tail of both ZO-1 and ZO-2.[38] Furthermore, in at least one study ZO-1 has also been demonstrated to interact with fodrin, another component of the cortical cytoskeleton.[4] This would appear to support the idea that proteins like ZO-1 provide a link to the cortical cytoskeleton, perhaps acting as an "anchor" to the plasma membrane.

However, this hypothesis seems too simplistic in the face of several more recent observations. First, many of the proteins that bind to ZO-1 such as cingulin, occludin and AF-6/afadin, can also bind independently to F-actin.[39-41] Thus, it is not clear why they would require ZO-1 simply for cytoskeletal attachment. Secondly, ZO-1 has also been demonstrated to interact with several very different proteins that are important regulators of cytoskeletal dynamics; these include cortactin, VASP (a.k.a. Vasodilator-Stimulated Phosphoprotein) and p120 catenin. Cortactin was initially identified by yeast 2 hybrid assay as a binding partner for the Drosophila protein PYD, a homolog of ZO-1,[42] and subsequent studies indicate that the human homolog of cortactin also binds to both ZO-1 and ZO-2[42] (and Fanning et al, unpublished). Cortactin is an F-actin binding protein that promotes the assembly of F-actin at the plasma membrane by the ARP2/3 complex.[43,44] VASP, like cortactin, is also an important regulator of F-actin dynamics, in part through effects on the ARP2/3 complex.[45] Colgan and colleagues found that VASP transiently localizes to the tight junction of PKA-stimulated endothelial and epithelial cells, and could be immunoprecipitated with ZO-1.[46,47]

The C-terminus of ZO-3 also binds directly to p120 catenin, a cortical membrane protein of the cadherin/catenin complex that also appears to have a second role as a regulator of the actin cytoskeleton.[14] Over-expression of a transgene encoding the C-terminus of ZO-3 in MDCK cells resulted in a dramatic reorganization of the actin cytoskeleton and an increase in cell migration.[14] These changes were accompanied by reduction of RhoA activity, a small GTPase that is an important cytoskeletal regulator. Interestingly, these effects were most dramatic in cells that were either in the process of assembling junctions (following calcium-switch) or which had been induced to undergo junction remodeling during wound healing. Polarized cells appeared relatively unaffected.

These observations are reminiscent of studies of ZO protein function in *Drosophila*, which is the only organismal model system that has been exploited to date. Mutations in the *Drosophila* ZO protein PYD can inhibit morphogenetic cell movements and cytoskeletal dynamics within the ectoderm during embryogenesis,[48,49] a process that appears very similar to wound healing at the molecular level. This effect appears to be mediated by direct interaction between PYD and Canoe, the *Drosophila* homolog of the tight junction protein AF-6/afadin. In fact, AF-6/afadin has also been implicated in cytoskeletal regulation in vertebrate cells.[39] This protein binds to profilin, which regulates the addition of actin monomers to growing filaments,[50] and disruption of AF-6/afadin in mice results in an early embryonic lethality that is associated with tight junction disruption and altered cellular polarity.[51,52]

Taken together, these observations suggest that the cytoskeletal activity of ZO proteins might be more dynamic than previously assumed. In fact, the observation that so many cytoskeletal proteins with regulatory activity interact with ZO proteins suggest that ZO proteins may be part of a regulatory mechanism that localizes the machinery of actin polymerization to the AJC and regulates assembly of this specialized cytoskeletal domain.

Organization of Signal Transduction Pathways and Transcription Factor Complexes That Regulate Junction Structure and Function

Tight junction structure and paracellular permeability are regulated by a large number of signaling pathways, although the molecular mechanisms that underlie these effects are still somewhat obscure (reviewed in ref. 53; see Chapters 7-9). Some of the components of these pathways localize to tight junctions, and it is possible to speculate that potential scaffolding proteins like ZO-1, ZO-2 and ZO-3 may integrate the readout of these pathways and translate them into functional or structural changes in tight junctions. For example, ZO-1 binds directly to the heterotrimeric G protein $G\alpha_{12}$, and modulation of $G\alpha_{12}$ activity can alter tight junction structure and permeability.[54,55] ZO-1 also binds netrin receptor-associated protein Neph1, and association between ZO-1 and Neph1 promotes signal transduction through the JNK signaling pathway[56] in cultured fibroblasts. More recently, Benais-Pont and colleagues have demonstrated that ZO-1 interacts directly with a guanine nucleotide exchange factor (GEF) for the Rho GTPase GEF-H1, and that modulation of ZO-1 expression can affect the activity of this cellular signaling protein.[57]

It has also been demonstrated that alteration of ZO protein expression can lead to disruption of cell growth and morphology in some cell types. For example, over-expression of an N-terminal fragment of ZO-1 in both MDCK I cells and a corneal epithelial cell line was associated with changes in cell morphology reminiscent of an epithelial to mesenchymal transition.[58,59] In MDCK I cells, this correlated with increased tumorigenicity and activation of beta catenin signaling.[59] There are also numerous indirect links between ZO proteins, signaling and cancer. Investigators have documented a loss of heterozygosity and a reduction in ZO-1 expression in one study of breast cancer biopsies,[60] and in a second study it was determined that a specific ZO-2 transcript is lost in most pancreatic adenocarcinomas.[61] Furthermore, interaction between ZO-2 and the E4-ORF1 gene product of Adenovirus type 9 promote transforming activity of this viral oncoprotein in fibroblasts.[62] Finally, as noted above genetic studies in *Drosophila* suggest Pyd, the homolog of ZO-1, is required for cell differentiation and morphogenetic movements of epithelial sheets during development.[49,63] These results are suggestive of a role for ZO proteins in cell signaling, although the molecular mechanism responsible for these changes have not been elucidated.

Curiously, ZO proteins also interact with several nuclear transcription factors. For example, ZO-1 binds directly to the Y-box transcription factor ZONAB.[21,64] ZONAB binds directly to the promoter sequences of the ErbB2 gene, a transmembrane receptor that regulates cell growth and differentiation, and inhibits expression of transgenes driven by ErbB2 promoter.[21] ZO-1 expression can reverse the inhibitory effect of ZONAB in reporter assays, and can also promote expression of ErbB2 gene in vivo. Similarly, ZO-2 binds directly to the Fos, Jun, and C/EBP transcription factors, and expression of ZO-2 in cultured cells can down-regulate the activity of a target promoter recognized by these proteins.[65] In *Drosophila*, mutations in Pyd reduce transcription of the hLh transcription factor *Extramachrochaeta*, resulting in ectopic bristle formation.[66] It is not yet clear exactly how ZO proteins regulate the activity of these transcription factors. ZO-1 and ZO-2 can be found in the nuclei of cells undergoing dynamic structural reorganizations, as is found in cells growing at low density, in cells adjacent to a wound margin, and during junction assembly following calcium switch.[67,68] Thus it is possible that ZO proteins might directly regulate transcription by association with transcription factor complexes, although there is currently no evidence to support this hypothesis in vertebrates. A more likely possibility is that ZO proteins regulate transcription indirectly by sequestering these transcription factors at the AJC. ZONAB, Fos and Jun all colocalize with ZO proteins at the tight junction in polarized cells. Furthermore, over-expression of ZO-1 can prevent accumulation of ZONAB and an associated cell cycle kinase, CDK4, in the nucleus.[21,64] It is tempting to speculate that the position of ZO proteins at the AJC allows them to transduce information on the state of cell-cell contact and modulate the appropriate cellular response by associated transcription factors. Many of the factors described above are known regulators of cell growth and differentiation. What still remains to be determined is what specific affect, if any, that the transcription factors associated with ZO-1 may have on the expression of genes involved in assembly and permeability of tight junctions.

Extension of the Paradigm: ZO Proteins and Global Organization of Cell-Cell Junctions

Given that ZO proteins were among the first tight junction proteins identified, it is perhaps surprising that the function of these molecules is still so poorly understood. This has been due, in part, to technical difficulties associated the study of large and ubiquitously distributed polypeptides. The ZO proteins have proven to be very difficult to work with in vitro, and until recently there were no cell lines available that lacked ZO-proteins or vertebrate model systems with targeted disruptions in these genes. This has meant that there were serious limitations on the type of structure-function analysis that could be applied. However, the current evidence clearly supports a role for ZO proteins in the organization of other tight junction proteins in the apical junction complex, in the cytoskeletal organization of the adjacent cortex, and in the signaling pathways that transect this cellular complex.

Importantly, there is also evidence that ZO proteins may have a similar functional role outside the tight junction. ZO proteins are components of cadherin-based cell-contacts in the adherens junctions of fibroblasts, the intercalated disks of cardiac myocytes,[4,69] and *puncta adherentes* that link neurons and dendrites within the hippocampus.[70] In fact, Imamura et al demonstrated that ZO-1 can bind directly to alpha catenin, a component of cadherin based cell-cell contacts, and that a large fragment of alpha catenin that contains this binding site (and perhaps several other binding domains) is necessary for cell-cell adhesion in fibroblasts[71]. ZO proteins also bind directly to the members of the p120 catenin family of proteins, which are cortical membrane proteins associated with cadherin at cell-cell junctions. In addition to the interaction between ZO-3 and p120 catenin described above, both ZO-1 and ZO-2 bind to ARVCF (Armadillo-repeat gene deleted in Velo-cardio-facial syndrome).[72] Both p120 catenin and ARVCF bind directly to the membrane proximal region of cadherins, and p120 has a clearly defined role in cell-cell adhesion.[73] Interestingly, the ZO-binding motif in ARVCF appears to be required for interaction with E-cadherin and localization to cell junctions.[72] These observations clearly suggest that ZO proteins have a distinct role in cadherin-mediated cell-cell adhesion.

ZO proteins are also associated with gap junctions in a large number of tissues, including retinal neurons,[74,75] oligodendrocytes,[76] epithelial cells,[77-79] vascular smooth muscle cells[79] and cardiac myocytes.[80] Although the functional role of this association is unclear (for review see ref. 81), preliminary studies suggest that ZO proteins may regulate the ultimate size of gap junction plaques.[82] It also cannot be ruled out that ZO proteins may have functional roles in the nervous system similar to other MAGUKs such as PSD-95 and SAP97/hDLG. ZO-1 has been specifically localized to synapses of pyramidal neurons by immunoelectron microscopy, where it is found in presynaptic axonal terminals.[83] Taken together, the current data suggest a more generalized function for ZO proteins at regions of cell-cell contact, and it is likely that the paradigms established in tight junctions may hold true in these structures as well. In other words, ZO proteins may serve as organizational scaffolds or cytoskeletal linkers at a variety of different specialized plasma membrane domains.

It is also highly likely that other tight junction proteins serve as scaffolding proteins with a key role in tight junction assembly. This may be particularly true of members of the PAR3/PAR6/aPKC and the CRB3/PATJ/PALS1 protein complexes. Like the ZO proteins, the polypeptides that compose these two complexes are all multi-domain proteins that localize to the tight junction in polarized cells and that are highly conserved through phylogeny. Disruption of the expression and/or localization of these polypeptides can disrupt cell polarity, tight junction structure and paracellular permeability (for review see refs. 84,85). For example, over-expression of PAR-3 promotes tight junction assembly following calcium switch,[86] whereas expression of dominant negative transgenes encoding PAR-6 or aPKC zeta disrupt tight junction assembly and the formation of the paracellular barrier.[17,87] Similarly, over-expression of CRB3, PATJ and PALS1 transgenes either disrupt localization of tight junction-associated proteins[88,89] or the formation of the paracellular barrier.[17,90,91] These observations suggest that these "polarity" proteins play a critical role in junction assembly, and imply that the formation of cell-cell contacts and the establishment of cell polarity are tightly linked. In addition, there are also at least two other multi-domain proteins, MAGI and MUPP1, that localize to the tight junction and bind to transmembrane proteins like JAM and the claudins,[92-94] although the functional roles of these proteins have not been defined.

Future Directions

It has become clear that the resolution of many of the issues relevant to the functional role of ZO proteins will require use of genetic systems in which expression of ZO proteins can be controlled in a tissue specific manner. The first steps in making cell lines lacking ZO proteins have been recently reported,[22] but these are laborious experiments and have some obvious limitations. For example, they do not allow the examination of the physiology of junctions or the dynamic cellular processes characteristic of development, and the issue of genetic redundancy between ZO proteins is difficult to circumvent. Mice or zebrafish provide attractive model systems, and the existence of a ZO-3 knockout in mice has been briefly mentioned in the literature.[95] The great benefit of these systems is their ability to address physiological and developmental questions and to serve as model systems of human disease. This is especially relevant given the recent identification of several genetic diseases arising from mutations in tight junction proteins,[96-100] including the identification of ZO-2 as a candidate gene in Familial Hypercholanemia.[101] Genetic redundancy between the function of different scaffolding proteins may pose an issue is these vertebrate systems, especially as they have several ZO proteins, but in these genetic systems this can be more easily circumvented through selective breeding of different lines. This problem of redundancy would not be an issue in invertebrate systems like *C.elegans* and *D. Melanogaster*, both of which have only one ZO gene. These systems have been generally overlooked, but both have adherens junctions, an epithelial barrier and defined organ systems (Reviewed in 102-104) and as such have a great potential to advance our understanding of the more fundamental cellular functions of ZO proteins.

Acknowledgements

The author would like to thank Dr. Christina Van Itallie for a critical reading of the manuscript. This work was supported by a grant from the National Institute of Diabetes and Digestive and Kidney Diseases (RO1 DK061397) and the State of North Carolina, USA.

References

1. Schneeberger EE, Lynch RD. The tight junction: A multifunctional complex. Am J Physiol Cell Physiol 2004; 286(6):C1213-C1228.
2. Stevenson BR, Siliciano JD, Mooseker MS et al. Identification of ZO-1: A high molecular weight polypeptide associated with the tight junction (zonula occludens) in a variety of epithelia. J Cell Biol 1986; 103(3):755-766.
3. Willott E, Balda MS, Fanning AS et al. The tight junction protein ZO-1 is homologous to the Drosophila discs-large tumor suppressor protein of septate junctions. Proc Natl Acad Sci USA 1993; 90(16):7834-7838.
4. Itoh M, Nagafuchi A, Yonemura S et al. The 220-kD protein colocalizing with cadherins in nonepithelial cells is identical to ZO-1, a tight junction-associated protein in epithelial cells: cDNA cloning and immunoelectron microscopy. J Cell Biol 1993; 121(3):491-502.
5. Gonzalez-Mariscal L, Betanzos A, Avila-Flores A. MAGUK proteins: Structure and role in the tight junction. Semin Cell Dev Biol 2000; 11(4):315-324.
6. Fanning AS, Anderson JM. Protein modules as organizers of membrane structure. Curr Opin Cell Biol 1999; 11(4):432-439.
7. Kim EJ, Sheng M. PDZ domain proteins of synapses. Nat Rev Neurosci 2004; 5(10):771-781.
8. Fei KY, Yan L, Zhang JS et al. Molecular and biological characterization of a zonula occludens-1 homologue in Hydra vulgaris, named HZO-1. Dev Genes Evol 2000; 210(12):611-616.
9. Leonardo ED, Hinck L, Masu M et al. Vertebrate homologues of C-elegans UNC-5 are candidate netrin receptors. Nature 1997; 386(6627):833-838.
10. Willott E, Balda MS, Heintzelman M et al. Localization and differential expression of 2 isoforms of the tight junction protein ZO-1. Am J Physiol 1992; 262(5):C1119-C1124.

11. Chlenski A, Ketels KV, Engeriser JL et al. ZO-2 gene alternative promoters in normal and neoplastic human pancreatic duct cells. Int J Cancer 1999; 83(3):349-358.
12. Tsukita S, Furuse M, Itoh M. Multifunctional strands in tight junctions. Nat Rev Mol Cell Biol 2001; 2(4):285-293.
13. Gonzalez-Mariscal L, Betanzos A, Nava P et al. Tight junction proteins. Prog Biophys Mol Biol 2003; 81(1):1-44.
14. Wittchen ES, Haskins J, Stevenson BR. NZO-3 expression causes global changes to actin cytoskeleton in Madin-Darby canine kidney cells: Linking a tight junction protein to Rho GTPases. Mol Biol Cell 2003; 14(5):1757-1768.
15. Roh MH, Liu CJ, Laurinec S et al. The carboxyl terminus of zona occludens-3 binds and recruits a mammalian homologue of discs lost to tight junctions. J Biol Chem 2002; 277(30):27501-27509.
16. Traweger A, Fuchs R, Krizbai IA et al. The tight junction protein ZO-2 localizes to the nucleus and interacts with the heterogeneous nuclear ribonucleoprotein scaffold attachment factor-B. J Biol Chem 2003; 278(4):2692-2700.
17. Suzuki A, Ishiyama C, Hashiba K et al. aPKC kinase activity is required for the asymmetric differentiation of the premature junctional complex during epithelial cell polarization. J Cell Sci 2002; 115(18):3565-3573.
18. Sheth B, Fontaine JJ, Ponza E et al. Differentiation of the epithelial apical junctional complex during mouse preimplantation development: A role for rab13 in the early maturation of the tight junction. Mech Dev 2000; 97(1-2):93-104.
19. Yonemura S, Itoh M, Nagafuchi A et al. Cell-to-cell adherens junction formation and actin filament organization: Similarities and differences between nonpolarized fibroblasts and polarized epithelial cells. J Cell Sci 1995; 108(Pt 1):127-142.
20. Ando-Akatsuka Y, Yonemura S, Itoh M et al. Differential behavior of E-cadherin and occludin in their colocalization with ZO-1 during the establishment of epithelial cell polarity. J Cell Physiol 1999; 179(2):115-125.
21. Balda MS, Matter K. The tight junction protein ZO-1 and an interacting transcription factor regulate ErbB-2 expression. EMBO J 2000; 19(9):2024-2033.
22. Umeda K, Matsui T, Nakayama M et al. Establishment and characterization of cultured epithelial cells lacking expression of ZO-1. J Biol Chem 2004; 279(43):44785-44794.
23. Mitic LL, Schneeberger EE, Fanning AS et al. Connexin-occludin chimeras containing the ZO-binding domain of occludin localize at MDCK tight junctions and NRK cell contacts. J Cell Biol 1999; 146(3):683-693.
24. Ebnet K, Aurrand-Lions M, Kuhn A et al. The junctional adhesion molecule (JAM) family members JAM-2 and JAM-3 associate with the cell polarity protein PAR-3: A possible role for JAMs in endothelial cell polarity. J Cell Sci 2003; 116(19):3879-3891.
25. Ebnet K, Schulz CU, Brickwedde MKMZ et al. Junctional adhesion molecule interacts with the PDZ domain-containing proteins AF-6 and ZO-1. J Biol Chem 2000; 275(36):27979-27988.
26. D'Atri F, Nadalutti F, Citi S. Evidence for a functional interaction between cingulin and ZO-1 in cultured cells. J Biol Chem 2002; 277(31):27757-27764.
27. Balda MS, Whitney JA, Flores C et al. Functional dissociation of paracellular permeability and transepithelial electrical resistance and disruption of the apical-basolateral intramembrane diffusion barrier by expression of a mutant tight junction membrane protein. J Cell Biol 1996; 134(4):1031-1049.
28. Chen YH, Merzdorf C, Paul DL et al. COOH terminus of occludin is required for tight junction barrier function in early Xenopus embryos. J Cell Biol 1997; 138(4):891-899.
29. Van Itallie CM, Colegio OR, Anderson JM. The cytoplasmic tails of claudius can influence tight junction barrier properties through effects on protein stability. J Membr Biol 2004; 199(1):29-38.
30. McCarthy KM, Francis SA, McCormack JM et al. Inducible expression of claudin-1-myc but not occludin-VSV-G results in aberrant tight junction strand formation in MDCK cells. J Cell Sci 2000; 113:3387-3398.
31. Colegio OR, Van Itallie CM, Mccrea HJ et al. Claudins create charge-selective channels in the paracellular pathway between epithelial cells. Am J Physiol Cell Physiol 2002; 283(1):C142-C147.
32. Furuse M, Sasaki H, Tsukita S. Manner of interaction of heterogeneous claudin species within and between tight junction strands. J Cell Biol 1999; 147(4):891-903.
33. Fanning AS, Jameson BJ, Jesaitis LA et al. The tight junction protein ZO-1 establishes a link between the transmembrane protein occludin and the actin cytoskeleton. J Biol Chem 1998; 273(45):29745-29753.
34. Itoh M, Morita K, Tsukita S. Characterization of ZO-2 as a MAGUK family member associated with tight as well as adherens junctions with a binding affinity to occludin and alpha catenin. J Biol Chem 1999; 274(9):5981-5986.

35. Itoh M, Nagafuchi A, Moroi S et al. Involvement of ZO-1 in cadherin-based cell adhesion through its direct binding to alpha catenin and actin filaments. J Cell Biol 1997; 138(1):181-192.
36. Wittchen ES, Haskins J, Stevenson BR. Exogenous expression of the amino-terminal half of the tight junction protein ZO-3 perturbs junctional complex assembly. J Cell Biol 2000; 151(4):825-836.
37. Fanning AS, Ma TY, Anderson JM. Isolation and functional characterization of the actin-binding region in the tight junction protein ZO-1. FASEB J 2002; 16(11).
38. Mattagajasingh SN, Huang SC, Hartenstein JS et al. Characterization of the interaction between protein 4.1R and ZO-2. A possible link between the tight junction and the actin cytoskeleton. J Biol Chem 2000; 275(39):30573-30585.
39. Mandai K, Nakanishi H, Satoh A et al. Afadin: A novel actin filament-binding protein with one PDZ domain localized at cadherin-based cell-to-cell adherens junction. J Cell Biol 1997; 139(2):517-528.
40. Wittchen ES, Haskins J, Stevenson BR. Protein interactions at the tight junction. Actin has multiple binding partners, and ZO-1 forms independent complexes with ZO-2 and ZO-3. J Biol Chem 1999; 274(49):35179-35185.
41. D'Atri F, Citi S. Cingulin interacts with F-actin in vitro. FEBS Lett 2001; 507(1):21-24.
42. Katsube T, Takahisa M, Ueda R et al. Cortactin associates with the cell-cell junction protein ZO-1 in both Drosophila and mouse. J Biol Chem 1998; 273(45):29672-29677.
43. Weaver AM, Karginov AV, Kinley AW et al. Cortactin promotes and stabilizes Arp2/3-induced actin filament network formation. Curr Biol 2001; 11(5):370-374.
44. Uruno T, Liu JL, Zhang PJ et al. Activation of Arp2/3 complex-mediated actin polymerization by cortactin. Nat Cell Biol 2001; 3(3):259-266.
45. Krause M, Dent EW, Bear JE et al. ENA/VASP proteins: Regulators of the actin cytoskeleton and cell migration. Annu Rev Cell Dev Biol 2003; 19:541-564.
46. Comerford KM, Lawrence DW, Synnestvedt K et al. Role of vasodilator-stimulated phosphoprotein in protein kinase A-induced changes in endothelial junctional permeability. FASEB J 2002; 16(2).
47. Lawrence DW, Comerford KM, Colgan SP. Role of VASP in reestablishment of epithelial tight junction assembly after Ca2+ switch. Am J Physiol Cell Physiol 2002; 282(6):C1235-C1245.
48. Takahashi K, Matsuo T, Katsube T et al. Direct binding between two PDZ domain proteins Canoe and ZO-1 and their roles in regulation of the jun N-terminal kinase pathway in Drosophila morphogenesis. Mech Dev 1998; 78(1-2):97-111.
49. Wei XR, Ellis HM. Localization of the Drosophila MAGUK protein Polychaetoid is controlled by alternative splicing. Mech Dev 2001; 100(2):217-231.
50. Boettner B, Govek EE, Cross J et al. The junctional multidomain protein AF-6 is a binding partner of the Rap1A GTPase and associates with the actin cytoskeletal regulator profilin. Proc Natl Acad Sci USA 2000; 97(16):9064-9069.
51. Zhadanov AB, Provance DWJ, Speer CA et al. Absence of the tight junctional protein AF-6 disrupts epithelial cell-cell junctions and cell polarity during mouse development. Curr Biol 1999; 9(16):880-888.
52. Ikeda W, Nakanishi H, Miyoshi J et al. Afadin: A key molecule essential for structural organization of cell-cell junctions of polarized epithelia during embryogenesis. J Cell Biol 1999; 146(5):1117-1131.
53. Matter K, Balda MS. Signalling to and from tight junctions. Nat Rev Mol Cell Biol 2003; 4(3):225-236.
54. Meyer TN, Hunt J, Schwesinger C et al. G alpha(12) regulates epithelial cell junctions through Src tyrosine kinases. Am J Physiol Cell Physiol 2003; 285(5):C1281-C1293.
55. Denker BM, Saha C, Khawaja S et al. Involvement of a heterotrimeric G protein alpha subunit in tight junction biogenesis. J Biol Chem 1996; 271(42):25750-25753.
56. Huber TB, Schmidts M, Gerke P et al. The carboxyl terminus of Neph family members binds to the PDZ domain protein zonula occludens-1. J Biol Chem 2003; 278(15):13417-13421.
57. Benais-Pont G, Punn A, Flores-Maldonada C et al. Identification of a tight junction-associated guanine nucleotide exchange factor that activates Rho and regulates paracellular permeability. J Cell Biol 2003; 160(5):729-740.
58. Ryeom SW, Paul D, Goodenough DA. Truncation mutants of the tight junction protein ZO-1 disrupt corneal epithelial cell morphology. Mol Biol Cell 2000; 11(5):1687-1696.
59. Reichert M, Muller T, Hunziker W. The PDZ domains of zonula occludens-1 induce an epithelial to mesenchymal transition of Madin-Darby canine kidney I cells - Evidence for a role of beta-catenin/Tcf/Lef signaling. J Biol Chem 2000; 275(13):9492-9500.
60. Hoover KB, Liao SY, Bryant PJ. Loss of the tight junction MAGUK ZO-1 in breast cancer - Relationship to glandular differentiation and loss of heterozygosity. Am J Pathol 1998; 153(6):1767-1773.

61. Chlenski A, Ketels KV, Tsao MS et al. Tight junction protein ZO-2 is differentially expressed in normal pancreatic ducts compared to human pancreatic adenocarcinoma. Int J Cancer 1999; 82(1):137-144.
62. Glaunsinger BA, Weiss RS, Lee SS et al. Link of the unique oncogenic properties of adenovirus type 9 E4-ORF1 to a select interaction with the candidate tumor suppressor protein ZO-2. EMBO J 2001; 20(20):5578-5586.
63. Takahashi K, Matsuo T, Katsube T et al. Direct binding between two PDZ domain proteins Canoe and ZO-1 and their roles in regulation of the Jun N-terminal kinase pathway in Drosophila morphogenesis. Mech Dev 1998; 78(1-2):97-111.
64. Balda MS, Garrett MD, Matter K. The ZO-1-associated Y-box factor ZONAB regulates epithelial cell proliferation and cell density. J Cell Biol 2003; 160(3):423-432.
65. Betanzos A, Huerta M, Lopez-Bayghen E et al. The tight junction protein ZO-2 associates with Jun, Fos and C/EBP transcription factors in epithelial cells. Exp Cell Res 2004; 292(1):51-66.
66. Takahisa M, Togashi S, Suzuki T et al. The Drosophila tamou gene, a component of the activating pathway of extramacrochaetae expression, encodes a protein homologous to mammalian cell-cell junction-associated protein ZO-1. Genes Dev 1996; 10(14):1783-1795.
67. Gottardi CJ, Arpin M, Fanning AS et al. The junction-associated protein, zonula occludens-1, localizes to the nucleus before the maturation and during the remodeling of cell-cell contacts. Proc Natl Acad Sci USA 1996; 93(20):10779-10784.
68. Jaramillo BE, Ponce A, Moreno J et al. Characterization of the tight junction protein ZO-2 localized at the nucleus of epithelial cells. Exp Cell Res 2004; 297(1):247-258.
69. Jesaitis LA, Goodenough DA. Molecular characterization and tissue distribution of ZO-2, a tight junction protein homologous to ZO-1 and the Drosophila discs-large tumor suppressor protein. J Cell Biol 1994; 124(6):949-961.
70. Inagaki M, Irie K, Deguchi-Tawarada M et al. Nectin-dependent localization of ZO-1 at puncta adhaerentia junctions between the mossy fiber terminals and the dendrites of the pyramidal cells in the CA3 area of adult mouse hippocampus. J Comp Neurol 2003; 460(4):514-524.
71. Imamura Y, Itoh M, Maeno Y et al. Functional domains of alpha-catenin required for the strong state of cadherin-based cell adhesion. J Cell Biol 1999; 144(6):1311-1322.
72. Kausalya PJ, Reichert M, Hunziker W. Connexin45 directly binds to ZO-1 and localizes to the tight junction region in epithelial MDCK cells. FEBS Lett 2001; 505(1):92-96.
73. Reynolds AB, Roczniak-Ferguson A. Emerging roles for p120-catenin in cell adhesion and cancer. Oncogene 2004; 23(48):7947-7956.
74. Li XB, Olson C, Lu SJ et al. Neuronal connexin36 association with zonula occludens-1 protein (ZO-1) in mouse brain and interaction with the first PDZ domain of ZO-1. Eur J Neurosci 2004; 19(8):2132-2146.
75. Rash JE, Pereda A, Kamasawa N et al. High-resolution proteomic mapping in the vertebrate central nervous system: Close proximity of connexin35 to NMDA glutamate receptor clusters and colocalization of connexin36 with immunoreactivity for zonula occludens protein-1 (ZO-1). J Neurocytol 2004; 33(1):131-151.
76. Li X, Ionescu AV, Lynn BD et al. Connexin47, connexin29 and connexin32 coexpression in oligodendrocytes and Cx47 association with zonula occludens-1 (Zo-1) in mouse brain. Neuroscience 2004; 126(3):611-630.
77. Giepmans BNG, Moolenaar WH. The gap junction protein connexin43 interacts with the second PDZ domain of the zona occludens-1 protein. Curr Biol 1998; 8(16):931-934.
78. Kojima T, Kokai Y, Chiba H et al. Cx32 but not Cx26 is associated with tight junctions in primary cultures of rat hepatocytes. Exp Cell Res 2001; 263(2):193-201.
79. Nielsen PA, Baruch A, Giepmans BNG et al. Characterization of the association of connexins and ZO-1 in the lens. Cell Commun Adhes 2001; 8(4-6):213-217.
80. Toyofuku T, Yabuki M, Otsu K et al. Direct association of the gap junction protein connexin-43 with ZO-1 in cardiac myocytes. J Biol Chem 1998; 273(21):12725-12731.
81. Giepmans BNG. Gap junctions and connexin-interacting proteins. Cardiovasc Res 2004; 62(2):233-245.
82. Hunter AW, Jourdan J, Gourdie RG. Fusion of GFP to the carboxyl terminus of connexin43 increases gap junction size in HeLa cells. Cell Commun Adhes 2003; 10(4-6):211-214.
83. Buchert M, Schneider S, Meskenaite V et al. The junction-associated protein AF-6 interacts and clusters with specific EPH receptor tyrosine kinases at specialized sites of cell-cell contact in the brain. J Cell Biol 1999; 144(2):361-371.
84. Roh MH, Margolis B. Composition and function of PDZ protein complexes during cell polarization. Am J Physiol Renal Physiol 2003; 285(3):F377-F387.

85. Nelson WJ. Adaptation of core mechanisms to generate cell polarity. Nature 2003; 422(6933):766-774.
86. Hirose T, Izumi Y, Nagashima Y et al. Involvement of ASIP/PAR-3 in the promotion of epithelial tight junction formation. J Cell Sci 2002; 115(12):2485-2495.
87. Gao L, Joberty G, Macara IG. Assembly of epithelial tight is negatively regulated by junctions Par6. Curr Biol 2002; 12(3):221-225.
88. Hurd TW, Gao L, Roh MH et al. Direct interaction of two polarity complexes implicated in epithelial tight junction assembly. Nat Cell Biol 2003; 5(2):137-142.
89. Lemmers C, Michel D, Lane-Guermonprez L et al. CRB3 binds directly to Par6 and regulates the morphogenesis of the tight junctions in mammalian epithelial cells. Mol Biol Cell 2004; 15(3):1324-1333.
90. Roh MH, Fan SL, Liu CJ et al. The Crumbs3-Pals1 complex participates in the establishment of polarity in mammalian epithelial cells. J Cell Sci 2003; 116(14):2895-2906.
91. Straight SW, Shin K, Fogg VC et al. Loss of PALS1 expression leads to tight junction and polarity defects. Mol Biol Cell 2004; 15(4):1981-1990.
92. Jeansonne B, Lu Q, Goodenough DA et al. Claudin-8 interacts with multi-PDZ domain protein 1 (MUPP1) and reduces paracellular conductance in epithelial cells. Cell Mol Biol 2003; 49(1):13-21.
93. Hamazaki Y, Itoh M, Sasaki H et al. Multi-PDZ domain protein 1 (MUPP1) is concentrated at tight junctions through its possible interaction with claudin-1 and junctional adhesion molecule. J Biol Chem 2002; 277(1):455-461.
94. Hirabayashi S, Tajima M, Yao I et al. JAM4, a junctional cell adhesion molecule interacting with a tight junction protein, MAGI-1. Mol Cell Biol 2003; 23(12):4267-4282.
95. Inoko A, Itoh M, Tamura A et al. Expression and distribution of ZO-3, a tight junction MAGUK protein, in mouse tissues. Genes Cells 2003; 8(11):837-845.
96. Simon DB, Lu Y, Choate KA et al. Paracellin-1, a renal tight junction protein required for paracellular Mg2+ resorption. Science 1999; 285(5424):103-106.
97. Wilcox ER, Burton QL, Naz S et al. Mutations in the gene encoding tight junction claudin-14 cause autosomal recessive deafness DFNB29. Cell 2001; 104(1):165-172.
98. Brancolini C, Edomi P, Marzinotto S et al. Exposure at the cell surface is required for Gas3/PMP22 to regulate both cell death and cell spreading: Implication for the Charcot-Marie-Tooth type 1A and Dejerine-Sottas diseases. Mol Biol Cell 2000; 11(9):2901-2914.
99. Wilson FH, Disse-Nicodeme S, Choate KA et al. Human hypertension caused by mutations in WNK kinases. Science 2001; 293(5532):1107-1112.
100. Hadj-Rabia S, Baala L, Vabres P et al. Claudin-1 gene mutation in Neonatal chthyosis-Sclerosing cholangitis. Gastroenterology 2004; 127(5):1386-1390.
101. Carlton VEH, Harris BZ, Puffenberger EG et al. Complex inheritance of familial hypercholanemia with associated mutations in TJP2 and BAAT. Nat Genet 2003; 34(1):91-96.
102. Cox EA, Hardin J. Sticky worms: Adhesion complexes in C-elegans. J Cell Sci 2004; 117(10):1885-1897.
103. Tepass U, Tanentzapf G. Epithelial cell polarity and cell junctions in Drosophila. Ann Rev Genet 2001; 35:747-784.
104. Tepass U. Claudin complexities at the apical junctional complex. Nat Cell Biol 2003; 5(7):595-597.
105. Kausalya PJ, Phua DC, Hunziker W. Association of ARVCF with ZO-1 and ZO-2: Binding to PDZ-domain proteins and cell-cell adhesion regulate plasma membrane and nuclear localization of ARVCF. Mol Biol Cell 2004; 15(12):5503-5515.
106. Mandai K, Nakanishi H, Satoh A et al. Afadin: A novel actin filament-binding protein with one PDZ domain localized at cadherin-based cell-to-cell adherens junction. J Cell Biol 1997; 139(2):517-528.
107. Itoh M, Furuse M, Morita K et al. Direct binding of three tight junction-associated MAGUKs, ZO-1, ZO-2, and ZO-3, with the COOH termini of claudins. J Cell Biol 1999; 147(6):1351-1363.
108. Nielsen PA, Baruch A, Shestopalov VI et al. Lens Connexins alpha 3Cx46 and alpha 8Cx50 interact with zonula occludens protein-1 (ZO-1). Mol Biol Cell 2003; 14(6):2470-2481.
109. Balda MS, Anderson JM, Matter K. The SH3 domain of the tight junction protein ZO-1 binds to a serine protein kinase that phosphorylates a region C-terminal to this domain. FEBS Lett 1996; 399(3):326-332.
110. Meyer TN, Schwesinger C, Denker BM. Zonula occludens-1 is a scaffolding protein for signaling molecules - G alpha(12) directly binds to the Src homology 3 domain and regulates paracellular permeability in epithelial cells. J Biol Chem 2002; 277(28):24855-24858.
111. Roh MH, Makarova O, Liu CJ et al. The Maguk protein, Pals1, functions as an adapter, linking mammalian homologues of Crumbs and Discs Lost. J Cell Biol 2002; 157(1):161-172.

CHAPTER 7

TJ Proteins That Make Round Trips to the Nucleus

Esther Lopez-Bayghen, Blanca Estela Jaramillo, Miriam Huerta, Abigail Betanzos and Lorenza Gonzalez-Mariscal

Abstract

The tight junction (TJ) located at the limit between the apical and basolateral plasma membranes, is a multiprotein complex integrated by both integral and cortical proteins. Through TJ epithelial cells establish a link with their neighbors that seals the paracellular pathway. Lately some TJ proteins like the MAGUK ZO-1 and ZO-2, MAGI 1c, as well as the unrelated proteins symplekin and ubinuclein, have been found to concentrate at the nucleus. In this chapter we describe such proteins and how their arrival to the nucleus is connected to the degree of cell-cell contact. We analyze the signals present in these TJ proteins that may be responsible for their movement from the membrane to the nucleus and vice-versa. We then detail, the interaction of these proteins to nuclear molecules involved in gene transcription, chromatin remodeling, RNA processing and polyadenylation.

In recent times, cell biologists have begun to recognize the dual location of certain proteins within the same cells. Such proteins appear to work as general constituents of two distant and different structures: they work as submembranous components of intercellular junctions and are also located in karyoplasms, Cajal bodies, or spliceosomes even in cells devoid of cell-cell junctions. Such proteins have recently been referred as NACos, for proteins that can localize to the nucleus and adhesion complexes.[1]

This chapter will deal with TJ proteins that shuttle between the plasma membrane and the nucleus. In all the cases so far studied, the subcellular distribution of the TJ NACo proteins is sensitive to the degree of cell-cell contact. Thus in epithelia cultured in a sparse condition, TJ NACos concentrate at the nucleus, whereas in a confluent state, they accumulate at the TJ and only a negligible proportion is maintained at the nucleus. Such behavior suggests that these proteins that mediate intercellular adhesion, also transmit information to the cell interior about the environment, such as the lack of neighboring cells. This information is crucial for determining epithelial behavior, especially for keeping the balance between proliferation and differentiation.

TJ Proteins Participate in Maintaining the Equilibrium between Proliferation and Differentiation

Epithelial cells constitute the boundary between the environment and the internal milieu of the organism. They are found lining the body cavities (e.g., stomach, uterus, intestine), the surface of the skin and forming the glandular tissues (mammary glands, liver, pancreas, etc.). In mammals the majority of cancers originate from epithelial cells. Benign tumors are generally well differentiated, whereas malignant ones display a wide range of un-differentiation, where

Tight Junctions, edited by Lorenza Gonzalez-Mariscal. ©2006 Landes Bioscience and Springer Science+Business Media.

loss of apicolateral polarity and intercellular junctions are frequently encountered. In recent years a number of components found in junctional complexes of polarized epithelial cells have been shown to have signaling functions involved in cell growth and differentiation.[2,3]

The first indication that TJ proteins could be related with the regulation of epithelial transformation, arose with the discovery that ZO-1 has a high homology with the product of the lethal discs-large (*dlg*) tumor suppressor gene of *Drosophila*.[4,5] ZO-1 and Dlg belong to the family of membrane associated guanylate kinase homologues (MAGUK), whose members are characterized for presenting a multidomain organization that includes PDZ repeats, SH3 and GK domains.[6] Considerable evidence has since accumulated, to confirm that genetic loss of ZO proteins is in fact coupled to cancer progression.[7-12] Furthermore, the tumorigenic potential of several viral proteins has been shown to depend on their ability to sequester junctional proteins MAGI-1, MUPP-1, Dlg, hScrib and ZO-2, thereby neutralizing the transformation repressive activity exerted by these junctional proteins (for detailed information on such issue refer to Chapter 9 by Ronald Javier).[13-19]

The recent discovery that TJ plaque proteins shuttle between the TJ and the nucleus, as well as the observation that these proteins associate with transcription factors and proteins involved in cell cycle regulation and chromatin remodeling, reinforces the idea that TJ plaque proteins, are necessary for maintaining the equilibrium between cell proliferation and differentiation.

Conditions Upon Which TJ Proteins Are Found at the Nucleus

The TJ plaque proteins symplekin, ZO-1 and ZO-2 shuttle between the plasma membrane and the nucleus. Other proteins, which have normally been identified as nuclear molecules, have in recent times been discovered to travel to the TJ under certain conditions. Such is the case of ubinuclein and the transcription factors ZONAB, huASH1, Jun, Fos and C/EBP. The results obtained studying these proteins indicate that their sub-cellular localization varies according to the degree of cell-cell contact in the epithelia. Thus in cultured epithelial monolayers, NACos concentrate at the nucleus of sparse cultures, while in a confluent condition they localize at the TJ. However, if these tissues are afflicted with cancer, TJ labeling diminishes drastically and nuclear staining becomes instead conspicuous. In multi-stratified tissues, the lowermost layers of the epithelium, which are the least differentiated and remain in a proliferative state, exhibit a strong nuclear distribution of NACos (Fig. 1). The following paragraphs of this section will describe in detail the distribution of different TJ NACos under variable culture conditions and in diverse epithelia.

ZO-1 and ZO-2

ZO-1 was the first TJ protein shown to be present at the nucleus of epithelial cells. In confluent cultures ZO-1 and ZO-2 are restricted to the plasma membrane, whereas in sparse monolayers they localize to the nucleus. This movement can be triggered in confluent monolayers by a mechanical injury. In such case, the cells that surround the wounded region begin to express these proteins at the nucleus. Furthermore, chemical stress induced by the addition of $CdCl_2$ and heat shock (42°C), have shown to increase nuclear ZO-2 levels.[20] The presence of Ca^{2+} at the surrounding media appears not to be determinant for the localization of ZO-2 at the nucleus. Since monolayers cultured at confluency under a low Ca^{2+} condition (LC; 1-5 μM Ca^{2+}) have ZO-2 diffusely distributed in punctae at the cytosol, while sparse cultures concentrate ZO-2 at the nucleus, independently of the presence or absence of Ca^{2+} in the bathing media (Fig. 2).[21] If inhibitors of protein synthesis are added during a switch of confluent monolayers from LC to normal Ca^{2+} media (NC; 1.8 mM), the transepithelial electrical resistance (TER) develops normally, thus indicating that TJ are being assembled with previously synthesized components, apparently stored in cytosolic vesicles, that arrive at the TJ region by an exocytic fusion process.[22]

The appearance of ZO-2 at the nucleus, upon monolayer injury, is again independent of protein synthesis, thus suggesting that nuclear ZO-2 originate from a preexisting pool of protein present at the TJ and/or at the cytosol. ZO-2 relocalization to the nucleus, requires

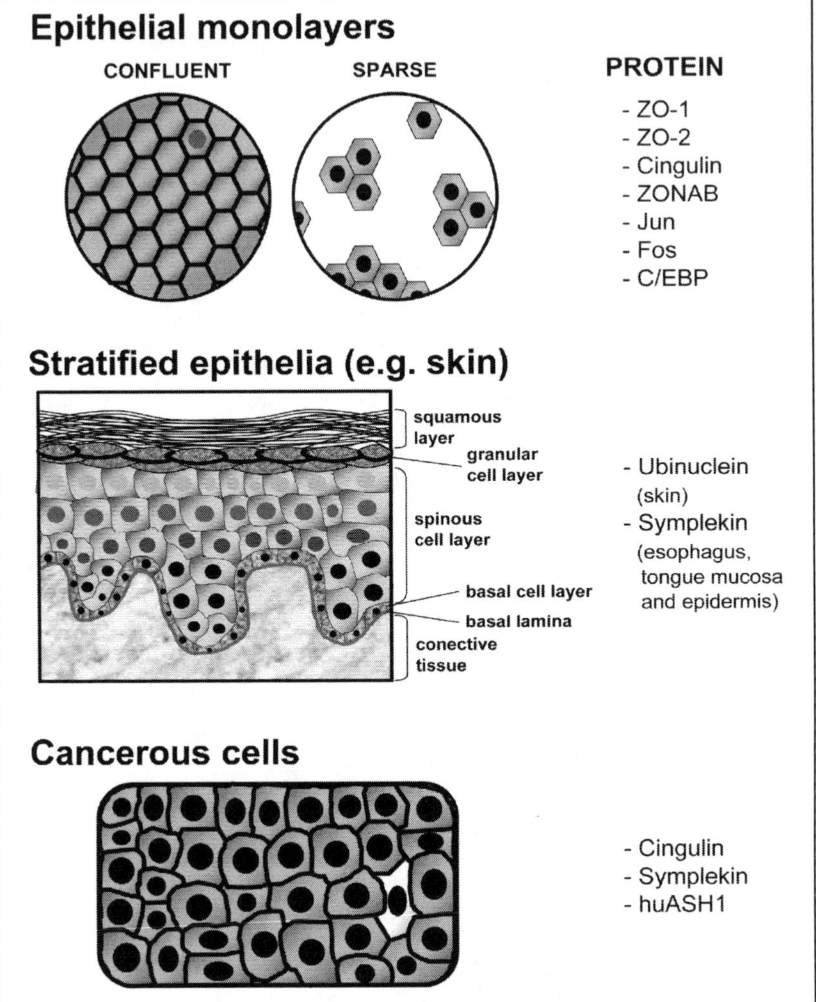

Figure 1. Schematic representation of the distribution of proteins that shuttle between the TJ and the nucleus, in different epithelia. In epithelial monolayers the indicated proteins concentrate at the cellular borders of confluent cultures, whereas in a sparse condition, they accumulate at the nucleus. In stratified epithelia, ubinuclein and symplekin display a nuclear concentration that is maximal at the basal cell layer and diminishes gradually as the cells approach the granular sheet. In the latter these proteins are only detected at the TJ. In contrast, in cancerous cells, cingulin, symplekin and huASH1 are present at the nucleus and cellular boundaries.

the participation of an intact actin cytoskeleton, since the addition of cytochalasin B, an agent that depolymerizes actin, inhibits the appearance of ZO-2 at the nucleus of mechanically injured cells.[21]

The accumulation of ZO proteins at the nucleus is not restricted to cultured cell lines. In kidney, purified nuclei have been shown to express considerable amounts of ZO-1 and ZO-2, and nuclear staining of both proteins has been observed in isolated mammalian renal tubules.[23] Nuclear ZO-1 has also been studied along distinct regions of the intestinal crypt-villus axis,

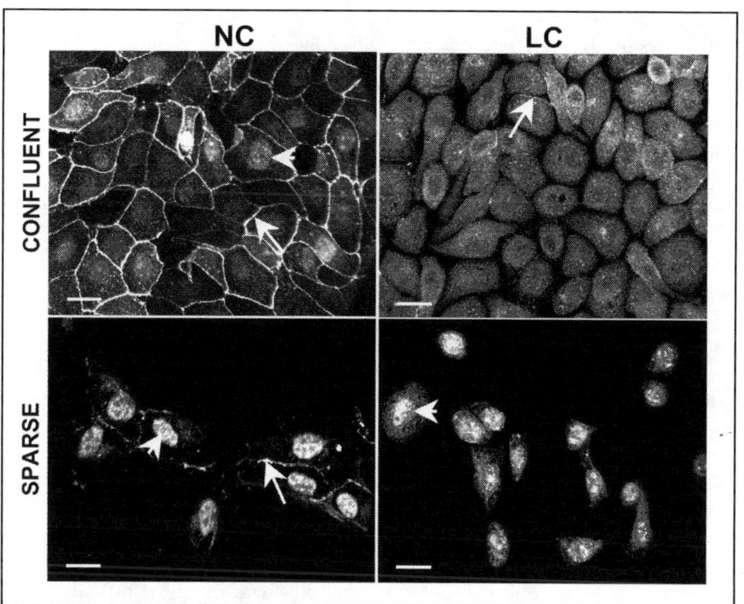

Figure 2. ZO-2 is conspicuously present in the nuclei of sparse cells cultured with and without Ca^{2+}. ZO-2 is detected in confluent and sparse monolayers cultured in normal (NC; 1.8 mM) and low Ca^{2+} (LC; 1-5 µM) condition. Arrows show ZO-2 at the cellular borders and arrowheads indicate the nuclear staining. Bar, 20 µm. (From: Islas S et al, Exp Cell Res 2002; 274:138,[21] ©2002 with permission from Elsevier.)

and could be detected in cells along the outer one fourth of the villus.[24] This corresponds to the region of the intestine where cells are committed to die and exfoliate. Instead at the other domains where the cells are proliferating (crypts), migrating and differentiating (crypt-villus and outer villus), no nuclear ZO-2 staining was found. These observations have related the nuclear localization of ZO-1 to changes in TJ structure accompanying cell death and extrusion events happening at the villus tips, and have argued that the nuclear accumulation of ZO-2 observed at subconfluency is more a reflection of the state of cell-cell contacts than the proliferative state of the cell.

MAGI-1c

MAGI-1, -2 and -3 constitute a family of MAGUK inverted proteins that localize at the TJ of epithelial cells.[25-27] The splicing variant MAGI-1c, has been found by subcellular fractionation assays to concentrate at the nucleus of confluent epithelial cells.[28] Whether the subcellular distribution of MAGI-1c is influenced by cell density remains unknown.

Par-6

Par-6, a protein with one PDZ repeat, forms a complex with Par-3 and aPKC that localizes to the TJ region. This complex is essential for the generation of epithelial polarity.

In MDCK cells, Par-6 is also present at the nuclei,[29] where it displays a speckled pattern.[30] If TJ are disrupted by the addition of hepatocyte growth factor, the nuclear localization of Par-6 remains unaffected.[29]

Cingulin

Cingulin, an actomyosin associated protein is a component of the cytoplasmic plaque of the TJ in vertebrates (see Chapter 5). In subconfluent epithelial cells[31] and in cancerous cell lines,[32] cingulin localizes at the nucleus and the cellular borders, while in confluent monolayers and in a variety of tissues including stratified epithelia, its presence is restricted to TJ.[33]

Symplekin

Symplekin is a 127 kDa protein, found at the TJ apical belt of multiple human tissues.[34] In the stratified epithelia of the esophagus, an exclusive nuclear localization of symplekin is observed in regions known to lack TJ, such as the proliferative compartment of the basal cell layer and at the nonproliferative strata of the suprabasal cells. A similar situation is found to occur in the stratified epithelia of tongue mucosa and epidermis. In human colon carcinoma, where only weak residual TJ staining is detected at the cell apex region, symplekin is strongly expressed at the nuclei of epithelial cells. In carcinoma cell lines, symplekin displays a dual residence at the nucleus and the TJ region. While in nonepithelial cells like endothelia and fibroblasts, and in cells that do not form stable junctions at all, such as lymphoma and erythroleukemia cells, only the nuclear reaction is consistently observed. These results would speculatively suggest symplekin sequestering to the plasma membrane by a set of TJ proteins that is absent in cells displaying poor intercellular contacts.

Ubinuclein

Ubinuclein is a ubiquitously expressed nuclear protein which interacts with cellular and viral transcription factors c-Jun and EB1.[35] Ubinuclein stains TJ in various tissues and in stratified epithelia such as skin, exhibits a nuclear localization throughout the basal and spinous cell layer, but localizes at TJ along the upper granular cell sheet. In epithelial cell lines, ubinuclein displays the same sensitivity to the sate of cell-cell contact as that of ZO proteins.[36]

Signals Contained in the Sequence of TJ Proteins That Facilitate Membrane-Nuclear Shuttling

Proteins can move into the nucleus if they contain nuclear localization signals (NLS).[37] Nevertheless, proteins without NLS, can still enter via cotransport with proteins that have their own NLS in a process known as "piggy-back".

The classical type of NLS is that of SV40 large T antigen.[38] Two rules are followed to detect it: (1) A pattern of 4 basic amino acids, lysine (K) or arginine (R), or three basic amino acids, K or R, and either histidine (H) or proline (P). (2) A sequence of 7 aminoacids, starting with P and followed within 3 residues by a basic segment containing 3 K/R residues out of 4.

Another type of NLS is the bipartite pattern, first found in *Xenopus* nucleoplasmin.[39] The pattern is: 2 basic residues, a 10 residue spacer, and another basic region consisting of at least 3 basic aminoacids out of 5 residues.

Table 1, presents the NLS we identified in all the MAGUK and MAGI proteins of the TJ as well as in PAR-6, cingulin, symplekin and ubinuclein, although we are aware that the presence of several of these proteins at the nucleus has not been so far reported (e.g., ZO-3, PALS1, MAGI-1a,1b, MAGI-2 and MAGI-3).

With regards to the functionality of the NLS, only those present in canine ZO-2 (c-ZO-2) have been tested. These signals concentrate on the region comprising the last amino acids of PDZ-1 and the first linker situated between PDZ-1 and PDZ-2. Upon comparing the NLS present in ZO-2 derived from different species, it becomes clear that the bipartite NLS located at the end of PDZ-1 is conserved in human, dog, mouse and chicken, while a second bipartite signal is only present in dog and mouse. Two monopartite signals are detected within the first bipartite signal in all the studied species, whereas only in dog, a monopartite NLS is detected between the bipartite NLSs.

In order to study the functionality of NLS present in c-ZO-2, the arrival at the nucleus of different amino ZO-2 constructs, lacking some or all the NLS was evaluated. The experiment

Table 1. Nuclear localization signals (NLS) found in MAGUK proteins of the TJ (A), MAGI family (B), Par-6 and cingulin (C), symplekin and ubinuclein (D) using the PSORT version 6.4

A. NLS: MAGUK Family

Scheme	Name		Sequence		Localization	
	h-ZO-1	96	RKSGKNAKITIRRKKKV	112	PDZ1 and 1st link	1st Bipartite
	c-ZO-1	96	RKSGKNAKITIRRKKKV	112	" " " "	
	m-ZO-1	96	RKSGKNAKITIRRKKKV	112	" " " "	
	h-ZO-2	106	RKSGKVAAIVVKRPRKV	122	PDZ1 and 1st link	
	c-ZO-2	83	RKSGKIAAIVVKRPRKV	99	" " " "	
	m-ZO-2	83	RKSGKIAAIVVKRPRKV	99	" " " "	
	ch-ZO-2	83	RKSGKVATIVVKRPRKV	99	" " " "	
	c-ZO-3	139	RRARTGRRNQAGSRGRR	155	1st link	
	c-ZO-2	246	RRTQPDARHAGSRSRSR	262	1st link	2nd Bipartite
	m-ZO-2	669	RRQRSRGGDKKTLRKSR	685	Between SH3 and GK	
	h-ZO-1	748	RKSARKLYERSHKLRKN	764	GK domain	
	c-ZO-1	747	RKSARKLYERSHKLRKN	763	" "	
	m-ZO-1	748	RKSARKLYERSHKLRKN	764	" "	
	hy-ZO-1	1141	KKPTDKEKTREKSKDRK	1157	Beyond GK	
	h-ZO-3	751	RRSTRRLYAQAQKLRKH	767	GK domain	
	c-ZO-3	707	RRSARRLYAQAQKLRKH	723	" "	
	m-ZO-3	866	RRQDSMRTYKHEALRKK	882	Acidic domain	
	hy-ZO-1	138	RKKPK	142	1st link	Monopartite
	h-ZO-1	307	PPRRSRSR	314	2nd link	
	hy-ZO-1	572	PSSGKKK	578	" "	
	c-ZO-2	185	RRPR	188	1st link	
	h-ZO-3	109	KRPRR	113	1st link	
	c-ZO-3	90	KRPRR	94	" "	
	m-ZO-3	90	KRPRR	94	" "	
	h-ZO-3	166	RRPR	169	" "	
	h-ZO-3	170	PGRRGRA	176	" "	
	h-ZO-3	365	PRLRRES	371	2nd link	
	m-ZO-3	324	PQPRRRERS	332	" "	
	h-PALS1	67	RRRR	70	U1 domain	
	m-PALS1	67	RRRR	70	" "	
	h-PALS1	442	KKKRKK	447	Between SH3 and GK	
	m-PALS1	442	KKKRKK	447	" " " "	
	h-PALS1	475	PANRKRP	481	GK	
	m-PALS1	475	PANRKRP	481	"	

continued on next page

Table 1. Continued

	B. NLS: MAGI Family				
Scheme	**Name**		**Sequence**		**Localization**
	h-MAGI-1a	383	RKTQYENPVLEAKRKKQ	399	WW and link to PDZ1
	h-MAGI-1b	383	RKTQYENPVLEAKRKKQ	399	" " " " "
	m-MAGI-1b	371	RKTQYENPVLEAKRKKQ	387	" " " " "
	h-MAGI-1c	383	RKTQYENPVLEAKRKKQ	399	" " " " "
	h-MAGI-2	372	RRTQFENPVLEAKRKLQ	388	" " " " "
	m-MAGI-2	371	RRTQFENPVLEAKRKLQ	387	" " " " "
	h-MAGI-1c	1326	RREGTRSADNTLERREK	1342	Beyond PDZ5
	h-MAGI-1c	1339	RREKHEKRRDVSPERRR	1355	" "
	h-MAGI-1c	1381	RRSPERRRGGSPERRAK	1397	" "
	h-MAGI-1c	1387	RRGGSPERRAKSTDRRR	1403	" "
	h-MAGI-1a	219	PKRTKSY	225	Between GK and WW
	h-MAGI-1b	219	PKRTKSY	225	" " " "
	m-MAGI-1b	219	PKRTKSY	225	" " " "
	h-MAGI-1c	219	PKRTKSY	225	" " " "
	h-MAGI-2	214	KRKR	217	" " " "
	m-MAGI-2	213	KRKR	216	" " " "
	h-MAGI-3	6	KKKHK	10	Before PDZ0
	m-MAGI-3	6	KKKHK	10	" "
	h-MAGI-3	400	KRKK	403	Between WW and PDZ1
	m-MAGI-3	378	KRKK	381	" " " "
	h-MAGI-2	1295	RKPK	1298	Beyond PDZ5
	h-MAGI-1c	1318	PKRRSPE	1324	" "
	h-MAGI-1c	1323	PEKRREG	1329	" "
	h-MAGI-1c	1351	PERRRER	1357	" "
	h-MAGI-1c	1359	PTRRRDG	1365	" "
	h-MAGI-1c	1367	PSRRRRS	1373	" "
	h-MAGI-1c	1407	PERRRER	1413	" "
	h-MAGI-1c	1445	PPEQRRRP	1451	" "

The first group of entries (RK…KRKKQ through RRGGSPERRAKSTDRRR) is marked **Bipartite**; the lower group (PKRTKSY through PPEQRRRP) is marked **Monopartite**.

continued on next page

revealed how even the construct lacking all NLS displays a high percentage of nuclear-cytosolic distribution. However, a higher number of cells displayed an exclusive nuclear localization pattern, when transfected with a construct with intact NLS. These results thus suggest that in order to enter the nucleus the amino ZO-2 segments can rely on their NLS as well as on their association to other proteins that are constantly traveling to the nucleus (piggy back mechanism).[40]

NACo proteins of the TJ should hypothetically count with a mechanism to allow them to export the nucleus. In this respect it has been observed, that proteins that exit the nucleus contain particular sequences named nuclear export signals (NES), that are sufficient to initiate exportation if placed on heterologous proteins.[41-43] A leucine rich type NES containing a characteristic spacing of leucine (L) or other hydrophobic residues was characterized in various

Table 1. Continued

C. NLS: Par6 and Cingulin

Name		Sequence		
h-Par6-γ	113	**RRRR**ALGALRDEG**PRRRA**H	131	⎫
m-Par6-γ	113	**RKKK**VLVTLRDDGL**RRR**PH	131	⎪
				⎬ Bipartite
h-Par6-β	112	**KK**NVLTNVLRPDNH**RKK**	128	⎪
m-Par6-β	112	**KK**NMLSNVLRPDNH**RKK**	128	⎪
ce-Par6	112	**KR**WKGISSLMAQK**PPKR**	128	⎭
h-Par6-α	112	**RRKK**	115	⎫ Monopartite
m-Par6-α	82	**RRKK**	85	⎭
h-cingulin	360	**RK**VEELQRKLDEEV**KKR**	376	⎫
m-cingulin	483	**RK**MEELQKKLDEEV**KKR**	499	⎬ Bipartite
xl-cingulin	957	**KR**DLESKLDEAQRS**LKR**	973	⎭
h-cingulin	199	PPEQ**RKR**	205	⎫
m-cingulin	328	PHEQ**RKR**	334	⎬ Monopartite
xl-cingulin	26	PRG**KRSK**	32	⎪
xl-cingulin	1319	**KRKP**	1325	⎭

continued on next page

proteins.[44] The receptor for such leucine rich NES is a protein called exportin/CRM1. These receptor associates in the nucleus with the NES signal in the presence of Ran-GTP, thus forming trimeric export complexes that are transferred to the cytoplasm. The anti-fungal, leptomycin B (LMB) specifically inhibits the nuclear export by blocking the interaction between NES and exportin/CRM1.[45-47]

Based on the NES consensus sequence detected for HIV-1 Rev,[48] PKI-α,[49] h-IκBα[50] and zyxin,[51] we previously searched for NES present in ZO proteins, and found several similar sequences.[21] Considering the NES of a broader number of proteins, including p53,[52] 4.1,[53] actin[54] and MAPKK,[55] we have now searched for the presence of the following NES consensus:

$$L/M/IX_{1-4}L/V/IX_{2-3}L/V/IX_{1-2}L/I$$

where X can be any amino acid.

Table 2 reveals that several sequences present in the MAGUK proteins of the TJ, the MAGI family, Par-6, cingulin, symplekin and ubinuclein, comply with such consensus. The number of NES present within each protein varies broadly. Thus, depending on the analyzed specie, the following number of putative NES is displayed: ZO-3, 1 or 2; ZO-2, none to 2 or 4; ZO-1, 5, 7 or 9; Pals-1, 7; MAGI proteins, 6 to 8; Par-6, 1-3 or 5; cingulin 8 or 9; ubinuclein 5 and symplekin 14 or 17. It is not known if the number of NES relates to the efficiency of these proteins to export the nucleus.

To the best of our knowledge, only the functionality of the NES present in ZO-2 has been explored. When MDCK cells were transfected with an amino ZO-2 construct containing all the NLS described above, and the first putative NES originally described for cZO-2 (aa 361-370),[21] the amino segment was exported from the nucleus.[20] Instead, the protein remained

Table 1. Continued

D. NLS: Symplekin and Ubinuclein

Name		Sequence		
h-Symplekin	214	**RKRPR**DDSDSTL**KKMKL**	230	} Bipartite
ce-Symplekin	1115	**KRL**RREEKKEKE**REKER**	1131	
h-Symplekin	388	PQA**KRRP**	394	
ce-Symplekin	360	PVA**KRPK**	366	} Monopartite
ce-Symplekin	604	H**KKR**	607	
h-Ubinuclein	84	**KK**DLSDPFNDEE**KERHK**	100	
m-Ubinuclein	84	**KK**DLSDPFNDEE**KERHK**	100	
h-Ubinuclein	106	**RK**FEEKYGG**KKRRK**DRI	122	
m-Ubinuclein	106	**RK**FEEKYGG**KKRRK**DRI	122	
h-Ubinuclein	576	**RR**GHGHLTSILA**KKK**VM	592	
m-Ubinuclein	576	**RR**GHGHLTSILA**KKK**VM	592	} Bipartite
h-Ubinuclein	189	**KKR**KLKEGGEKI**KKKKK**D	205	
m-Ubinuclein	189	**KKR**KLKEGGEKI**KKKKK**D	205	
h-Ubinuclein	201	**KKKKK**DDTYDKE**KK**S**KKS**	218	
m-Ubinuclein	201	**KKKKK**DDTYDKE**KK**S**KKS**	218	
h-Ubinuclein	203	**KKK**DDTYDKE**KKS**KK**SKF**	220	
m-Ubinuclein	203	**KKK**DDTYDKE**KKS**KK**SKF**	220	
h-Ubinuclein	80	PGD**KKK**D	86	
m-Ubinuclein	80	PGD**KKK**D	86	
h-Ubinuclein	188	**PKKRK**LK	194	
m-Ubinuclein	188	**PKKRK**LK	194	
h-Ubinuclein	509	**PRKKF**QW	515	
m-Ubinuclein	509	**PRKKF**QW	515	} Monopartite
h-Ubinuclein	360	PL**EKRVK**	366	
m-Ubinuclein	360	PL**EKRVK**	366	
h-Ubinuclein	183	**KKK**D	186	
m-Ubinuclein	183	**KKK**D	186	
h-Ubinuclein	234	**KKKKK**	238	
m-Ubinuclein	234	**KKKKK**	238	

h, human; c, canine; m, mouse; ch, chicken; ce, *Caenorhabditis elegans*; hy, *Hydra vulgaris*; xl, *Xenopus laevis*. Letters in bold, highlight the basic residues comprising the NLS. The overlapping monopartite and bipartite signals are respectively indicated with continuous and discontinuous underlining. Links refer to the residues located between the PDZ domains of MAGUK and MAGI proteins.

at the nucleus, when this NES was deleted from the construct. To test whether this leucine rich sequence could function as an autonomous NES, a corresponding peptide was chemically conjugated to ovalbumin (NES-1-OVA) and injected into the nucleus of MDCK cells. Since NES-1-OVA remained within the nucleus,[40] it is unlikely for the sequence 361-370 of cZO-2 to constitute a functional NES. In contrast, when an ovalbumin conjugated peptide corresponding to the second NES of cZO-2 (aa 728-738), was injected into the nucleus of epithelial

Table 2. *Nuclear export signals (NES) found in MAGUK proteins of the TJ (A), MAGI family (B), Par-6 and ubinuclein (C), cingulin (D) and symplekin (E)*

A. NES: MAGUK Family

Scheme	Name	Sequence		Localization
	hy-ZO-1	67	LKVGDILI	74	PDZ1 domain
	h-PALS1	302	LEINGIEI	309	" "
	m-PALS1	302	LEINGIEI	309	" "
	h-PALS1	309	IRGKDVNEVFDL	320	" "
	m-PALS1	309	IRGKDVNEVFDL	320	" "
	h-PALS1	317	INSGKICLLSL	327	" "
	m-PALS1	317	INSGKICLLSL	327	" "
	h-PALS1	324	MHGTLTFVLI	333	" "
	m-PALS1	324	MHGTLTFVLI	333	" "
	c-ZO-2	305	LRLGSQIFI	313	PDZ2 domain
	m-ZO-2	301	LRLGSQIFI	309	" "
	ch-ZO-2	288	LRLGSQIFI	296	" "
	h-ZO-1	384	MWIYLSVHL	392	2nd link
	h-ZO-1	388	LSVHLMVSYL	397	" "
	hy-ZO-1	514	INGTPLDNLKI	524	" "
	hy-ZO-1	519	LDNLKISECIELI	531	" "
	hy-ZO-1	736	LILLALPDDVSL	747	PDZ3 domain
	h-ZO-1	479	IREEAVLFLLDL	490	Between PDZ3 and SH3
	c-ZO-1	478	IREEAVLFLLDL	489	" " " "
	m-ZO-1	479	IREEAVLFLLDL	490	" " " "
	h-ZO-1	548	LYNGKLGSWLAI	559	SH3 domain
	c-ZO-1	537	LYNGKLGSWLAI	548	" "
	m-ZO-1	549	LYNGKLGSWLAI	560	" "
	hy-ZO-1	736	LILLALPDDVSL	747	" "
	h-PALS1	335	LRNSDLKPYIIFI	347	" "
	m-PALS1	335	LRNSDLKPYIIFI	347	" "
	h-ZO-1	688	IIRLHTIKQI	697	GK domain
	c-ZO-1	687	IIRLHTIKQI	696	" "
	m-ZO-1	688	IIRLHTIKQI	697	" "

continued on next page

cells, it became rapidly exported to the cytosol, suggesting it is a functional NES.[40] When MDCK cells were transfected with ZO-2 constructs lacking both NES, but containing the NLS, the protein was exported from the nucleus, suggesting that yet unidentified NES domains should be active in ZO-2.[20] This could in fact be the case, since a new search performed

Table 2. Continued

Scheme	Name	Sequence			Localization
	hy-ZO-1	897	MRPVVVLGPLADL	909	GK domain
	hy-ZO-1	906	LADLARIKL	914	" "
	hy-ZO-1	955	LLDIVPEGIEML	966	" "
	hy-ZO-1	966	LMYAQLCPIVVML	978	" "
	c-ZO-2	719	LFGPIADIAL	728	" "
	c-ZO-2	728	LEKLANELPDL	738	" "
	m-ZO-2	726	MERLANELPDL	736	" "
	ch-ZO-2	713	MEKLSTDLPHL	723	" "
	h-ZO-3	691	IIKLDTVRVI	700	" "
	c-ZO-3	647	IIKLDTVRVI	656	" "
	m-ZO-3	651	IIKLDTVRVI	660	" "
	h-ZO-3	808	LDGSLEDNLDL	818	" "
	hy-ZO-1	1259	IKDTVYEPVSL	1269	Beyond GK
	h-ZO-1	1635	LSSIETGVSII	1645	Prolin-rich domain
	c-ZO-1	1648	LSSIETGVSII	1658	" " "
	m-ZO-1	1644	LSSIETGVSII	1654	" " "
	h-ZO-1	1693	LKFLKPVEL	1701	" " "
	m-ZO-1	1702	LKFLKPVEL	1710	" " "
	c-ZO-1	1706	LKFLKPVEL	1714	" " "
	h-PALS1	121	LEIEDLFSSLKHI	133	Between U1 and L27N
	m-PALS1	121	LEIEDLFSSLKHI	133	" " " "
	h-PALS1	212	LNTPHIQALLL	222	Between L27C and PDZ1
	m-PALS1	212	LNAPHIQALLL	222	" " " "

continued on next page

with the modified NES consensus, described above, revealed the existence two more putative NES in cZO-2: one located at the beginning of the GK domain and the other at the start of the PDZ-2 module (Table 2).

No NES could be detected for hZO-2, following either the previous[21] or the new NES consensus, and shuttling between the TJ and the nucleus has not yet been described for this protein. Nevertheless, the amino acid identity for dog and human ZO-2 is so high,[56] that we expected them to display a similar behavior. Therefore hZO-2 is proposed to leave the nucleus either employing NES sequences with other consensus or by association to proteins that export the nucleus.

A Protein Involved in Chromatin Remodeling Colocalizes with ZO-1 at the TJ

In eukaryotes the genome is partitioned into discrete chromatin domains that represent structural units where the mode of DNA packing can be changed without affecting neighboring

Table 2. Continued

B. NES: MAGI Family

Scheme	Name		Sequence		Localization
	h-MAGI-2	77	**LTIRDVLAVI**	86	PDZ0 domain
	m-MAGI-2	77	**LTIRDVLAVI**	86	" "
	m-MAGI-2	103	**IVDKDLRHYLNL**	114	Between PDZ0 and GK
	h-MAGI-2	125	**LQQIIRDNLYL**	135	GK domain
	m-MAGI-2	125	**LQQIIRDNLYL**	135	" "
	h-MAGI-3	109	**INKDLRHYLSL**	119	" "
	m-MAGI-3	111	**INKDLRHYLSL**	121	" "
	h-MAGI-3	130	**LQQVIRDNLYL**	140	" "
	m-MAGI-3	132	**LQQVIRDNLYL**	142	" "
	h-MAGI-1a	162	**LTVKEFLDL**	170	" "
	h-MAGI-1b	162	**LTVKEFLDL**	170	" "
	m-MAGI-1b	162	**LTVKEFLDL**	170	" "
	h-MAGI-1c	162	**LTVKEFLDL**	170	" "
	h-MAGI-2	473	**IVYINEVCVL**	482	PDZ1 domain
	m-MAGI-2	472	**IVYINEVCVL**	481	" "
	h-MAGI-3	504	**LVPVNQYVNL**	513	" "
	m-MAGI-3	482	**LVPVNQYVNL**	491	" "
	h-MAGI-1a	496	**LQIKSLVL**	503	" "
	h-MAGI-1b	496	**LQIKSLVL**	503	" "
	m-MAGI-1b	476	**LQIKSLVL**	583	" "
	h-MAGI-1c	496	**LQIKSLVL**	503	" "
	h-MAGI-1a	542	**IPIGASVDL**	550	1st link
	h-MAGI-1b	542	**IPIGASVDL**	550	" "
	m-MAGI-1b	522	**IPIGASVDL**	530	" "
	h-MAGI-1c	542	**IPIGASVDL**	550	" "
	m-MAGI-3	625	**IYHQNVQNLTHL**	636	PDZ2 domain
	h-MAGI-3	647	**IYHQNVQNLTHL**	658	" "
	m-MAGI-3	642	**LKQFPVGADVPL**	653	" "
	h-MAGI-3	655	**LTHLQVVEVL**	664	" "
	m-MAGI-3	633	**LTHLQVVEVL**	642	" "
	h-MAGI-3	775	**IYIGAIIPL**	783	PDZ3 domain
	m-MAGI-3	753	**IYIGAIIPL**	761	" "

continued on next page

regions. In differentiated cells, only a subset of these domains is active. In zones characterized by highly condensed chromatin (heterochromatin), repressive signals are generated that cause gene silencing, while in areas of less tight DNA packing (euchromatin), actives genes are preferentially located. The configuration of chromatin can be modulated in specific and regulated ways by

Table 2. Continued

Name		Sequence		Localization
h–MAGI–1a	837	**IYIGHIVPL**	845	PDZ3 domain
h–MAGI–1b	864	**IYIGHIVPL**	872	" "
m–MAGI–1b	816	**IYIGHIVPL**	824	" "
h–MAGI–1c	837	**IYIGHIVPL**	845	" "
h–MAGI–2	948	**ITVPHKIGRI**	957	PDZ4 domain
m–MAGI–2	946	**ITVPHKIGRI**	955	" "
m–MAGI–2	994	**IKDAGLSVTLRI**	1105	" "
h–MAGI–1a	1078	**MDLYVLRL**	1085	PDZ5 domain
h–MAGI–1b	1173	**MDLYVLRL**	1180	" "
h–MAGI–1b	1057	**MDLYVLRL**	1064	" "
h–MAGI–1c	1145	**MDLYVLRL**	1152	" "
h–MAGI–2	1169	**MDLYVLRL**	1176	" "
m–MAGI–2	1160	**MDLYVLRL**	1167	" "
h–MAGI–3	1068	**MGLFILRL**	1075	" "
m–MAGI–3	1044	**MGLFILRL**	1051	" "
h–MAGI–2	1190	**MRVGDQIIEI**	1199	" "
m–MAGI–2	1180	**MRVGDQIIEI**	1189	" "
h–MAGI–3	1087	**IHVGDQIVEI**	1096	" "
m–MAGI–3	1163	**IHVGDQIVEI**	1172	" "
h–MAGI–1a	1167	**MRIGDEILEI**	1176	Beyond PDZ5
h–MAGI–1b	1192	**MRIGDEILEI**	1201	" "
m–MAGI–1b	1076	**MRIGDEILEI**	1085	" "
h–MAGI–1c	1164	**MRIGDEILEI**	1173	" "

continued on next page

relocation of genomic domains in the cell nucleus and through chromatin packing.[57] The latter mechanism includes DNA methylation that in general, represses transcription and histone post-translational modifications like acetylation and deacetylation that stabilizes the active/nonactive status of a chromatin domain.

The human homologue of *Drosophila ash1* (*huASH1*) belongs to the *trithorax* group of genes (*trxG*), which act as transcriptional regulators for the activities of homeotic genes.[58,59] When the human ASH1 protein was studied, it was unexpectedly found distributed in the nucleus and the TJ region of cancerous cell lines.[32]

ASH1 contains a SET domain characteristic of a number of chromatin-associated proteins. The SET domain has a histone methyl transferase activity (HMTase), implicated in conferring both active and repressed transcriptional states to chromatin.[60,61] ASH1 also has a PHD finger, which is a Cys-rich Zn finger-like motif, implicated in protein-protein interaction, and four AT hooks for DNA binding.[62]

The human homologous huASH1, presents a region with high homology to bromodomains that is absent in the *Drosophila* gene.[32] Chromatin remodeling complexes always include bromodomain-containing proteins that recognize different patterns of acetylated histones.[63,64]

Table 2. Continued

C. NES: Par6 and Ubinuclein

Name		Sequence	
h-Par6-γ	62	**IGYADVHGDLLPI**	74
m-Par6-γ	62	**IGYADVHGDLLPI**	74
m-Par6-α	87	**LLLRPVAPL**	95
h-Par6-α	117	**LLLRPVAPL**	125
m-Par6-α	147	**MSVRVALQGL**	156
m-Par6-α	183	**LQGLERVPGI**	192
h-Par6-γ	187	**LEKVPGIFI**	195
m-Par6-γ	187	**LEKVPGIFI**	195
h-Par6-β	186	**LEKVPGIFI**	194
m-Par6-β	186	**LEKVPGIFI**	194
h-Par6-α	186	**LERVPGIFI**	194
m-Par6-α	156	**LERVPGIFI**	164
ce-Par6	215	**LEVNGIEVL**	223
h-Par6-β	317	**LESLTQIEL**	325
m-Par6-β	316	**LESLTQIEL**	324
h-Par6-γ	331	**LAQRLQRDLAL**	341
h-Par6-γ	339	**LALDGGLQRL**	348
h-Par6-γ	341	**LDGGLQRLLSSL**	352
h-Par6-α	45	**LLRAVHQIPGL**	55
m-Par6-α	45	**LLCVVHQIPGL**	55
h-Ubinuclein	67	**LVKNIRGKVKGL**	78
m-Ubinuclein	67	**LVKNIRGKVKGL**	78
h-Ubinuclein	308	**MDSLTDLDL**	316
m-Ubinuclein	308	**MDSLTDLDL**	316
h-Ubinuclein	357	**LPAPLEKRVKEL**	368
m-Ubinuclein	357	**LPIPLEKRVKEL**	368
h-Ubinuclein	519	**IRELLCQVVKI**	529
m-Ubinuclein	519	**IRELLCQVVKI**	529
h-Ubinuclein	522	**LLCQVVKIKL**	531
m-Ubinuclein	522	**LLCQVVKIKL**	531

continued on next page

Rather than proposing a particular role for huASH1 in the organization and function of TJ, we hypothesize that the colocalization of this factor with ZO-1 at the TJ may allow huASH1 to be retained away from the nucleus in order to attenuate its role in gene transcription.

Nuclear Matrix and TJ Proteins

The nuclear matrix is a structural and functional entity[65] obtained after a serial extraction of lipids, soluble proteins, intermediate filaments, DNA and most of the RNA.[66] In this nuclear

Table 2. Continued

D. NES: Cingulin

Name		Sequence	
xl-cingulin	6	IPVGQGVQI	14
h-cingulin	164	IDTAPLSSVDSL	175
m-cingulin	293	IDTAPLSSVDSL	304
xl-cingulin	251	IDTKPLSSVDSL	262
xl-cingulin	415	ILLEKLPSL	423
h-cingulin	431	LRHGLETQVMEL	442
m-cingulin	554	LRHGLEAQVKEL	565
h-cingulin	593	MEEELGEKIEVL	604
m-cingulin	558	LEAQVKELQL	567
h-cingulin	778	LQLQKTLQQL	787
m-cingulin	801	LQLQKTLQQL	810
xl-cingulin	1216	LQEIQERLQL	1225
h-cingulin	902	LSRLQDEIQRL	912
m-cingulin	1025	LSRLQDELQRL	1035
h-cingulin	958	LKGLEEKVSRL	968
m-cingulin	1081	LKSLEEKVSRL	1091
h-cingulin	928	LDKELLAQRLQGL	940
m-cingulin	1051	LDKELLAQRLQGL	1063
xl-cingulin	1096	LDKQMISQRLQSL	1108
h-cingulin	1072	LERKVKELSI	1081
m-cingulin	1195	LERRVKELSI	1204
m-cingulin	15	LETLKILDL	23
xl-cingulin	798	IQDQLKQVLL	807
xl-cingulin	1073	MQEEVQKLKL	1082
xl-cingulin	1240	LERKLKELNI	1249

continued on next page

matrix a large number of proteins are present, including structural proteins such as F-actin, and lamins, proteins involved in replication, transcription, splicing complexes as well as transcriptional regulators.

The nuclear matrix maintains the nuclear structure and serves to localize gene sequences and to keep the distribution of regulatory factors throughout the nuclear space.[67]

Since TJ sub-membranous molecules participate in the link between transmembrane proteins and the underlying cortical cytoskeleton, it was reasonable to study if at the nucleus a similar situation occurred. To explore this issue relative extractability assays have been performed. Concerning ZO-1 and symplekin, both are completely Triton X-100 extractable from the nucleus, while the fraction present at the junction remains largely un-removable.[24,34] ZO-1 also disappears from the nucleus upon methanol fixation,[24] in lieu, nuclear ZO-2 is resistant to treatment with methanol or ethanol.[21]

Table 2. Continued

E. NES: Symplekin

Name		Sequence	
h-Symplekin	1	**MTQLYKVAL**	9
ce-Symplekin	42	**MHLLIDPSLSI**	52
h-Symplekin	47	**IRTHAIKFVEGL**	58
h-Symplekin	52	**IKFVEGLIVTL**	62
h-Symplekin	108	**LKFMVHPAISSI**	119
h-Symplekin	118	**INLTTALGSL**	127
h-Symplekin	111	**MVHPAISSINL**	121
h-Symplekin	121	**LTTALGSLANI**	131
ce-Symplekin	250	**LITVIESLCMI**	260
ce-Symplekin	271	**LPRVFDVIKAL**	281
h-Symplekin	274	**LTPDNVANLVLI**	285
h-Symplekin	282	**LVLISMVYL**	290
ce-Symplekin	309	**LKLPASVPL**	317
ce-Symplekin	410	**LNHETVMNLVKI**	421
h-Symplekin	411	**IFRLSDVLKPL**	421
ce-Symplekin	416	**MNLVKISLYTL**	426
h-Symplekin	476	**ILEDVRARLDL**	486
h-Symplekin	477	**LEDVRARLDL**	486
ce-Symplekin	512	**IVKQSLQEINTL**	523
ce-Symplekin	517	**LQEINTLPVI**	526
ce-Symplekin	688	**IERLKQVCL**	696
ce-Symplekin	735	**LRSSCLEVVKEL**	746
ce-Symplekin	740	**LEVVKELCYL**	749
h-Symplekin	748	**LYHKRLPDVRFL**	759
h-Symplekin	753	**LPDVRFLIPVL**	763
h-Symplekin	772	**IQALPKLIKL**	781
h-Symplekin	876	**MYPRLGGFVMNI**	887
ce-Symplekin	888	**MDIRALLPII**	897
ce-Symplekin	890	**IRALLPIIGGL**	900
ce-Symplekin	802	**MPSDPLLLIPL**	812
h-Symplekin	984	**LEEDDLEPLTL**	994

h, human; c, canine; m, mouse; ch, chicken; ce, *Caenorhabditis elegans*; hy, *Hydra vulgaris*; xl, *Xenopus laevis*. Letters in bold, highlight the hydrophobic residues comprising the NES.

Furthermore, upon studying the nuclear characteristics of ZO-2, we found several evidences that strongly suggest that nuclear ZO-2 is linked to the nuclear matrix: (1) ZO-2 is present in a nuclear matrix preparation derived from sparse epithelial cells. (2) ZO-2 remains located at the nucleus of sparse MDCK monolayers subjected to a sequential extraction of soluble proteins, the salt labile cytoskeleton and of chromatin associated proteins and (3) lamin B_1 and actin coinmunoprecipitate with ZO-2 derived from the nucleus of sparse epithelial cells.[40]

Interaction of TJ Cortical Proteins with Molecules Involved in RNA Processing and Transcriptosome Assembly

Speckles or splicing factor compartments occupy ~20% of total nuclear volume. They are characterized for containing high concentrations of pre-mRNA splicing factors as well as transcription factors, 3' processing factors and ribosomal proteins. The central region of speckles corresponds to clusters of 20 nm granules called interchromatin granule clusters, which are thought to represent transcriptosomes containing the entire transcription and RNA processing machineries (for reviews see refs. 69,70).

The proteins described below play a role in RNA processing and/or transcriptosome assembly, and have been found to associate at the nucleus to cortical TJ proteins.

The Essential Pre-mRNA Splicing Protein SC35, Colocalizes with ZO-2 at the Nucleus

In epithelial cells, nuclear ZO-2 concentrates in speckles, where it colocalizes with splicing factor SC-35 (Fig. 3).[21] This factor, required in the first splicing step of precursor mRNAs,[68] belongs to the family of Serine-Arginine (SR) phosphoproteins[69] that contain multiple RS domains. The latter have shown to be motifs necessary and sufficient for targeting to the nuclear speckles.[70]

ZO-2 Interacts with SAF-B, a Chromatin Component Proposed to Participate in the Assembly of Transcriptosome Complexes

Double hybrid screening identified the scaffold attachment factor B (SAF-B), as a molecule that interacts through its carboxyl terminal domain with the PDZ-1 region of ZO-2.[20]

SAF-B is an abundant component of chromatin but not of the nuclear matrix. The factor binds to scaffold or matrix attachment regions (S/MAR) of DNA.[71,72] The latter are chromatin domains associated with the nuclear matrix that maintain chromatin structure, and that sometimes also harbor tissue specific sites for transcription factors.

SAF-B colocalizes with SC-35 in nuclear speckles, and interacts directly with various splicing factors and RNA polymerase II via its C-terminal domain. Therefore, SAF-B has been proposed to serve as a molecular link that keeps assembled transcriptosome complexes in the vicinity of actively transcribed genes.[73]

In breast cancer cells, SAF-B functions as estrogen receptor corepressor and growth inhibitor. SAF-B is lost in 20% of breast cancers.[74]

SAF-B colocalizes and coimmunoprecipitates with nuclear ZO-2 in epithelial cells. The significance of such interaction is still unknown, however it reinforces the idea that ZO-2 somehow participates in the maintenance and/or nuclear matrix anchoring of transcriptosome complexes.

Symplekin, a TJ Cortical Protein Involved in RNA Polyadenylation

Symplekin plays a critical role in the assembly of the polyadenylation machinery. It acts as a molecular scaffold that brings together in the same complex, a cleavage and polyadenylation specificity factor (CPSF) and a cleavage stimulatory factor (CstF).[75,76]

In stressed cells symplekin colocalizes in punctuate structures of the nuclei with the heat shock factor (HSF-1). The latter is a transcription factor that binds to HSP promoters to

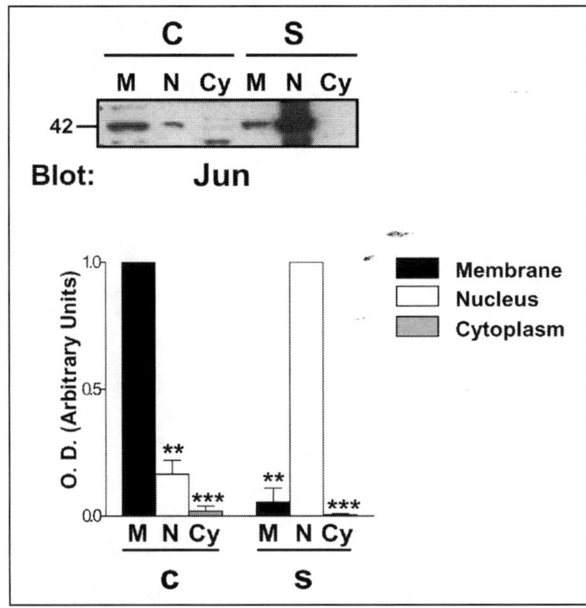

Figure 6. Jun is differentially expressed in the cytosol the membrane and the nuclear extracts from both sparse and confluent MDCK cells. Membrane (M), nuclear (N) and cytoplasm (Cy) fractions obtained from sparse (S) and confluent (C) monolayers were run on a SDS PAGE and blotted with an antibody against Jun. (From: Betanzos A et al, Exp Cell Res 2004; 292:51,[80] ©2004 with permission from Elsevier.)

In pull down assays, the carboxyl terminal domain of ZO-2 interacts with C/EBPα- and in gel mobility shift experiments, the addition of the middle or carboxyl domains of ZO-2, generates the appearance of new complexes associated to the C/EBP oligo probe. The association of C/EBP with ZO-2 was confirmed in vivo in MDCK monolayers by coimmunoprecipitation and colocalization. The subcellular localization of C/EBPα displays the same pattern as that observed in epithelial monolayers for Jun and Fos.[80]

ZO-2 and KyoT2

Double hybrid and pull down assays have identified KyoT2 as a protein associated to a new spliced isoform of human ZO-2.[85] This TF belongs to the LIM superfamily of proteins defined by the presence of one or more LIM domains.[86]

The LIM zinc finger does not interact with DNA but functions as a protein-protein binding module.[87] LIM proteins at the nucleus regulate transcription by acting as scaffolds for the assembly of multiple transcription factors. In the cytoplasm, LIM proteins associate signaling proteins with the actin cytoskeleton. KyoT2 interacts by its LIM domain with the RBP-Jκ transcription factor (recombination signal binding protein-J kappa),[88] a 60-kDa protein recognizing the core sequence C/TGTGGGAA.[89] RBP-Jκ mediates transcriptional activation that follows Notch signalling. The Notch system is responsible for neurogenesis and ectodermal specification in the fruit fly.[90] Once the Notch receptor is activated by a ligand from a neighboring cell, proteolytic cleavage results in the liberation of the intracytoplasmic domain of Notch. This domain is then translocated into the nucleus, where it associates with the transcription factor RBP-Jk, to activate the expression of target genes.[91]

KyoT2 inhibits the expression of Notch regulated genes, by recruiting through its LIM 2 domain, the corepressor RING1 to RPB-J.[92] The same LIM domain of KyoT2 interacts with ZO-2.[85] Therefore under confluency conditions, ZO-2 could speculatively retain KyoT2 at the

TJ, promoting as a consequence the activation of the Notch pathway which leads at least in the case of epidermal stem cells to differentiation.[93] In contrast in proliferating cultures, the association between ZO-2 and KyoT2 might allow this TF that lacks NLS,[88] to travel to the nucleus.

Concluding Remarks

The inverse correlation between the expression of TJ proteins and cancer development raised the idea that TJ proteins may be involved in the control of cell growth and differentiation. This hypothesis was reinforced with the observation that the cortical TJ proteins ZO-1, ZO-2, cingulin, symplekin and ubinuclein distribute to the nucleus and the TJ, in a manner sensitive to the degree of cell-cell contact.

To our knowledge, none of the TJ proteins found at the nucleus have so far been found to bind to DNA. Therefore their role at the nucleus might be exerted through association to nuclear proteins. Since MAGUK proteins have been described as scaffold at the TJ, it might not result surprising that at the nucleus they act as platforms for the establishment of multiprotein complexes probably involved in the transcription of genes that regulate cell proliferation (Fig. 7).

At the TJ, cortical proteins may regulate proliferation and differentiation by sequestration of transcription factors and cell-cycle regulators.

Finally, it remains to be explored whether the TJ cortical proteins are present in the same complexes at the nucleus. Furthermore, it is not yet known if such proteins shuttle together between the membrane and the nucleus, although their movement appears to be triggered by the same changes in the intensity of cell-cell contact.

In summary, the observation of the presence of cortical TJ proteins at the nucleus suggests that proteins shuttle between the two subcellular locations, in order to assemble different protein complexes involved either in cell proliferation or in the maintenance of cell-cell contact at the TJ.

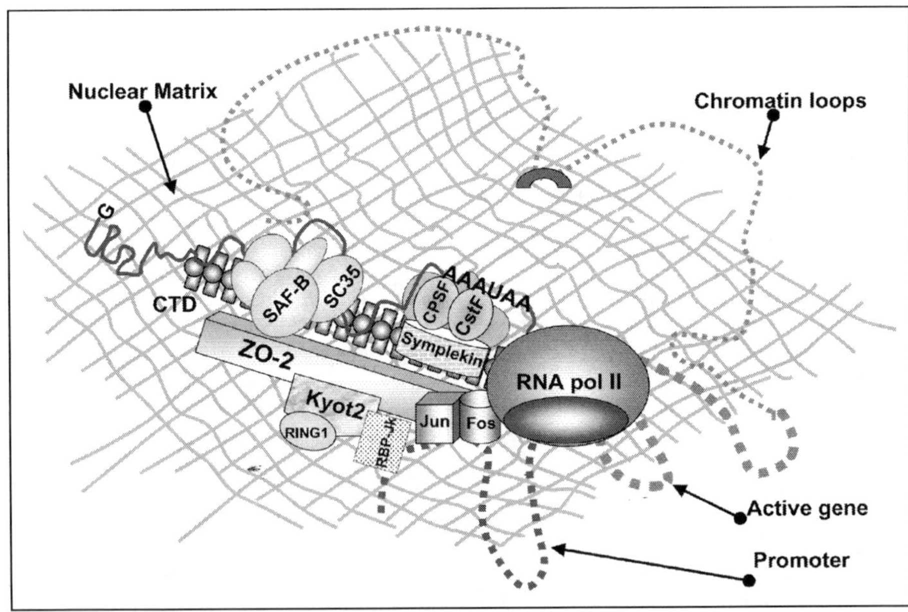

Figure 7. Schematic representation of the participation of TJ cortical proteins within an active transcriptosome. ZO-2 associated to the nuclear matrix, interacts with transcription factors Jun, Fos, and Kyot 2, with splicing factor SC-35 and with SAF-B, a molecule involved in the assembly of transcriptosomes. Symplekin participates in the polyadenylation process.

References

1. Balda MS, Matter K. Epithelial cell adhesion and the regulation of gene expression. Trends Cell Biol 2003; 13(6):310-318.
2. Birchmeier W, Hulsken J, Behrens J. Adherens junction proteins in tumour progression. Cancer Surv 1995; 24:129-140.
3. Gumbiner BM. Cell adhesion: The molecular basis of tissue architecture and morphogenesis. Cell 1996; 84(3):345-357.
4. Woods DF, Bryant PJ. The discs-large tumor suppressor gene of Drosophila encodes a guanylate kinase homolog localized at septate junctions. Cell 1991; 66(3):451-464.
5. Willott E, Balda MS, Fanning AS et al. The tight junction protein ZO-1 is homologous to the Drosophila discs-large tumor suppressor protein of septate junctions. Proc Natl Acad Sci USA 1993; 90(16):7834-7838.
6. Gonzalez-Mariscal L, Betanzos A, Avila-Flores A. MAGUK proteins: Structure and role in the tight junction. Semin Cell Dev Biol 2000; 11(4):315-324.
7. Hoover KB, Liao SY, Bryant PJ. Loss of the tight junction MAGUK ZO-1 in breast cancer: Relationship to glandular differentiation and loss of heterozygosity. Am J Pathol 1998; 153(6):1767-1773.
8. Mauro L, Bartucci M, Morelli C et al. IGF-I receptor-induced cell-cell adhesion of MCF-7 breast cancer cells requires the expression of junction protein ZO-1. J Biol Chem 2001; 276(43):39892-39897.
9. Palmer HG, Gonzalez-Sancho JM, Espada J et al. Vitamin D(3) promotes the differentiation of colon carcinoma cells by the induction of E-cadherin and the inhibition of beta-catenin signaling. J Cell Biol 2001; 154(2):369-387.
10. Chlenski A, Ketels KV, Korovaitseva GI et al. Organization and expression of the human zo-2 gene (tjp-2) in normal and neoplastic tissues. Biochim Biophys Acta 2000; 1493(3):319-324.
11. Chlenski A, Ketels KV, Engeriser JL et al. zo-2 gene alternative promoters in normal and neoplastic human pancreatic duct cells. Int J Cancer 1999; 83(3):349-358.
12. Chlenski A, Ketels KV, Tsao MS et al. Tight junction protein ZO-2 is differentially expressed in normal pancreatic ducts compared to human pancreatic adenocarcinoma. Int J Cancer 1999; 82(1):137-144.
13. Glaunsinger BA, Weiss RS, Lee SS et al. Link of the unique oncogenic properties of adenovirus type 9 E4-ORF1 to a select interaction with the candidate tumor suppressor protein ZO-2. EMBO J 2001; 20(20):5578-5586.
14. Glaunsinger BA, Lee SS, Thomas M et al. Interactions of the PDZ-protein MAGI-1 with adenovirus E4-ORF1 and high-risk papillomavirus E6 oncoproteins. Oncogene 2000; 19(46):5270-5280.
15. Lee SS, Glaunsinger B, Mantovani F et al. Multi-PDZ domain protein MUPP1 is a cellular target for both adenovirus E4-ORF1 and high-risk papillomavirus type 18 E6 oncoproteins. J Virol 2000; 74(20):9680-9693.
16. Thomas M, Glaunsinger B, Pim D et al. HPV E6 and MAGUK protein interactions: Determination of the molecular basis for specific protein recognition and degradation. Oncogene 2001; 20(39):5431-5439.
17. Pim D, Thomas M, Javier R et al. HPV E6 targeted degradation of the discs large protein: Evidence for the involvement of a novel ubiquitin ligase. Oncogene 2000; 19(6):719-725.
18. Thomas M, Laura R, Hepner K et al. Oncogenic human papillomavirus E6 proteins target the MAGI-2 and MAGI-3 proteins for degradation. Oncogene 2002; 21(33):5088-5096.
19. Nakagawa S, Huibregtse JM. Human scribble (Vartul) is targeted for ubiquitin-mediated degradation by the high-risk papillomavirus E6 proteins and the E6AP ubiquitin-protein ligase. Mol Cell Biol 2000; 20(21):8244-8253.
20. Traweger A, Fuchs R, Krizbai IA et al. The tight junction protein ZO-2 localizes to the nucleus and interacts with the heterogeneous nuclear ribonucleoprotein scaffold attachment factor-B. J Biol Chem 2003; 278(4):2692-2700.
21. Islas S, Vega J, Ponce L et al. Nuclear localization of the tight junction protein ZO-2 in epithelial cells. Exp Cell Res 2002; 274(1):138-148.
22. Gonzalez-Mariscal L, Chavez dR, Cereijido M. Tight junction formation in cultured epithelial cells (MDCK). J Membr Biol 1985; 86(2):113-125.
23. Gonzalez-Mariscal L, Namorado MC, Martin D et al. Tight junction proteins ZO-1, ZO-2, and occludin along isolated renal tubules. Kidney Int 2000; 57(6):2386-2402.
24. Gottardi CJ, Arpin M, Fanning AS et al. The junction-associated protein, zonula occludens-1, localizes to the nucleus before the maturation and during the remodeling of cell-cell contacts. Proc Natl Acad Sci USA 1996; 93(20):10779-10784.
25. Ide N, Hata Y, Nishioka H et al. Localization of membrane-associated guanylate kinase (MAGI)-1/BAI-associated protein (BAP) 1 at tight junctions of epithelial cells. Oncogene 1999; 18(54):7810-7815.

26. Wu X, Hepner K, Castelino-Prabhu S et al. Evidence for regulation of the PTEN tumor suppressor by a membrane-localized multi-PDZ domain containing scaffold protein MAGI-2. Proc Natl Acad Sci USA 2000; 97(8):4233-4238.
27. Wu Y, Dowbenko D, Spencer S et al. Interaction of the tumor suppressor PTEN/MMAC with a PDZ domain of MAGI3, a novel membrane-associated guanylate kinase. J Biol Chem 2000; 275(28):21477-21485.
28. Dobrosotskaya I, Guy RK, James GL. MAGI-1, a membrane-associated guanylate kinase with a unique arrangement of protein-protein interaction domains. J Biol Chem 1997; 272(50):31589-31597.
29. Johansson A, Driessens M, Aspenstrom P. The mammalian homologue of the Caenorhabditis elegans polarity protein PAR-6 is a binding partner for the Rho GTPases Cdc42 and Rac1. J Cell Sci 2000; 113(Pt 18):3267-3275.
30. Cline EG, Nelson WJ. Nuclear localization of mammalian Par 6. Mol Biol Cell 2003; 14S:244a.
31. Citi S, Cordenonsi M. The molecular basis for the structure, function and regulation of tight junctions. In: Garrod DR, North AJ, Chidgey MAJ, eds. Adhesive Interactions of Cells. Greenwich: JAI Press, Inc., 1999:203-233.
32. Nakamura T, Blechman J, Tada S et al. huASH1 protein, a putative transcription factor encoded by a human homologue of the Drosophila ash1 gene, localizes to both nuclei and cell-cell tight junctions. Proc Natl Acad Sci USA 2000; 97(13):7284-7289.
33. Citi S, Sabanay H, Kendrick-Jones J et al. Cingulin: Characterization and localization. J Cell Sci 1989; 93(Pt 1):107-122.
34. Keon BH, Schafer S, Kuhn C et al. Symplekin, a novel type of tight junction plaque protein. J Cell Biol 1996; 134(4):1003-1018.
35. Aho S, Buisson M, Pajunen T et al. Ubinuclein, a novel nuclear protein interacting with cellular and viral transcription factors. J Cell Biol 2000; 148(6):1165-1176.
36. Aho S. Ubinuclein a nuclear protein intracting with cellular and viral transcription factors associates with tight junctions. Mol Biol Cell 2002; 13(suppl):494a.
37. Hicks GR, Raikhel NV. Protein import into the nucleus: An integrated view. Annu Rev Cell Dev Biol 1995; 11:155-188.
38. Kalderon D, Richardson WD, Markham AF et al. Sequence requirements for nuclear location of simian virus 40 large-T antigen. Nature 1984; 311(5981):33-38.
39. Robbins J, Dilworth SM, Laskey RA et al. Two interdependent basic domains in nucleoplasmin nuclear targeting sequence: Identification of a class of bipartite nuclear targeting sequence. Cell 1991; 64(3):615-623.
40. Jaramillo BE, Ponce A, Moreno J et al. Characterization of the tight junction protein ZO-2 localized at the nucleus of epithelial cells. Exp Cell Res, 2004; 297:247-258.
41. Gorlich D, Mattaj IW. Nucleocytoplasmic transport. Science 1996; 271(5255):1513-1518.
42. Nigg EA. Nucleocytoplasmic transport: Signals, mechanisms and regulation. Nature 1997; 386(6627):779-787.
43. Nakielny S, Dreyfuss G. Nuclear export of proteins and RNAs. Curr Opin Cell Biol 1997; 9(3):420-429.
44. Izaurralde E, Adam S. Transport of macromolecules between the nucleus and the cytoplasm. RNA 1998; 4(4):351-364.
45. Yoshida M, Nishikawa M, Nishi K et al. Effects of leptomycin B on the cell cycle of fibroblasts and fission yeast cells. Exp Cell Res 1990; 187(1):150-156.
46. Fukuda M, Asano S, Nakamura T et al. CRM1 is responsible for intracellular transport mediated by the nuclear export signal. Nature 1997; 390(6657):308-311.
47. Ossareh-Nazari B, Bachelerie F, Dargemont C. Evidence for a role of CRM1 in signal-mediated nuclear protein export. Science 1997; 278(5335):141-144.
48. Fischer U, Huber J, Boelens WC et al. The HIV-1 rev activation domain is a nuclear export signal that accesses an export pathway used by specific cellular RNAs. Cell 1995; 82(3):475-483.
49. Wen W, Meinkoth JL, Tsien RY et al. Identification of a signal for rapid export of proteins from the nucleus. Cell 1995; 82(3):463-473.
50. Johnson C, Van Antwerp D, Hope TJ. An N-terminal nuclear export signal is required for the nucleocytoplasmic shuttling of IkappaBalpha. EMBO J 1999; 18(23):6682-6693.
51. Nix DA, Beckerle MC. Nuclear-cytoplasmic shuttling of the focal contact protein, zyxin: A potential mechanism for communication between sites of cell adhesion and the nucleus. J Cell Biol 1997; 138(5):1139-1147.
52. Stommel JM, Marchenko ND, Jimenez GS et al. A leucine-rich nuclear export signal in the p53 tetramerization domain: Regulation of subcellular localization and p53 activity by NES masking. EMBO J 1999; 18(6):1660-1672.

53. Luque CM, Perez-Ferreiro CM, Perez-Gonzalez A et al. An alternative domain containing a leucine-rich sequence regulates nuclear cytoplasmic localization of protein 4.1R. J Biol Chem 2003; 278(4):2686-2691.
54. Wada A, Fukuda M, Mishima M et al. Nuclear export of actin: A novel mechanism regulating the subcellular localization of a major cytoskeletal protein. EMBO J 1998; 17(6):1635-1641.
55. Fukuda M, Gotoh I, Gotoh Y et al. Cytoplasmic localization of mitogen-activated protein kinase kinase directed by its NH2-terminal, leucine-rich short amino acid sequence, which acts as a nuclear export signal. J Biol Chem 1996; 271(33):20024-20028.
56. Beatch M, Jesaitis LA, Gallin WJ et al. The tight junction protein ZO-2 contains three PDZ (PSD-95/Discs-Large/ZO-1) domains and an alternatively spliced region. J Biol Chem 1996; 271(42):25723-25726.
57. Recillas-Targa F, Razin SV. Chromatin domains and regulation of gene expression: Familiar and enigmatic clusters of chicken globin genes. Crit Rev Eukaryot Gene Expr 2001; 11(1-3):227-242.
58. Mazo AM, Huang DH, Mozer BA et al. The trithorax gene, a trans-acting regulator of the bithorax complex in Drosophila, encodes a protein with zinc-binding domains. Proc Natl Acad Sci USA 1990; 87(6):2112-2116.
59. Shearn A. The ash-1, ash-2 and trithorax genes of Drosophila melanogaster are functionally related. Genetics 1989; 121(3):517-525.
60. Beisel C, Imhof A, Greene J et al. Histone methylation by the Drosophila epigenetic transcriptional regulator Ash1. Nature 2002; 419(6909):857-862.
61. Rea S, Eisenhaber F, O'Carroll D et al. Regulation of chromatin structure by site-specific histone H3 methyltransferases. Nature 2000; 406(6796):593-599.
62. Tripoulas N, LaJeunesse D, Gildea J et al. The Drosophila ash1 gene product, which is localized at specific sites on polytene chromosomes, contains a SET domain and a PHD finger. Genetics 1996; 143(2):913-928.
63. Kanno T, Kanno Y, Siegel RM et al. Selective recognition of acetylated histones by bromodomain proteins visualized in living cells. Mol Cell 2004; 13(1):33-43.
64. Zeng L, Zhou MM. Bromodomain: An acetyl-lysine binding domain. FEBS Lett 2002; 513(1):124-128.
65. Berezney R, Coffey DS. Identification of a nuclear protein matrix. Biochem Biophys Res Commun 1974; 60(4):1410-1417.
66. Fey EG, Bangs P, Sparks C et al. The nuclear matrix: Defining structural and functional roles. Crit Rev Eukaryot Gene Expr 1991; 1(2):127-143.
67. Konety BR, Getzenberg RH. Nuclear structural proteins as biomarkers of cancer. J Cell Biochem 1999; (Suppl)32-33:183-191.
68. Fu XD, Maniatis T. Factor required for mammalian spliceosome assembly is localized to discrete regions in the nucleus. Nature 1990; 343(6257):437-441.
69. Fu XD. The superfamily of arginine/serine-rich splicing factors. RNA 1995; 1(7):663-680.
70. Hedley ML, Amrein H, Maniatis T. An amino acid sequence motif sufficient for subnuclear localization of an arginine/serine-rich splicing factor. Proc Natl Acad Sci USA 1995; 92(25):11524-11528.
71. Renz A, Fackelmayer FO. Purification and molecular cloning of the scaffold attachment factor B (SAF-B), a novel human nuclear protein that specifically binds to S/MAR-DNA. Nucleic Acids Res 1996; 24(5):843-849.
72. Romig H, Ruff J, Fackelmayer FO et al. Characterisation of two intronic nuclear-matrix-attachment regions in the human DNA topoisomerase I gene. Eur J Biochem 1994; 221(1):411-419.
73. Nayler O, Stratling W, Bourquin JP et al. SAF-B protein couples transcription and premRNA splicing to SAR/MAR elements. Nucleic Acids Res 1998; 26(15):3542-3549.
74. Oesterreich S. Scaffold attachment factors SAFB1 and SAFB2: Innocent bystanders or critical players in breast tumorigenesis? J Cell Biochem 2003; 90(4):653-661.
75. Hofmann I, Schnolzer M, Kaufmann I et al. Symplekin, a constitutive protein of karyo- and cytoplasmic particles involved in mRNA biogenesis in Xenopus laevis oocytes. Mol Biol Cell 2002; 13(5):1665-1676.
76. Takagaki Y, Manley JL. Complex protein interactions within the human polyadenylation machinery identify a novel component. Mol Cell Biol 2000; 20(5):1515-1525.
77. Xing H, Mayhew CN, Cullen KE et al. HSF1 modulation of Hsp70 mRNA polyadenylation via interaction with symplekin. J Biol Chem 2004; 279(11):10551-10555.
78. Balda MS, Matter K. The tight junction protein ZO-1 and an interacting transcription factor regulate ErbB-2 expression. EMBO J 2000; 19(9):2024-2033.
79. Balda MS, Garrett MD, Matter K. The ZO-1-associated Y-box factor ZONAB regulates epithelial cell proliferation and cell density. J Cell Biol 2003; 160(3):423-432.

80. Betanzos A, Huerta M, Lopez-Bayghen E et al. The tight junction protein ZO-2 associates with Jun, Fos and C/EBP transcription factors in epithelial cells. Exp Cell Res 2004; 292(1):51-66.
81. Wisdom R. AP-1: One switch for many signals. Exp Cell Res 1999; 253(1):180-185.
82. Croniger C, Leahy P, Reshef L et al. C/EBP and the control of phosphoenolpyruvate carboxykinase gene transcription in the liver. J Biol Chem 1998; 273(48):31629-31632.
83. Darlington GJ, Ross SE, MacDougald OA. The role of C/EBP genes in adipocyte differentiation. J Biol Chem 1998; 273(46):30057-30060.
84. Poli V. The role of C/EBP isoforms in the control of inflammatory and native immunity functions. J Biol Chem 1998; 273(45):29279-29282.
85. Huang HY, Li R, Sun Q et al. [LIM protein KyoT2 interacts with human tight junction protein ZO-2-i3]. Yi Chuan Xue Bao 2002; 29(11):953-958.
86. Schmeichel KL, Beckerle MC. The LIM domain is a modular protein-binding interface. Cell 1994; 79(2):211-219.
87. Dawid IB. LIM protein interactions: Drosophila enters the stage. Trends Genet 1998; 14(12):480-482.
88. Taniguchi Y, Furukawa T, Tun T et al. LIM protein KyoT2 negatively regulates transcription by association with the RBP-J DNA-binding protein. Mol Cell Biol 1998; 18(1):644-654.
89. Tun T, Hamaguchi Y, Matsunami N et al. Recognition sequence of a highly conserved DNA binding protein RBP-J kappa. Nucleic Acids Res 1994; 22(6):965-971.
90. Artavanis-Tsakonas S, Rand MD, Lake RJ. Notch signaling: Cell fate control and signal integration in development. Science 1999; 284(5415):770-776.
91. Panin VM, Irvine KD. Modulators of notch signaling. Semin Cell Dev Biol 1998; 9(6):609-617.
92. Qin H, Wang J, Liang Y et al. Ring1 inhibits transactivation of RBP-J by Notch through interaction with LIM protein KyoT2. Nucleic Acids Res 2004; 32(4):1492-1501.
93. Thelu J, Rossio P, Favier B. Notch signalling is linked to epidermal cell differentiation level in basal cell carcinoma, psoriasis and wound healing. BMC Dermatol 2002; 2(1):7.

CHAPTER 8

Tight Junctions and the Regulation of Epithelial Cell Proliferation and Gene Expression

Emma Kavanagh,* Anna Tsapara,* Karl Matter and Maria S. Balda

Abstract

Cell interactions with the extracellular matrix and neighboring cells regulate epithelial cell proliferation and differentiation. Tight Junctions are the most apical cell-cell junctions of epithelial cells and evidence indicates that they participate in the suppression of cell proliferation and stimulation of differentiation. Some tight junction components are expressed at increased levels in differentiated versus proliferating cells; and their down-regulation has been linked to epithelial-mesenchymal transition as well as cancer. Other tight junction proteins are also found in the nucleus of proliferating cells and have been linked to the regulation of cell proliferation, transcription and RNA processing. Therefore, it seems that the accumulation of proteins at forming tight junctions is related to signaling pathways that control cell growth arrest and differentiation. We propose a model in which the assembly state of tight junctions is used as a sensor for cell proliferation and density. When cell density increases, the expression levels of tight junction proteins that are inhibitors of proliferation increase and they become stabilized at the forming junctions, resulting in the suppression of proliferation promoting signaling pathways. Furthermore, the tight junction signaling pathways in connection with signals from other cell-cell and extracellular matrix interactions ensure that epithelial cells stop to proliferate and begin to differentiate.

Introduction

Epithelial cells interact with the extracellular matrix and with each other via specialized adhesive structures that are critical for the development and function of epithelial tissues. These structures not only mediate adhesion but also regulate cell proliferation and differentiation. In order to understand the molecular mechanisms that regulate epithelial cell differentiation and cell proliferation, it is necessary to understand how epithelial cells interact with their neighbors and the extracellular matrix, and how this affects intracellular signaling pathways that control cell proliferation, differentiation and gene expression.

The epithelial intercellular junctional complex consists of tight junctions (TJs), adherens junctions (AJs), and desmosomes. AJs and desmosomes are adhesive junctions that are linked to the actin and intermediate filament cytoskeleton, respectively.[1-3] TJs function as barriers that restrict paracellular diffusion as well as apical/basolateral intramembrane diffusion and also interact with the actin cytoskeleton.[4-7] At the basal membrane, epithelial cells adhere to

---

* E. Kavanagh and A. Tsapara contributed equality to this chapter.

Tight Junctions, edited by Lorenza Gonzalez-Mariscal. ©2006 Landes Bioscience and Springer Science+Business Media.

the extracellular matrix mainly via integrins and syndecans.[8,10] All of these adhesion complexes consist of a set of transmembrane components that interact extracellularly with ligands and in the cytoplasm with large multimeric protein complexes consisting of cytoskeletal linkers, signal transduction proteins as well as factors involved in DNA transcription or RNA processing. These cell adhesion protein complexes also transmit information to the cell interior about the environment, which influence various aspects of cell behavior and phenotype, including proliferation, differentiation and gene expression. These signals are transmitted to the cell interior either by sequential modulation of signal transduction pathways and cascades or, alternatively, by specific proteins that shuttle between sites of adhesion at the plasma membrane and the nucleus.[9-14] In this review, we will highlight the role of tight junctions in regulating cell proliferation and gene expression and speculate on how tight junctions and other types of cell adhesion might interact to regulate cell differentiation.

Junctional Membrane Proteins

Tight junctions are composed of four types of membrane proteins: claudins, JAMs, occludin, and Crumbs.[5,15-19] There is evidence that suggests that all four proteins are involved in the regulation of epithelial cell proliferation or differentiation. Occludin and claudins have been linked to the regulation of epithelial proliferation. In a salivary gland-derived epithelial cell line, occludin overexpression is able to reverse Raf-1-induced transformation and occludin deficient mice exhibit hyperplasia and epithelial dedifferentiation in different tissues.[20,21]

Surprisingly, overexpression of claudin-11 induces cell proliferation of an oligodendrocyte cell line; and claudin-1, claudin-3 and claudin-4 are overexpressed in certain tumors.[22-24] However, overexpression of claudin-4 was associated with a significantly reduced invasive potential of a pancreatic cancer cell line, suggesting that claudin expression does not directly correlate with the proliferative phenotype of a cell or its differentiation state.[25] This is further supported by the observation that EGF inhibits the expression of claudin-2 but upregulates the expression of claudin-1, claudin-3 and claudin-4 in MDCK cells.[26] Furthermore, the claudin-2 and claudin-1 promoters were shown to be stimulated by the growth factor pathways of the TCF/β-catenin transcription complex in mouse mammary epithelial and colon cancer cells,[23] but dedifferentiation of mammary epithelial cells induced by the transcription factor Snail downregulates the promoters of claudin-3, claudin-4 and claudin-7.[27,28] All together, these conflicting results do not permit the association of claudins with a particular signalling pathway.

The other two types of junctional membrane proteins, JAMs and Crumbs3, have not been linked to the regulation of proliferation but both interact with evolutionarily conserved signaling complexes that control epithelial polarization and, therefore, differentiation. These pathways are out of the scope of this review and have been described elsewhere (summary in Table 1).[29,30]

Junctional Plaque Proteins

The junctional membrane proteins interact with a complex cytoplasmic plaque that contains different types of proteins which are linked to the actin cytoskeleton. The cytoplasmic plaque consists of multiple adaptor proteins as well as different types of signaling components such as protein kinases and phosphatases, small and heterotrimeric GTP binding proteins, and transcriptional regulators.[6,31-34]

Several of the adaptor proteins belong to the membrane-associated guanylate kinases (MAGUK) protein family. They contain different domains such as one homologous to yeast guanylate kinase, an SH3 and one or more PDZ domains.[35-37] However, none of the MAGUKs appears to function as a guanylate kinase. Since the Drosophila tumor suppressor Discs large A was one of the first members described of the MAGUK family, it was suggested that TJs might be involved in processes regulating cell proliferation and differentiation.[38-40] This hypothesis was further supported by the finding that the Drosophila homologue of the TJ-associated ZO-1, Tamou/Polychaetoid, participates in the regulation of dorsal closure, epithelial migration, and cell fate determination in sensory organs.[41,42]

Table 1. Tight junction components and the regulation of cell proliferation and differentiation*

Tight Junction Proteins	Effect on Cell Proliferation and/or Differentiation	Key Features
Transmembrane proteins		
Occludin	Reverses transformation and proliferation induced by Raf-1	Interaction with ZO-1/2/3, PI3K, aPKCζ, Itch
Claudins	Induces proliferation in oligodendrocytes	Upregulated in certain cancers, transcriptionally regulated by the LEF/catenin and SNAIL pathways, interaction with integrins, MMPs, ZO-1/2/3, MUPP1
JAMs	Regulation of junction assembly	Interaction with integrins, reovirus, adenovirus, CASK, ZO-1, cingulin, afadin/AF-6, PAR3, MUPP1
Crumbs3	Cell polarity and apical membrane biogenesis	Interaction with Pals1
Cytoplamic proteins		
ZO-1	Regulation of gene expression, cell proliferation and cell density	Reduction of the nuclear accumulation of ZONAB/cdk4 and cyclin D1
ZO-2	Unknown	Interaction with ZO-1, adenovirus E4-ORF1, SAF-B, Fos, Jun, C/EBP
ZO-3	Unknown	Interaction with claudins, ZO-1
PATJ/Pals1 complex	Cell polarity and apical membrane biogenesis	Interaction with ZO-3, PAR6, Crumbs3
aPKCζ/PAR3/6 complex	Regulation of junction assembly and cell polarity	Interaction with cdc42, Pals1
ZONAB	Regulation of gene expression, cell proliferation and cell density	Regulation of nuclear cdk4/cyclin D1 and gene expression
Symplekin	Regulation of gene expression and RNA polyadenylation	Interaction with HSF-1
GEF-H1	Unknown	Activation of RhoA and mitotic spindle localization
MAGI-1	Unknown	Interaction with JAM4, adenovirus E4-ORF1, papillomavirus E6
MAGI-2 and 3	Unknown	Interaction with PTEN and regulation of PKB/AKT
MUPP1	Unknown	Interaction with claudins, JAMs, adenovirus E4-ORF1, papillomavirus E6

* Only TJ components with evidence for a possible role on cell proliferation and/or differentiation are listed.

Experimental work published over the last few years indeed provided evidence for several of the TJ-associated MAGUKs to be involved in the regulation of epithelial proliferation. ZO-1 is downregulated in proliferative cells such as during corneal wound repair and in colorectal epithelial cells transformed by overexpression of β-catenin, suggesting that reduced expression of ZO-1 is related to increased proliferation of epithelial cells.[43,44] This is supported by the finding that ZO-1 is downregulated in breast cancer tissues and increased expression of ZO-1 in MDCK cells reduces proliferation.[45,46] Similarly, expression of ZO-2, which is another MAGUK that binds to ZO-1, is also reduced in certain tumors.[47] ZO-1 and ZO-2 have been reported to accumulate transiently in the nucleus of proliferating cells; however, there is some disagreement about the nuclear localization of both proteins, which suggests that additional unknown parameters might affect their nuclear distribution.[32,33,48-53] In fact, ZO-1 inhibits cell proliferation outside of the nucleus by cytoplasmic sequestration of the transcription factor ZONAB and the cell cycle regulator CDK4.[46] Strikingly, ZO-2, MAGI-1 (which is an inverted MAGUK) and MUPP1 (a multi-PDZ domain protein) have been shown to bind and inactivate oncogenic viral proteins.[51,54,55] Furthermore, ZO-2 binds to DNA scaffolding factor SAF-B[53] and the transcription factors Fos, Jun and C/EBP;[56] but the functional relevance of these ZO-2 interactions is unknown. Moreover, the tumor suppressor and lipid phosphatase PTEN binds to a PDZ domain of MAGI-3, another TJ-associated MAGUK, and this interaction seems to be important for modulating the activity of PKB/AKT, a protein kinase involved in cell survival.[57] Nevertheless, although most of the molecular mechanisms are not completely understood, the combined accumulated evidence suggests that TJ-associated MAGUKs and related proteins function as proliferation suppressors (summary in Table 1).

Several other signaling proteins have been localized to TJs but there is yet no direct evidence of their function in the regulation of cell proliferation. A likely candidate for a protein that regulates proliferation is GEF-H1/Lfc, a guanine nucleotide exchange factor specific for Rho[33,58] It is well known that Rho activation stimulates epithelial cell proliferation and is important during different steps of the cell cycle.[59-61] GEF-H1/Lfc also associates with the mitotic spindle, further supporting a possible role of this protein in cell proliferation but it is not known whether the dual localization of GEF-H1 reflects a regulatory function of TJs in mitosis or is part of a mechanism that regulates cytokinesis and junction formation.

TJs have been linked to the regulation of gene expression via mechanisms that rely on proteins involved in DNA transcription and RNA processing. These proteins can shuttle between the nucleus and cell adhesion complexes, and are therefore called NACos (Nucleus and Adhesion Complexes).[9] One such protein is ZONAB, a transcription factor that interacts with ZO-1. ZONAB localizes to the nucleus, where it functions as a transcriptional repressor, and to intercellular junctions, where it associates with ZO-1. ZONAB and ZO-1 functionally interact to regulate cell proliferation and gene expression.[46,50]

HuASH1 is the human homologue of *Drosophila* ASH1 (absent, small or homeotic discs 1), a transcription factor that belongs to the trithorax group and that participates in the maintenance of expression of segment-specific homeotic genes. HuASH1 was found to localize at cell-cell junctions and colocalizes with TJ markers by immunofluorescence.[62] Drosophila ASH1 functions as a histone methyl-transferase, a catalytic activity important for chromatin remodeling and gene expression.[63] However, how huASH1 becomes recruited to intercellular junctions, whether it is also involved in chromatin remodeling, and the functional relevance of its junctional association are unknown.

The TJ and nucleus-associated protein symplekin has been linked to the machinery involved in 3'-end processing of pre-mRNA and polyadenylation.[64-66] In agreement with this, symplekin was recently shown to interact with HSF-1 (heat shock inducible transcription factor one) to regulate hsp70 mRNA polyadenylation in stressed cells.[67] Although it will be necessary to characterise other mRNA transcripts that are regulated by symplekin, this might represent a mechanism by which intercellular junctions participate in the regulation of mRNA processing (summary in Table 1).

ZO-1 and ZONAB in the Regulation of Cell Proliferation

We recently identified ZONAB, a Y-box transcription factor that specifically binds to the SH3 domain of ZO-1 and regulates the expression of the growth factor coreceptor erbB-2 and cell proliferation in a cell density dependent manner.[9,46,50] Y-box transcription factors are multifunctional regulators of gene expression and have been proposed to play a general role in promoting proliferation.[68,69] ZONAB binds to inverted CCAAT box sequences of the promoters of erbB-2 and several cell cycle regulators.[50]

ZONAB localizes to the nucleus, where it participates in the regulation of gene expression, and to TJs, where it binds to ZO-1. In MDCK cells, this dual distribution is regulated by the expression levels of ZO-1, which, in turn, are determined by cell density. In cells that are at a high density and therefore not proliferating, ZO-1 levels are high and ZONAB expression is low. ZONAB consequently localizes primarily to TJs. In contrast, in proliferating low density cells, which express only low levels of ZO-1 but have high ZONAB expression levels, ZONAB accumulates in the nucleus as well as TJs. The importance of the expression levels of ZO-1 for the distribution of ZONAB is supported by the finding that overexpression of ZO-1 reduces the nuclear pool of ZONAB.[46,50]

If the nuclear ZONAB pool is reduced either by overexpression of ZO-1 or by depletion of total ZONAB expression using antisense or RNA-interference techniques, proliferation of MDCK cells is reduced.[46] This suggests that the proliferation suppressive function of ZO-1 is based on the junctional sequestration of ZONAB. This is further supported by the finding that the SH3 domain of ZO-1, which binds to ZONAB, is necessary and sufficient to reduce proliferation. These data indicate that the cell-density dependent accumulation of ZO-1 at TJs results in the inactivation of the proliferation-promoting ZONAB pathway.[46] This model is further supported by the finding that changes in the relative expression levels of ZO-1 and ZONAB by overexpression or RNA interference not only resulted in changes in cell proliferation but also in the final cell densities of mature MDCK cell monolayers.[46]

ZO-1 and ZONAB appear to regulate G1/S phase transition of the cell cycle (see Fig. 1). This is mediated at least in part by an interaction of ZONAB with CDK4, a regulator of G1/S phase transition, which also colocalizes with ZO-1 at TJs. A reduction in the nuclear ZONAB accumulation was found to reduce also the nuclear pool of CDK4 and cyclin D1 and reduced hyperphosphorylation of the retinoblastoma protein, a nuclear CDK4 substrate.[46] However, as ZONAB is a transcription factor, it is likely that it may also affect proliferation by regulating the expression of cell cycle genes. Potential ZONAB binding sites were indeed found in the genes of several regulators of G1/S-phase transition;[50] however, there is currently no direct evidence for a regulation of their expression by ZONAB (see Fig. 2).

CDK4 is a central regulator of cellular proliferation and its activity is tightly controlled by several interacting proteins.[70-73] Inhibitory proteins can be transcription factors that are upregulated during terminal differentiation.[74-76] The ZONAB and ZO-1 mechanism of regulating CDK4 is different in two respects. Firstly, ZONAB does not directly affect the protein kinase activity but links CDK4 to a mechanism that controls the nuclear accumulation of the kinase. Secondly, ZONAB is upregulated during proliferation and downregulated in differentiated cells, resulting in nuclear accumulation at times of high nuclear CDK4 activity. Because ZONAB does not affect CDK4 activity, it will be interesting to determine how the ZONAB/ CDK4 interaction is regulated and whether CDK4 regulates the transcriptional activity of ZONAB.

The observation that the ZO-1/ZONAB pathway regulates cell density in MDCK cells suggests that this pathway might be important for the regulation of cell numbers in epithelial tissues. Interestingly, experiments in mice linked the ZONAB-interacting kinase CDK4 to the regulation of cell numbers in certain tissues,[77] suggesting that regulation of CDK4 function by ZONAB and ZO-1 might be important for the regulation of cell density. This is further supported by the observation that inhibition of CDK4 by expression of p16/INK4a, a CDK4

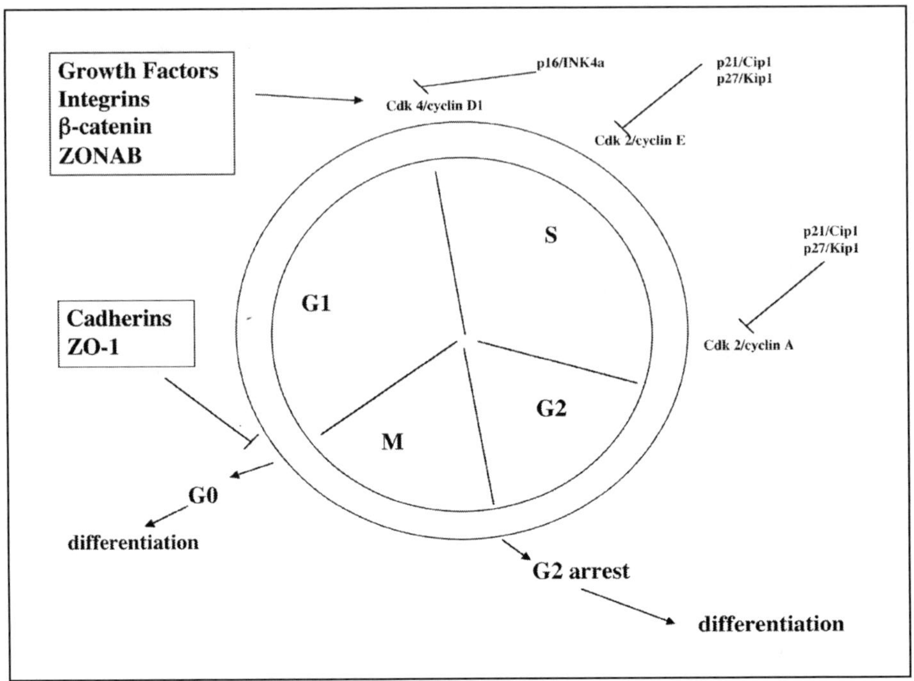

Figure 1. Scheme of the different phases of the cell cycle and points where it might be affected by growth factors, as well as integrin, adherens junction and tight junction associated proteins. Points of arrest for differentiation are also indicated. Only cell cycle kinases, cyclins and inhibitors involved in G1/S phase transition are shown.

inhibitor, was sufficient to reduce cell density in mature MDCK monolayers similarly to overexpression of ZO-1.[46]

Since the level of ZO-1 expression can affect the activity of ZONAB, it is important to understand how expression of ZO-1 is regulated. ZO-1 is expressed at low levels in proliferating low density MDCK cells and becomes upregulated with increasing cell density.[50] At least in part, this appears to be mediated by stabilization of the protein at the forming junctions since its half-life increases with cell density.[78] Because ZO-1 interacts with several junctional membrane proteins, it is likely that stabilization is mediated by the junctional accumulation of transmembrane components. Nevertheless, it is likely that the contribution of transcriptional mechanisms is also important for the regulation of ZO-1 expression. This is suggested by the observation that ZO-1 mRNA is downregulated during corneal wound repair[43] and by overexpression of β-catenin.[44] The observation that a truncated protein consisting only of the HA-tagged SH3 domain of ZO-1 accumulated in the cytosol and was sufficient to reduce proliferation and nuclear accumulation of ZONAB and CDK4 suggests that the interaction of ZO-1 with tight junctions may only be important for the stabilization of ZO-1 but not for preventing the nuclear accumulation of ZONAB and associated proteins.[46] Hence, binding of the SH3 domain to ZONAB may compete with another interaction required for nuclear transport by, for example, restricting access to a nuclear localization signal. Therefore, it will be necessary to identify ZO-1 and ZONAB interacting proteins and to analyze their functions as well as to determine the transcriptional regulation of ZO-1 and ZONAB expression to understand the role of ZO-1 and ZONAB in cell proliferation.

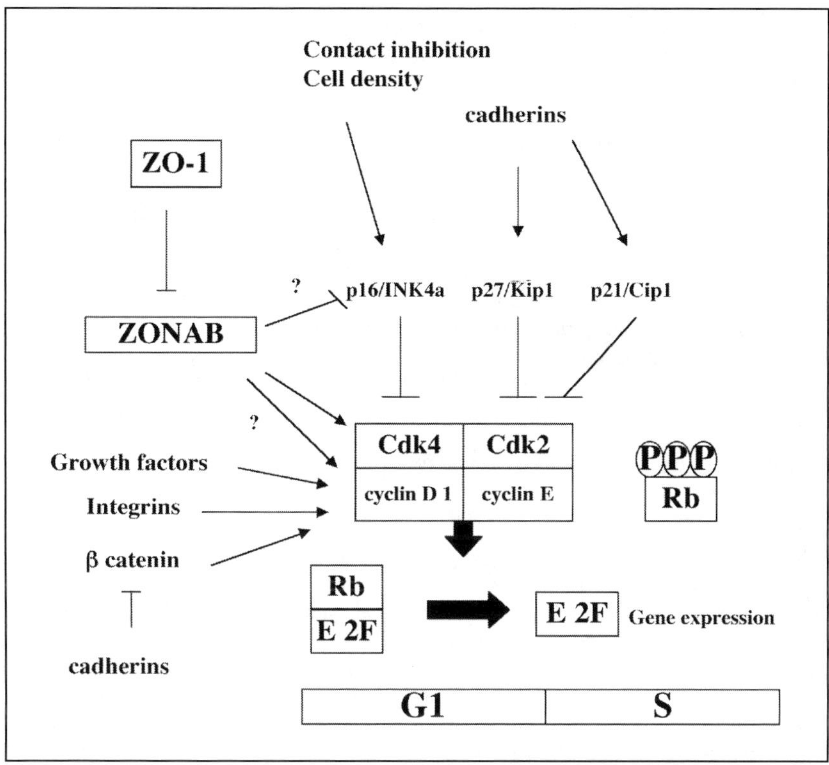

Figure 2. Scheme of regulators and targets of G1/S phase transition. Integrins, growth factors and β-catenin can induce G1/S phase transition by up-regulation of cyclin D1 expression. Cadherins were shown to arrest cells in G1/G0 by upregulation of the cell cycle inhibitor p27/Kip1 and p21/Cip1 and sequestration of β-catenin to the plasma membrane what reduces cyclin D1 expression. ZO-1 was shown to arrest cells in G1/G0 by reduction of the nuclear concentration of ZONAB/cdk4 and cyclin D1. Active cdk4/cyclin D1 phosphorylates retinoblastoma which in turn releases E2F to activate transcription of genes involved in proliferation.

ZONAB and the Regulation of Gene Expression

ZONAB binds to an inverted CCAAT box sequence in the promoter of erbB-2. ZO-1 and ZONAB functionally interact to regulate the expression of endogenous erbB-2.[50] In low density cells, overexpression of ZO-1 increases the expression of endogenous erbB-2, and cotransfection of ZONAB reverses the ZO-1 effect. Based on gene reporter assays, ZONAB functions as a repressor of the erbB-2 promoter and ZO-1 stimulates erbB-2 expression by sequestering nuclear ZONAB. The increased expression of endogenous erbB-2 induced by ZO-1 is not related to reduced proliferation because neither overexpression of erbB-2 nor downregulation with a KDEL-tagged single chain antibody affected the proliferation rates of MDCK cells (Matter and Balda, unpublished). This indicates that regulation of erbB-2 expression is not part of the mechanism by which ZONAB and ZO-1 regulate proliferation of MDCK cells.[46,50] Regulation of erbB-2 expression as well as reduction of the ZONAB/CDK4 complex does not require detectable nuclear accumulation of ZO-1, which indicates that ZO-1 does not directly participate in the regulation of the transcriptional activity of ZONAB but controls the nuclear pool by cytoplasmic sequestration. ZO-1 and ZONAB thus provide epithelial cells with a mechanism to regulate cell proliferation and gene expression in a cell density-dependent manner.

Upregulation of erbB-2 has been linked to organ development and cell differentiation, as well as in certain cellular contexts to tumorigenesis.[79-81] Since overexpression of erbB-2 has been associated with tumorigenesis, it was surprising that overexpression of ZO-1 reduced cell proliferation but increased endogenous erbB-2 expression. However, the levels of erbB-2 overexpression in cancer cells is around 10-100 times higher than in their nontransformed counterparts and correlate with gene amplification. ZO-1 only increases endogenous erbB-2 expression 2-4 times in low confluent cells, which corresponds to the normal expression levels in high density MDCK cells. Furthermore, increased levels of erbB2 expression have been observed when other cells reach confluence and differentiate[82,83] and during mouse development.[79-81] Moreover, overexpression of erbB-2 does not always correlate with increased cell proliferation or transformation, and the activity of constitutively active alleles depends on the cellular background and environment.[84,85] Therefore, by regulating the expression of erbB-2, ZO-1 and ZONAB could be part of a TJ-associated signal transduction pathway important for differentiation of certain tissues as well as a possible cross talk with growth factor pathways.

Tight Junctions and the Ras Pathway in the Regulation of Cell Proliferation

Overexpression of ZONAB resulted in increased cell density but did not cause overgrowth of MDCK cells, suggesting that ZONAB becomes either inactivated or overridden by other signaling pathways. Such mechanisms may be cell type-dependent as ZONAB-overexpression accelerates proliferation of a mammary epithelial cell line (Balda and Matter, unpublished). Therefore, it will be crucial to determine how the ZO-1/ZONAB pathway is regulated as well as cross talk with other signaling systems that influence cell proliferation. We will discuss here the evidence of cell-adhesion-associated signaling pathways that have been implicated in proliferation and that might be related to TJs.

Cell proliferation is controlled by soluble mitogens, components of the extracellular matrix to which cells adhere mainly via integrins, as well as cell-cell junctions. It is clear that there is an extensive cross talk between these pathways.[9,86-88] Two of the most-studied signaling pathways originating from integrins involve the focal adhesion kinase, FAK, and integrin linked kinase, ILK. Activation of FAK triggers signaling cascades such as the extracellular signal-regulated kinase (ERK) of the Ras pathway, which is important for the regulation of cell proliferation and survival.[89] Via this pathway as well as independently, FAK controls cell proliferation by regulation of the expression of the cell cycle regulator cyclin D1.[90,91] Activation of ILK influences a number of signaling components such as GSK3β, resulting in modulation of TCF/LEF and AP-1 activities, which are known to regulate the expression of cyclin D1 on the transcriptional level.[92,93] Thus, extracellular matrix and cell-cell adhesion can cross talk to regulate cell proliferation and differentiation via at least two signaling pathways, the TCF/LEF and the Ras pathway. The cross talk of extracellular matrix, growth factors and cell-cell adhesion by the TCF/LEF pathway has been discussed recently;[9,88,92] so we will focus here only on possible cross talk for the regulation of cell proliferation by the Ras pathway (see Fig. 3).

The loss of epithelial differentiation in cancers often correlates with mutations in Ras, affecting the cellular response to growth factors, extracellular matrix signaling, and cell–cell adhesion.[86] In nontransformed cells, the main mitogenic signal transduction pathways are the ones downstream of Ras such as Raf kinases, Rho family small GTPases and the phosphoinositide 3-OH kinase-PKB/AKT pathway, and they are regulated by integrin-mediated cell adhesion, resulting in anchorage-dependent growth. In contrast, in cancer cells, constitutive activation of these pathways makes cancer cells to grow independently of adhesion.[94] Similarly, during epithelial mesenchymal transition, there is cell-cell junction dissociation and a massive cytoskeletal reorganization with activation of signaling proteins such as Ras, Rho, Src and β-catenin,[95-97] which are known to regulate cell cycle progression at different levels.[60,93,98,99] It is thus striking that Ras, Rho and Src affect cell adhesion at different levels, and

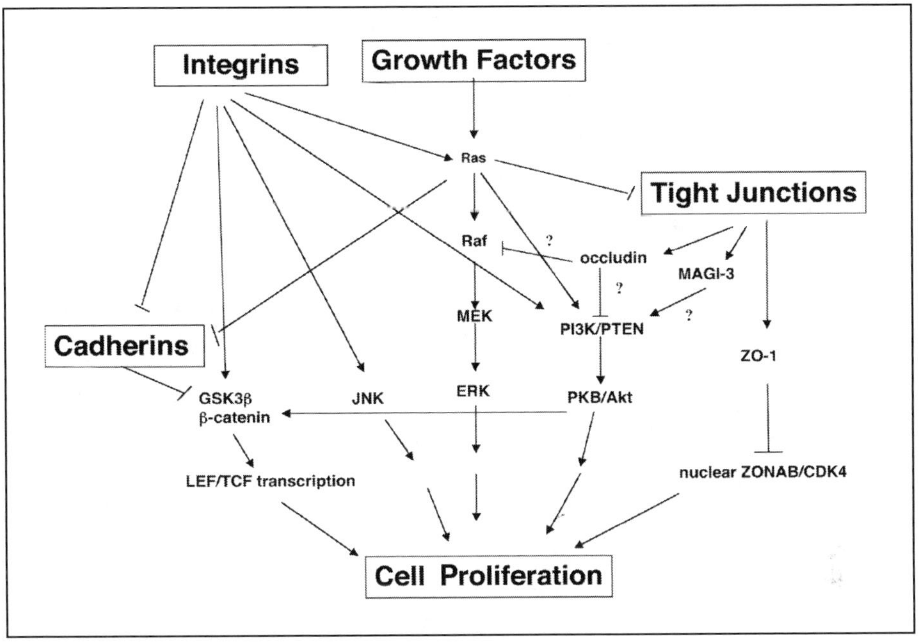

Figure 3. The Ras/Raf and Ras/PI3K signal transduction pathways are possible points of cross talk or self-regulation between signals that favour and inhibit proliferation. Integrins and growth factors activate the Ras pathway to induce cell proliferation. In normal cells there is a tight control of this pathway, in transformed cells this control is lost and many adherens and tight junction proteins are downregulated. Overexpression of E-cadherin as well as occludin were shown to reverse Ras-induced transformation in certain cell types. Only possible cross talk for the regulation of cell proliferation via the Ras pathway is shown (see *Note Added in Proof*). For reason of simplicity cross talk via the TCF/LEF pathway that has been discussed recently[9,88,92] has been omitted.

that multiple pathways originating at structures mediating cell adhesion can regulate the activity of the Ras pathway, which in turn can regulate cell proliferation.

TJs have been linked to different effectors of the Ras pathway. Raf kinases are common Ras effectors, the activity of which leads to the sequential activation of the pathway of the mitogen-activated protein kinases of extracellular-signal regulated kinases, MEKs and ERKs, which in turn regulates transcription. Expression of constitutively active Raf-1 or MEK1 is sufficient to transform immortalized epithelial cells.[20,100] Consequently, treatment of H-Ras transformed MDCK cells with the selective MEK1 inhibitor PD98059 resulted in increased expression of adherens junction proteins and assembly of functional adherens junctions and TJs.[101,102] TJs seem to be intimately connected with Raf-1 signalling, as overexpression of the TJ-associated membrane protein occludin is sufficient to reverse Raf-1-mediated transformation of a salivary gland epithelial cell line.[20] The molecular mechanism by which TJs can counteract Raf signalling has not been elucidated, and it is not clear whether inhibition of Raf-1 mediated transformation by occludin is direct or involves other signalling pathways.

AF-6/afadin is a Ras target that was proposed to serve as a peripheral component of TJs in epithelial cells because it interacts with ZO-1 in a Ras dependent manner and also binds to JAM.[103,104] It was also demonstrated that AF-6/afadin is recruited to nectin-based adherens junctions, suggesting that AF-6/afadin is not a specific component of TJs.[105,106] Furthermore, AF-6/afadin is crucial for the establishment of intercellular junctions during mouse development.[107] The physiological relevance for the interaction of AF-6/afadin with ZO-1 and how this contributes to Ras signaling is unknown, the fact that AF-6/afadin can associate with components of more than one intercellular junction might be of importance for signaling cross talk.

The Ras effector phosphatidylinositol-3 kinase (PI3K) is important for the transduction of signals from growth factors, integrins and cadherins.[16,108,109] PI3K is an upstream regulator of protein kinase B, PKB/AKT, and this pathway plays a central role in regulating cell survival, adhesion, migration, proliferation and differentiation. The action of PI3K is counteracted by the lipid phosphatase and tumor suppressor PTEN, which converts PI-3,4,5-P(3) to PI-4,5-P(2). Inactivation of the PI3K pathway inhibits G1/S phase transition.[110] PI3K and PTEN have been linked to TJs. PI3K associates with the C-terminal cytoplasmic domain of occludin in vitro[111] and seems to be involved in junctional assembly as well as response to oxidative stress.[112] PTEN binds to a PDZ domain of a junctional MAGUK, MAGI-3, which is thought to be responsible for junctional recruitment of the lipid phosphatase. Moreover, MAGI-3 and PTEN also seem to cooperate to modulate the kinase activity of PKB/AKT.[57] Whether the interactions of PI3K and PTEN with TJ proteins are related to the role of the PI3K pathway in cell proliferation observed in certain cancers is unknown.[113] (see *Note Added in Proof*) Nevertheless, it is tempting to speculate that TJs serve as a platform to anchor multiple proliferation-regulating pathways to the plasma membrane to regulate them in a cell density-dependent manner as well as to facilitate cross talk between these pathways to ensure proper regulation of proliferation and differentiation.

Conclusions and Perspectives

Many TJ proteins have been identified, sequenced and characterized since 1986, when ZO-1 was first isolated.[114] Several of these TJ associated proteins have been linked not only to the regulation of paracellular permeability and restriction of apical/basolateral intramembrane diffusion of lipids, but have also been connected to different types of signaling pathways that modulate cell proliferation and differentiation. The exciting recent results that indicate that TJs play a role in signal transduction processes that regulate epithelial proliferation and differentiation, suggest that TJs suppress pathways that stimulate proliferation and inhibit differentiation. For example, upregulation of occludin expression suppresses Raf-1-mediated transformation; overexpression of ZO-2 can counteract oncogenic viral proteins; and upregulation of ZO-1 expression results in cytoplasmic sequestration of the transcription factor ZONAB and reduces cell proliferation and cell density. However, most of these signaling systems are incompletely characterized and little is known about their physiological functions during development and in tissue homeostasis.

Our current knowledge has been obtained either by studying single molecules or linear signaling pathways, and it is not clear whether and how these pathways interact between them as well as with the ones originating from other sites of cell adhesion such as adherens junctions and integrins. For example, we are eager to know which is (are) the mechanism(s) by which occludin inhibits Raf-1-mediated transformation and how this influences ZO-1 associated signaling molecules such as ZONAB; whether and how the ZO-1/ZONAB pathway of cell proliferation is regulated by association with other tight junction proteins; and whether ZO-2, symplekin, huASH1and GEF-H1 regulate cell proliferation or any other nuclear function.

TJs assemble when epithelial cells reach high cell densities, stop proliferating and differentiate, resulting in increased expression levels of TJ-associated proteins. Therefore, it is tempting to speculate that TJs function in the sensing of cell density. In agreement, manipulation of specific TJ proteins, such as ZO-1 and ZONAB, affects the cell density at which cell culture monolayers stop to proliferate. It will be necessary to identify the tissues that are sensitive to such manipulations in vivo. Furthermore, TJ-associated signaling systems are likely to be affected by and influence other pathways that control proliferation and differentiation; so it will be important to find out which are the cross talk between TJs and better characterized cell proliferation regulatory systems such as cadherin-based adhesion and signaling from the extracellular matrix and growth factors.

Note Added in Proof

A target of the Ras pathway, RalA, has recently been linked to tight junction signalling. RalA interacts with ZONAB in a GTP-dependent manner. The amount of the RalA-ZONAB complex increases as epithelial cells become more dense and increase cell contacts. The RalA-ZONAB interaction results in relief of transcriptional repression of a ZONAB regulated promoter. Additionally, expression of oncogenic Ras alleviates transcriptional repression by ZONAB in a RalA dependent manner. These data implicate the RalA/ZONAB can control ZONAB function (Frankel P, Aronheim A, Kavanagh E et al. RalA interacts with ZONAB in a cell density dependent manner and regulates its transcriptional activity. EMBO J; in press).

Acknowledgements

We thank Drs. S. Aijaz and S. Goodrick for critical reading of the manuscript. The research in our laboratories is supported by The Wellcome Trust, Cancer Research UK and BBSRC.

References

1. Jamora C, Fuchs E. Intercellular adhesion, signalling and the cytoskeleton. Nat Cell Biol 2002; 4(4):E101-108.
2. Green KJ, Gaudry CA. Are desmosomes more than tethers for intermediate filaments? Nat Rev Mol Cell Biol 2000; 1(3):208-216.
3. Garrod DR, Merritt AJ, Nie Z. Desmosomal cadherins. Curr Opin Cell Biol 2002; 14(5):537-545.
4. Anderson JM. Molecular structure of tight junctions and their role in epithelial transport. News Physiol Sci 2001; 16:126-130.
5. Tsukita S, Furuse M, Itoh M. Multifunctional strands in tight junctions. Nat Rev Mol Cell Biol 2001; 2:286-293.
6. Cereijido M, Shoshani L, Contreras RG. Molecular physiology and pathophysiology of tight junctions. I. Biogenesis of tight junctions and epithelial polarity. Am J Physiol 2000; 279(3):G477-482.
7. Fanning AS, Ma TY, Anderson JM. Isolation and functional characterization of the actin binding region in the tight junction protein ZO-1. FASEB J 2002; 16(13):1835-1837.
8. Woods A, Couchman JR. Syndecan-4 and focal adhesion function. Curr Opin Cell Biol 2001; 13(5):578-583.
9. Balda MS, Matter K. Epithelial cell adhesion and the regulation of gene expression. Trends Cell Biol 2003; 13(6):310-318.
10. Damsky CH, Ilic D. Integrin signaling: It's where the action is. Curr Opin Cell Biol 2002; 14(5):594-602.
11. Ben-Ze'ev A, Geiger B. Differential molecular interactions of β-catenin and plakoglobin in adhesion, signaling and cancer. Curr Opin Cell Biol1998; 10(5):629-639.
12. Assoian RK, Schwartz MA. Coordinate signaling by integrins and receptor tyrosine kinases in the regulation of G1 phase cell-cycle progression. Curr Opin Genet Dev 2001; 11(1):48-53.
13. Gottardi CJ, Gumbiner BM. Adhesion signaling: How β-catenin interacts with its partners. Curr Biol 2001; 11(19):R792-794.
14. Seidensticker MJ, Behrens J. Biochemical interactions in the wnt pathway. Biochim Biophys Acta 2000; 1495(2):168-182.
15. Anderson JM, Van Itallie CM, Fanning AS. Setting up a selective barrier at the apical junction complex. Curr Opin Cell Biol 2004; 16(2):140-145.
16. Dejana E, Lampugnani MG, Martinez-Estrada O et al. The molecular organization of endothelial junctions and their functional role in vascular morphogenesis and permeability. Int J Dev Biol 2000; 44(6):743-748.
17. Bazzoni G. The JAM family of junctional adhesion molecules. Curr Opin Cell Biol 2003; 15(5):525-530.
18. Lemmers C, Michel D, Lane-Guermonprez L et al. CRB3 binds directly to Par6 and regulates the morphogenesis of the tight junctions in mammalian epithelial cells. Mol Biol Cell 2004; 15(3):1324-1333.
19. Roh MH, Fan S, Liu CJ et al. The Crumbs3-Pals1 complex participates in the establishment of polarity in mammalian epithelial cells. J Cell Sci 2003; 116(Pt 14):2895-2906.
20. Li D, Mrsny RJ. Oncogenic Raf-1 disrupts epithelial tight junctions via downregulation of occludin. J Cell Biol 2000; 148(4):791-800.

21. Saitou M, Furuse M, Sasaki H et al. Complex phenotype of mice lacking occludin, a component of tight junction strands. Mol Biol Cell 2000; 11(12):4131-4142.
22. Tiwari-Woodruff SK, Buznikov AG, Vu TQ et al. OSP/claudin-11 forms a complex with a novel member of the tetraspanin super family and β1 integrin and regulates proliferation and migration of oligodendrocytes. J Cell Biol 2001; 153(2):295-305.
23. Miwa N, Furuse M, Tsukita S et al. Involvement of claudin-1 in the β-catenin/Tcf signaling pathway and its frequent upregulation in human colorectal cancers. Oncol Res 2000; 12(11-12):469-476.
24. Long H, Crean CD, Lee WH et al. Expression of Clostridium perfringens enterotoxin receptors claudin-3 and claudin-4 in prostate cancer epithelium. Cancer Res 2001; 61(21):7878-7881.
25. Michl P, Barth C, Buchholz M et al. Claudin-4 expression decreases invasiveness and metastatic potential of pancreatic cancer. Cancer Res 2003; 63(19):6265-6271.
26. Singh AB, Harris RC. Epidermal growth factor receptor activation differentially regulates claudin expression and enhances transepithelial resistance in Madin-Darby canine kidney cells. J Biol Chem 2004; 279(5):3543-3552.
27. Ohkubo T, Ozawa M. The transcription factor Snail downregulates the tight junction components independently of E-cadherin downregulation. J Cell Sci 2004; 117(Pt 9):1675-1685.
28. Ikenouchi J, Matsuda M, Furuse M et al. Regulation of tight junctions during the epithelium-mesenchyme transition: Direct repression of the gene expression of claudins/occludin by Snail. J Cell Sci 2003; 116(Pt 10):1959-1967.
29. Roh MH, Margolis B. Composition and function of PDZ protein complexes during cell polarization. Am J Physiol Renal Physiol 2003; 285(3):F377-387.
30. Ebnet K, Aurrand-Lions M, Kuhn A et al. The junctional adhesion molecule (JAM) family members JAM-2 and JAM-3 associate with the cell polarity protein PAR-3: A possible role for JAMs in endothelial cell polarity. J Cell Sci 2003; 116(19):3879-3891.
31. D'Atri F, Citi S. Molecular complexity of vertebrate tight junctions. Mol Membrane Biol 2002; 19:103-112.
32. Gonzalez-Mariscal L, Betanzos A, Nava P et al. Tight junction proteins. Prog Biophys Mol Biol 2003; 81(1):1-44.
33. Matter K, Balda MS. Signalling to and from tight junctions. Nat Rev Mol Cell Biol 2003; 4(3):225-236.
34. Schneeberger EE, Lynch RD. The tight junction: A multifunctional complex. Am J Physiol Cell Physiol 2004; 286(6):C1213-1228.
35. Itoh M, Nagafuchi A, Yonemura S et al. The 220-kD protein colocalizing with cadherins in nonepithelial cells is identical to ZO-1, a tight junction-associated protein in epithelial cells: cDNA cloning and immunoelectron microscopy. J Cell Biol 1993; 121(3):491-502.
36. Willott E, Balda MS, Fanning AS et al. The tight junction protein ZO-1 is homologous to the Drosophila discs-large tumor suppressor protein of septate junctions. Proc Natl Acad Sci USA 1993; 90:7834-7838.
37. Fanning AS, Anderson JM. Protein modules as organizers of membrane structure. Curr Opin Cell Biol 1999; 11(4):432-439.
38. Anderson JM, Balda MS, Faming A. The structure and regulation of tight junctions. Curr Opin Cell Biol 1993; 5:772-778.
39. Tsukita S, Furuse M, Itoh M. Structural and signalling molecules come together at tight junctions. Curr Opin Cell Biol 1999; 11:628-633.
40. Dimitratos SD, Woods DF, Stathakis DG et al. Signaling pathways are focused at specialized regions of the plasma membrane by scaffolding proteins of the MAGUK family. Bioessays 1999; 21(11):912-921.
41. Takahisa M, Togashi S, Suzuki T et al. The Drosophila tamou gene, a component of the activating pathway of extramacrochaetae expression, encodes a protein homologous to mammalian cell-cell junction-associated protein ZO-1. Genes Dev 1996; 10(14):1783-1795.
42. Chen CM, Freedman JA, Bettler Jr DR et al. Polychaetoid is required to restrict segregation of sensory organ precursors from proneural clusters in Drosophila. Mech Dev 1996; 57(2):215-227.
43. Cao Z, Wu HK, Bruce A et al. Detection of differentially expressed genes in healing mouse corneas using cDNA Microarrays. Invest Ophthalmol Vis Sci 2002; 43:2897-2904.
44. Mann B, Gelos M, Siedow A et al. Target genes of β-catenin-T cell-factor/lymphoid-enhancer-factor signaling in human colorectal carcinomas. Proc Natl Acad Sci USA 1999; 96(4):1603-1608.
45. Hoover KB, Liao SY, Bryant PJ. Loss of the tight junction MAGUK ZO-1 in breast cancer: Relationship to glandular differentiation and loss of heterozygosity. Am J Pathol 1998; 153:1767-1773.
46. Balda MS, Garrett MD, Matter K. The ZO-1-associated Y-box factor ZONAB regulates epithelial cell proliferation and cell density. J Cell Biol 2003; 160(3):423-432.

47. Chlenski A, Ketels KV, Korovaitseva GI et al. Organization and expression of the human zo-2 gene (tjp-2) in normal and neoplastic tissues. Biochim Biophys Acta 2000; 1493(3):319-324.

48. Gottardi CJ, Arpin M, Fanning AS et al. The junction-associated protein, zonula occludens-1, localizes to the nucleus before the maturation and during the remodeling of cell-cell contacts. Proc Natl Acad Sci USA 1996; 93:10779-10784.

49. Islas S, Vega J, Ponce L et al. Nuclear localization of the tight junction protein ZO-2 in epithelial cells. Exp Cell Res 2002; 274(1):138-148.

50. Balda MS, Matter K. The tight junction protein ZO-1 and an interacting transcription factor regulate ErbB-2 expression. EMBO J 2000; 19(9):2024-2033.

51. Glaunsinger BA, Weiss RS, Lee SS et al. Link of the unique oncogenic properties of adenovirus type 9 E4-ORF1 to a select interaction with the candidate tumor suppressor protein ZO-2. EMBO J 2001; 20(20):5578-5586.

52. Reichert M, Muller T, Hunziker W. The PDZ domains of zonula occludens-1 induce an epithelial to mesenchymal transition of Madin-Darby canine kidney I cells. Evidence for a role of β-catenin/Tcf/Lef signaling. J Biol Chem 2000; 275(13):9492-9500.

53. Traweger A, Fuchs R, Krizbai IA et al. The tight junction protein ZO-2 localizes to the nucleus and interacts with the heterogeneous nuclear ribonucleoprotein scaffold attachment factor-B. J Biol Chem 2003; 278(4):2692-2700.

54. Lee SS, Glaunsinger B, Mantovani F et al. Multi-PDZ domain protein MUPP1 is a cellular target for both adenovirus E4-ORF1 and high-risk papillomavirus type 18 E6 oncoproteins. J Virol 2000; 74(20):9680-9693.

55. Glaunsinger BA, Lee SS, Thomas M et al. Interactions of the PDZ-protein MAGI-1 with adenovirus E4-ORF1 and high- risk papillomavirus E6 oncoproteins. Oncogene 2000; 19(46):5270-5280.

56. Betanzos A, Huerta M, Lopez-Bayghen E et al. The tight junction protein ZO-2 associates with Jun, Fos and C/EBP transcription factors in epithelial cells. Exp Cell Res 2004; 292(1):51-66.

57. Wu Y, Dowbenko D, Spencer S et al. Interaction of the tumor suppressor PTEN/MMAC with a PDZ domain of MAGI3, a novel membrane-associated guanylate kinase. J Biol Chem 2000; 275(28):21477-21485.

58. Benais-Pont G, Punn A, Flores-Maldonado C et al. Identification of a tight junction-associated guanine nucleotide exchange factor that activates Rho and regulates paracellular permeability. J Cell Biol 2003; 160(5):729-740.

59. Sahai E, Marshall CJ. RHO-GTPases and cancer. Nat Rev Cancer 2002; 2(2):133-142.

60. Pruitt K, Der CJ. Ras and Rho regulation of the cell cycle and oncogenesis. Cancer Lett 2001; 171(1):1-10.

61. Jaffe AB, Hall A. Rho GTPases in transformation and metastasis. Adv Cancer Res 2002; 84:57-80.

62. Nakamura T, Blechman J, Tada S et al. huASH1 protein, a putative transcription factor encoded by a human homologue of the Drosophila ash1 gene, localizes to both nuclei and cell-cell tight junctions. Proc Natl Acad Sci USA 2000; 97(13):7284-7289.

63. Beisel C, Imhof A, Greene J et al. Histone methylation by the Drosophila epigenetic transcriptional regulator Ash1. Nature 2002; 419(6909):857-862.

64. Keon BH, Schäfer S, Kuhn C et al. Symplekin, a novel type of tight junction plaque protein. J Cell Biol 1996; 134:1003-1018.

65. Takagaki Y, Manley JL. Complex protein interactions within the human polyadenylation machinery identify a novel component. Mol Cell Biol 2000; 20(5):1515-1525.

66. Hofmann I, Schnolzer M, Kaufmann I et al. Symplekin, a constitutive protein of karyo- and cytoplasmic particles involved in mRNA biogenesis in Xenopus laevis oocytes. Mol Biol Cell 2002; 13(5):1665-1676.

67. Xing H, Mayhew CN, Cullen KE et al. HSF1 modulation of Hsp70 mRNA polyadenylation via interaction with symplekin. J Biol Chem 2004; 279(11):10551-10555.

68. Kohno K, Izumi H, Uchiumi T et al. The pleiotropic functions of the Y-box-binding protein, YB-1. Bioessays 2003; 25(7):691-698.

69. Gallia GL, Johnson EM, Khalili K. Puralpha: A multifunctional single-stranded DNA- and RNA-binding protein. Nucleic Acids Res 2000; 28(17):3197-3205.

70. Ekholm SV, Reed SI. Regulation of G(1) cyclin-dependent kinases in the mammalian cell cycle. Curr Opin Cell Biol 2000; 12(6):676-684.

71. Ortega S, Malumbres M, Barbacid M. Cyclin D-dependent kinases, INK4 inhibitors and cancer. Biochim Biophys Acta 2002; 1602(1):73-87.

72. Bringold F, Serrano M. Tumor suppressors and oncogenes in cellular senescence. Exp Gerontol 2000; 35(3):317-329.

73. Sherr CJ. The Pezcoller lecture: Cancer cell cycles revisited. Cancer Res 2000; 60(14):3689-3695.

74. Wang H, Iakova P, Wilde M et al. C/EBPα arrests cell proliferation through direct inhibition of Cdk2 and Cdk4. Mol Cell 2001; 8(4):817-828.
75. Zhang JM, Zhao X, Wei Q et al. Direct inhibition of G(1) cdk kinase activity by MyoD promotes myoblast cell cycle withdrawal and terminal differentiation. EMBO J 1999; 18(24):6983-6993.
76. Li J, Joo SH, Tsai MD. An NF-κB-specific inhibitor, IκBα, binds to and inhibits cyclin-dependent kinase 4. Biochemistry 2003; 42(46):13476-13483.
77. Malumbres M, Barbacid M. To cycle or not to cycle: A critical decision in cancer. Nat Rev Cancer 2001; 1:222-231.
78. Gumbiner B, Lowenkopf T, Apatira D. Identification of a 160-kDa polypeptide that binds to the tight junction protein ZO-1. Proc Natl Acad Sci USA 1991; 88(8):3460-3464.
79. Troyer KL, Lee DC. Regulation of mouse mammary gland development and tumorigenesis by the ERBB signaling network. J Mammary Gland Biol Neoplasia 2001; 6(1):7-21.
80. Citri A, Skaria KB, Yarden Y. The deaf and the dumb: The biology of ErbB-2 and ErbB-3. Exp Cell Res 2003; 284(1):54-65.
81. Garratt AN, Ozcelik C, Birchmeier C. ErbB2 pathways in heart and neural diseases. Trends Cardiovasc Med 2003; 13(2):80-86.
82. Kornilova ES, Taverna D, Hoeck W et al. Surface expression of erbB-2 protein is post-transcriptionally regulated in mammary epithelial cells by epidermal growth factor and by the culture density. Oncogene1992; 7(3):511-519.
83. Park SK, Miller R, Krane I et al. The erbB2 gene is required for the development of terminally differentiated spinal cord oligodendrocytes. J Cell Biol 2001; 154(6):1245-1258.
84. Khoury H, Dankort DL, Sadekova S et al. Distinct tyrosine autophosphorylation sites mediate induction of epithelial mesenchymal like transition by an activated ErbB-2/Neu receptor. Oncogene 2001; 20(7):788-799.
85. Janda E, Litos G, Grunert S et al. Oncogenic Ras/Her-2 mediate hyperproliferation of polarized epithelial cells in 3D cultures and rapid tumor growth via the PI3K pathway. Oncogene 2002; 21(33):5148-5159.
86. Schwartz MA, Ginsberg MH. Networks and crosstalk: Integrin signalling spreads. Nat Cell Biol 2002; 4(4):E65-68.
87. Giancotti FG, Tarone G. Positional control of cell fate through joint integrin/receptor protein kinase signaling. Annu Rev Cell Dev Biol 2003; 19:173-206.
88. Nelson WJ, Nusse R. Convergence of Wnt, β-catenin, and cadherin pathways. Science 2004; 303(5663):1483-1487.
89. Howe AK, Aplin AE, Juliano RL. Anchorage-dependent ERK signaling—mechanisms and consequences. Curr Opin Genet Dev 2002; 12(1):30-35.
90. Zhao J, Pestell R, Guan JL. Transcriptional activation of cyclin D1 promoter by FAK contributes to cell cycle progression. Mol Biol Cell 2001; 12(12):4066-4077.
91. Zhao J, Bian ZC, Yee K et al. Identification of transcription factor KLF8 as a downstream target of focal adhesion kinase in its regulation of cyclin D1 and cell cycle progression. Mol Cell 2003; 11(6):1503-1515.
92. Yoganathan N, Yee A, Zhang Z et al. Integrin-linked kinase, a promising cancer therapeutic target: Biochemical and biological properties. Pharmacol Ther 2002; 93(2-3):233-242.
93. Behrens J. Control of β-catenin signaling in tumor development. Ann NY Acad Sci 2000; 910:21-33, discussion 33-25.
94. Danen EH, Yamada KM. Fibronectin, integrins, and growth control. J Cell Physiol 2001; 189(1):1-13.
95. Savagner P. Leaving the neighborhood: Molecular mechanisms involved during epithelial-mesenchymal transition. Bioessays 2001; 10:912-923.
96. Mercer JA. Intercellular junctions: Downstream and upstream of Ras? Semin Cell Dev Biol 2000; 11(4):309-314.
97. Vincent-Salomon A, Thiery JP. Host microenvironment in breast cancer development: Epithelial-mesenchymal transition in breast cancer development. Breast Cancer Res 2003; 5(2):101-106.
98. Coleman ML, Marshall CJ, Olson MF. RAS and RHO GTPases in G1-phase cell-cycle regulation. Nat Rev Mol Cell Biol 2004; 5(5):355-366.
99. Amanatullah DF, Zafonte BT, Albanese C et al. Ras regulation of cyclin D1 promoter. Methods Enzymol 2001; 333:116-127.
100. Montesano R, Soriano JV, Hosseini G et al. Constitutively active mitogen-activated protein kinase kinase MEK1 disrupts morphogenesis and induces an invasive phenotype in Madin-Darby canine kidney epithelial cells. Cell Growth Differ 1999; 10(5):317-332.

101. Lu Q, Paredes M, Zhang J et al. Basal extracellular signal-regulated kinase activity modulates cell-cell and cell-matrix interactions. Mol Cell Biol 1998; 18(6):3257-3265.
102. Chen Y, Lu Q, Schneeberger EE et al. Restoration of tight junction structure and barrier function by down-regulation of the mitogen-activated protein kinase pathway in ras-transformed Madin-Darby canine kidney cells. Mol Biol Cell 2000; 11(3):849-862.
103. Yamamoto T, Harada N, Kano K et al. The ras target AF-6 interacts with ZO-1 and serves as a peripheral component of tight junctions in epithelial cells. J Cell Biol 1997; 139:785-795.
104. Ebnet K, Schulz CU, Meyer Zu Brickwedde MK et al. Junctional adhesion molecule interacts with the PDZ domain-containing proteins AF-6 and ZO-1. J Biol Chem 2000; 275(36):27979-27988.
105. Mandai K, Nakanishi H, Satoh A et al. Afadin: A novel actin filament-binding protein with one PDZ domain localized at cadherin-based cell-to-cell adherens junction. J Cell Biol 1997; 139(2):517-528.
106. Takai Y, Nakanishi H. Nectin and afadin: Novel organizers of intercellular junctions. J Cell Sci 2003; 116(Pt 1):17-27.
107. Zhadanov AB, Provance Jr DW, Speer CA et al. Absence of the tight junctional protein AF-6 disrupts epithelial cell-cell junctions and cell polarity during mouse development. Curr Biol 1999; 9(16):880-888.
108. Yart A, Chap H, Raynal P. Phosphoinositide 3-kinases in lysophosphatidic acid signaling: Regulation and cross-talk with the Ras/mitogen-activated protein kinase pathway. Biochim Biophys Acta 2002; 1582(1-3):107-111.
109. Bakin AV, Tomlinson AK, Bhowmick NA et al. Phosphatidylinositol 3-kinase function is required for transforming growth factor β-mediated epithelial to mesenchymal transition and cell migration. J Biol Chem 2000; 275(47):36803-36810.
110. Burgering BM, Kops GJ. Cell cycle and death control: Long live Forkheads. Trends Biochem Sci 2002; 27(7):352-360.
111. Nusrat A, Chen JA, Foley CS et al. The coiled-coil domain of occludin can act to organize structural and functional elements of the epithelial tight junction. J Biol Chem 2000; 275(38):29816-29822.
112. Sheth P, Basuroy S, Li C et al. Role of phosphatidylinositol 3-kinase in oxidative stress-induced disruption of tight junctions. J Biol Chem 2003; 278(49):49239-49245.
113. Mills GB, Lu Y, Fang X et al. The role of genetic abnormalities of PTEN and the phosphatidylinositol 3-kinase pathway in breast and ovarian tumorigenesis, prognosis, and therapy. Semin Oncol 2001; 28(5 Suppl 16):125-141.
114. Stevenson BR, Siliciano JD, Mooseker MS et al. Identification of ZO-1: A high molecular weight polypeptide associated with the tight junction (zonula occludens) in a variety of epithelia. J Cell Biol 1986; 103:755-766.

CHAPTER 9

Tight Junction Proteins and Cancer

Isabel J. Latorre, Kristopher K. Frese and Ronald T. Javier

Abstract

The tight junction (TJ) has been the subject of intense investigations since its discovery in the early 1960s. It has long been recognized as a key determinant of the epithelial cell barrier function, but only recently has the role of the TJ in the control of cellular proliferation and differentiation been appreciated. Identification of the molecular components of the TJ has shown that, in addition to their structural functions, these proteins also regulate both signal transduction emanating from the plasma membrane and gene expression in the nucleus. In addition, studies of human tumors reveal a direct correlation between the loss of functional TJs in cancer progression and metastasis, and several different mutant mice lacking specific TJ proteins develop hyperproliferative disorders. Work conducted in vitro with cell lines has provided additional evidence that the loss of TJ proteins can promote transformation, as well as increased metastatic potential. Intriguingly, the tumorigenic potential of several different viral oncoproteins is also linked to their ability to disrupt TJs. In summary, studies directed at determining the mechanisms whereby the TJ controls cellular proliferation and motility are expected to aid in the development of new effective strategies for treating and preventing human cancers.

Introduction

Despite the fact that approximately 90% of all human malignancies are epithelial in origin,[1,2] much of our knowledge about mechanisms responsible for the tumorigenic transformation of cells stems from studies of fibroblasts. A growing body of evidence, however, suggests that features unique to epithelial cells, particularly specialized cell-cell junctions and apico-basal cell polarity, play central roles in regulating cellular proliferation and differentiation and, when disrupted, contribute to the development of cancer.

In multicellular organisms, proper functioning of many types of epithelial cells depends on the partitioning of their plasma membranes into distinct apical and basal compartments, where apical membranes face a luminal space or the environment and basal membranes contact the underlying body tissues. This polarization of epithelial cells mainly involves two types of cell-cell membrane structures, the adherens junction and the zonula occludens or tight junction (TJ). Whereas adherens junctions are responsible for strong cell-cell adhesion,[3] TJs create a barrier that separates not only the apical and basal membrane compartments within an individual epithelial cell but also the external environment from internal body tissues within a sheet of epithelial cells.

In 1963, Farquhar and Palade first described the TJs of epithelial cells, and also provided evidence that these apically-located, belt-like structures form an impermeable barrier that restricts macromolecules from diffusing out of luminal spaces.[4] Although they incorrectly hypothesized that the TJ resulted from "fusion of the adjacent cell membranes", these investigators described a "diffuse band of dense cytoplasmic material...associated with this junction",

Tight Junctions, edited by Lorenza Gonzalez-Mariscal. ©2006 Landes Bioscience and Springer Science+Business Media.

which we now know corresponds to a variety of TJ-associated proteins. In recent years, there have been intense efforts to identify and determine functions for the polypeptides present at the TJ, and today more than 40 different such proteins have been identified. For most of these factors, however, their precise functions and contributions to TJ establishment and maintenance are not yet fully understood.

Central to the subject of this chapter, the disruption of cell-cell junctions and the concomitant loss of polarity represent hallmark phenotypes for many different cancer cells. Moreover, a variety of evidence supports the argument that these two common phenotypes of cancer cells directly contribute to tumorigenesis rather than simply representing indirect consequences of the transformed state of these cells. Two recent studies highlight this important point by reporting compelling data indicating that disruption of adherens junctions per se in cells promotes the development of some cancers.[5,6]

In this chapter, we discuss findings that point to a similar important role for TJ disruption in carcinogenesis. While previous studies primarily linked loss of TJs to late stages of tumor progression such as metastasis, a burgeoning body of evidence now argues that some TJ-associated proteins are tumor suppressors and/or regulators of signal transduction, suggesting that TJ disruption may likewise contribute to early stages of carcinogenesis. Consistent with this proposal, work from several different groups has shown that the transforming potentials of oncoproteins encoded by human adenovirus, human papillomavirus, and human T-cell leukemia virus type 1 depend in part on their ability to interact with cellular PDZ domain-containing proteins found at the TJs in epithelial cells.

TJ Proteins

TJ Transmembrane Proteins

The formation of functional TJs relies in part on interactions between the extracellular domains of integral membrane proteins present at these specialized membrane sites. Three types of such proteins, occludin,[7] junctional adhesion molecules (JAMs),[8,9] and claudins,[10,11] are known to localize to the TJ.

Occludin, which was the first integral membrane protein identified at the TJ,[7] specifically localizes to this site,[12] and its protein expression levels correlate with the number of TJ strands present in cells.[13] Nevertheless, expression of occludin alone is insufficient to induce TJ formation in cells.[14] Instead, this protein appears to provide both the barrier (paracellular sealing) and fence (membrane domain differentiation) functions of the TJ.[15] The carboxyl-terminal cytoplasmic domain of occludin interacts with several TJ-associated proteins, including ZO-1[16], ZO-2,[17] and ZO-3,[18] as well as with F-actin[19] and the regulatory subunit of phosphoinositide 3-kinase (PI3K).[20,21] Although dispensable for maintaining TER, the cytoplasmic domain of occludin is important for maintaining the paracellular barrier.[22]

A second family of integral membrane proteins found at the TJ is the JAMs, which are members of the immunoglobulin superfamily.[8,23] Although the four different known JAM proteins can be detected in circulating leukocytes and platelets,[24] these proteins are predominantly expressed in epithelial and endothelial cells, where they localize to the TJ and, to a lesser extent, along the basolateral membrane.[25] Similar to occludin, JAM expression alone fails to promote TJ formation in cells.[26] While the function of JAMs at the TJ is not fully understood, several lines of evidence indicate their requirement for TJ function and integrity.[25,27] In addition, their cytoplasmic domains mediate interactions with several TJ-associated proteins, including occludin, cingulin, ASIP/Par-3,[26] MUPP1,[28] ZO-1,[29] MAGI-1,[30] and AF-6.[31,32] These interactions are thought to contribute to the proper TJ localization of both JAMs and their partners.

Claudins are a third family of integral membrane proteins found at the TJ. The claudin family comprises more than 20 different members,[14,33] which vary widely in their tissue distribution patterns.[11] Contrary to occludin and JAMs, claudin expression is both necessary and sufficient to induce TJ formation in cells,[14,34] indicating that this protein represents a key structural element of TJs required to initiate formation of these cell-cell junctions. In addition, similar to both occludin and JAMs, claudins likewise bind to several other TJ-associated proteins (e.g., MUPP1,[28] PATJ,[35] ZO-1, ZO-2, and ZO-3[36]), which could be envisioned to regulate the capacity of claudins to initiate TJ formation and/or to transmit signals from these cell-cell contact sites.

TJ Cytoplasmic Proteins

In addition to transmembrane proteins, a number of cytoplasmic proteins also associate with TJs, as was eluded to above. These proteins can be divided into two general groups, PDZ domain-containing adaptor proteins and a collection of other adaptor and signaling proteins.

A large group of related TJ-associated cytoplasmic proteins consists of members from the diverse family of PDZ domain-containing proteins, including the membrane-associated guanylate kinase-homology (MAGUK) proteins ZO-1,[37] ZO-2,[38] ZO-3,[18] PALS1,[39] MAGI-1, -2, and -3,[40-42] the multi-PDZ domain proteins PAR3,[43] PAR6,[44] MUPP1,[28] and PATJ,[39] and the Ras effector protein AF-6.[31] PDZ domains are protein-binding modules named after the first three founding member proteins: post-synaptic density protein *PSD95*, *Drosophila* tumor suppressor Discs-large protein *Dlg*, and zonula occludens protein *ZO-1*. In addition to having several tandem PDZ domains, PDZ proteins also typically contain other protein-binding modules, such as SH3, WW, and L27 domains. The presence of a guanylate kinase-homology domain, which likewise functions as a protein-binding module, serves to distinguish MAGUK-family proteins from other PDZ proteins. Consistent with having multiple protein-binding domains, PDZ proteins typically function as scaffolds to assemble transmembrane and cytosolic proteins into supramolecular signaling complexes, as well as to tether such complexes to the actin cytoskeleton and to localize them at specialized membrane sites of cell-cell contact,[45-47] such as TJs.

There is also some evidence indicating that at least some of the TJ-associated PDZ proteins are required for both TJ establishment and cell polarity. For example, two different TJ-associated protein complexes, Crb/PALS1/PATJ and Par3/Par6/aPKC, form a functional complex required for proper epithelial cell polarity.[48] Consistent with the fact that an interaction between PALS1 and Par6 mediates formation of this important functional complex, it was recently shown that siRNA knockdown of PALS1 expression leads to defects in TJ formation and polarity in epithelial cells.[49] Numerous other studies have also shown that over-expression of dominant-negative PDZ-protein mutants adversely affects TJ function, structure, and assembly.[44,48,50-54]

Another group of TJ-associated cytoplasmic proteins consists of otherwise unrelated adaptor (cingulin,[55] 7H6[56]) and signaling proteins (PTEN,[40] heterotrimeric G-proteins,[57,58] atypical protein kinase C [aPKC][43] protein phosphatase 2A [PP2A][59]). Interestingly, aPKC and PP2A are believed to phosphorylate or dephosphorylate certain TJ proteins,[59-61] respectively, and in so doing, regulate cell polarity,[62] as well as TJ assembly[59] and function.[63] Thus, an intriguing possible scenario is that, in some cases, the loss of TJs and polarity observed in cancer cells stems from dysregulation of aPKC and/or PP2A.

Importantly, in vitro and in vivo studies of multiple TJ-associated proteins have revealed that changes in their expression and localization can promote tumorigenesis and metastasis (Fig. 1).

TJ Proteins and Cancer

Some epithelial cells (e.g., those lining the respiratory and gastrointestinal tracts) have constant exposure to the external environment and, therefore, frequently become injured or killed by toxins, microbes, and viruses. At sites of epithelial cell damage, loss of TJ integrity represents a cue for cells to launch a repair program involving cellular proliferative and migratory activities at the wound edge, and ending with the reassembly of TJs to reform an intact epithelial layer.[64] An elegant molecular mechanism regulating this repair process has been elucidated in human airway epithelia. Under normal conditions, TJs in these cells act to segregate a growth

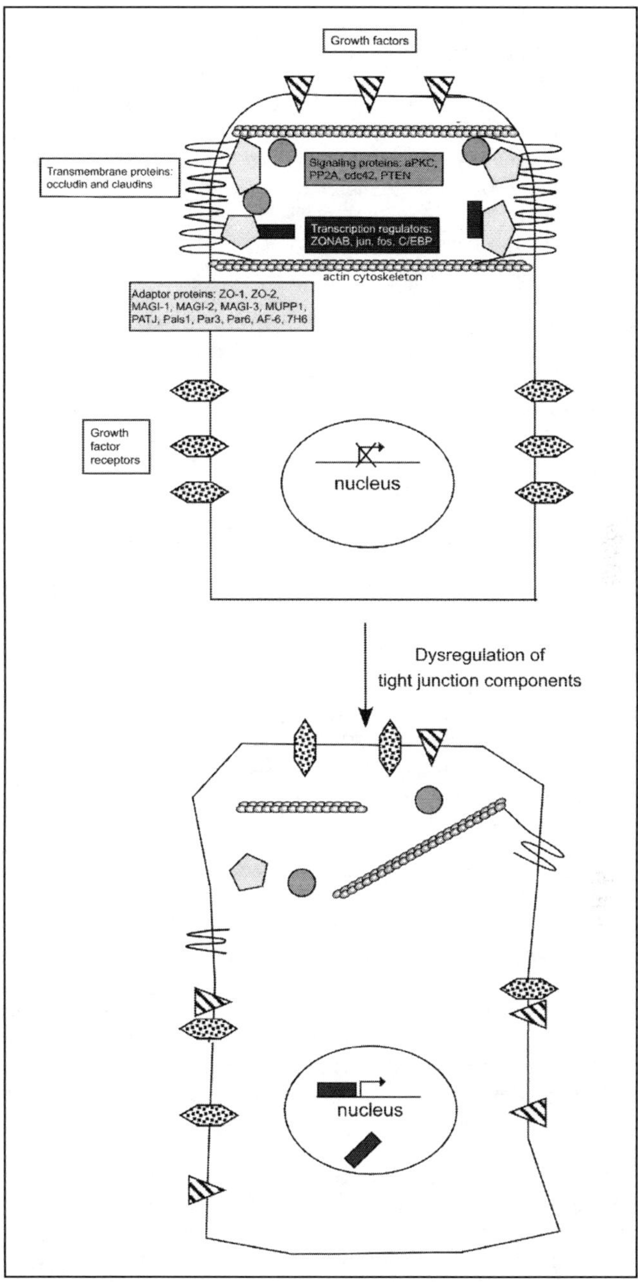

Figure 1. Mechanisms of TJ dysregulation that may promote tumorigenesis and metastasis. TJ-associated proteins involved in controlling cellular proliferation can be divided into four different groups: transmembrane proteins, adaptor proteins, signaling proteins, and transcription factors. Altered expression levels or localization of these proteins could adversely affect other TJ components, resulting in increased intercellular permeability, loss of polarity and contact inhibition, dysregulated signaling from the TJ, and elevated expression of growth-stimulatory genes in the nucleus.

factor within the apical membrane compartment away from its receptor in the basolateral membrane compartment, thereby precluding receptor activation.[65] Upon injury or death of a cell within the epithelial cell layer, the disruption of TJs in adjacent healthy cells permits the growth factor to bind and activate its receptor, thereby inducing cellular proliferation and migration to repair the wound. In light of this model, one can also easily envision that, in epithelial cells, the persistent loss of TJs and polarity caused by mutations in genes encoding TJ proteins or their regulators could initiate neoplastic transformation. Supporting this idea, disrupted TJs, as measured by freeze-fracture analyses or by decreased transepithelial resistance (TER) of tissues, represent a common alteration of epithelial cell-derived tumor cells, some examples of which include hepatocellular carcinomas,[66] thyroid tumors,[67] colon carcinomas, and adenomatous polyps.[68]

Identification and characterization of the many protein components of TJs should provide important insights into the molecular mechanisms whereby disruption of TJs can trigger malignant transformation in a wide variety of epithelial cell types. This point can be highlighted by studies of the two closely related TJ proteins, ZO-1 and ZO-2. The fact that these two proteins showed homology to the *Drosophila* tumor suppressor protein Dlg, mutation of which leads to loss of cell polarity and neoplastic growth,[69] initially pointed to their suspected roles as tumor suppressors.[70] Additional useful insights were gleaned from the demonstration that conditions such as cell subconfluency induce TJ-associated ZO-1 and ZO-2 to translocate into the nucleus,[71,72] reminiscent of the ability of Wnt pathway activation to induce the β-catenin oncoprotein[73] to move from adherens junctions into the nucleus and, thereby, to stimulate transcription of the c-myc,[74] cyclin D1,[75,76] and c-jun[77] genes. Collectively, these observations suggested that ZO-1 and ZO-2 might suppress cellular proliferation by regulating the activity of nuclear factors, a hypothesis for which compelling evidence now exists (see below).

To metastasize, cancer cells must have the capacities to invade surrounding normal tissues, to move into the circulation, and to transmigrate through vascular endothelial cells into surrounding tissues. Because reduced intercellular adhesion promotes increased cellular motility and invasiveness, disruption of TJs likewise plays a central role in conferring a metastatic phenotype to cancer cells. Interestingly, in addition to showing that TJ proteins are generally downregulated in metastatic cancer cells,[1,78] findings also indicate that microvessels feeding tumors often display reduced occludin and claudins expression and increased permeability.[79-81]

Dysregulated TJ Protein Expression in Cancer

Numerous reports demonstrate that TJ proteins are downregulated in both human (Table 1) and experimental cancers.[82,83] In most instances, the expression level of these proteins directly correlates with the degree of differentiation of the cancer cells. For instance, ZO-1 expression in infiltrating breast ductal carcinomas was shown to be reduced in 42%, 83%, or 93% of highly, moderately, or poorly differentiated tumors, respectively.[84] The downregulation of ZO-1 in these carcinomas was also proposed to represent a late event in cancer progression.

The upregulation of TJ proteins represents an additional alteration observed in some cancers (Table 1). One potential explanation for this somewhat surprising finding may be that over-expression of TJ proteins can dominantly negatively interfere with proper TJ formation and function, thereby potentiating malignant transformation and metastasis. Alternatively, while TJ-protein upregulation and increased TJ function may decrease the tumorigenic and metastatic potentials of most cancer cells, it is possible that such effects on the TJ fail to hinder the ability of certain cancer cells to thrive. In this regard, the functional consequences of TJ-protein upregulation in cancer cells likely depend on variables such as the cell type, the presence of specific oncogenic mutations, and the subcellular distribution of the TJ protein.[85] Nonetheless, whereas downregulation of TJ proteins clearly promotes cancer progression, more work is needed to clarify the role of TJ-protein upregulation in this process.

Table 1. Changes in TJ protein expression reported in human cancers

Tight Junction Protein	Expression Status	Type of Cancer	Refs.
Occludin	Reduced	Poorly differentiated gastrointestinal adenocarcinomas	168
	Reduced	Gastric cancer	169
	Lost	High-grade prostate cancer (except in cells facing the lumen)	170
	Lost	Colorectal cancer	171
	Reduced	Uterine adenocarcinoma	172
Claudin 1	Increased	Colorectal cancer	173
Claudin 3	Increased	Prostate adenocarcinoma, prostatic intraepithelial neoplasia	174
	Increased	Ovarian cancer	85,175
Claudin 4	Increased	Prostate adenocarcinoma, prostatic intraepithelial neoplasia	174
	Increased	Ovarian cancer	85,175
	Increased	Pancreatic cancer	87,88
	Increased	Intraductal papillary-mucinous tumors of the pancreas	176
Claudin 7	Reduced	Invasive ductal carcinomas of the breast	177
	Reduced	Head and neck squamous cell carcinomas	178
Claudin 16	Increased	Ovarian cancer	179
Claudin 23	Reduced	Intestinal-type gastric cancer	180
ZO-1	Reduced	Infiltrating ductal carcinomas	84
	Reduced	Poorly differentiated gastrointestinal adenocarcinomas	168
	Reduced	Colorectal cancer with liver metastasis	181,182
	Increased	Primary pancreatic cancer cells	183
ZO-2	ZO-2-A lost	Pancreatic duct carcinomas	184
	Reduced	Breast adenocarcinoma	185
ASIP/Par3	Par3 Δexon 17b reduced	Liver cancer cell lines, hepatocellular carcinomas	186

Animal Studies That Reveal Potential Roles for TJ Proteins in Cancer

To examine the direct consequences of dysregulating expression of TJ proteins in vivo, several groups have generated knockout mice unable to express TJ proteins. While such mutant animals typically display various developmental or pathological defects, a predisposition for developing hyperproliferative disorders has been observed less frequently. The studies detailed below represent examples in which dysregulated expression of TJ proteins promotes abnormal cellular proliferation, neoplasia, or metastasis in experimental animal model systems.

TJ Transmembrane Proteins

Mice lacking both isoforms of occludin are born at Mendelian ratios with no gross anatomical defects.[86] While these mutant mice appear to develop TJs morphologically and functionally indistinguishable from those observed in wild-type mice, demonstrating the dispensability of occludin for TJ establishment and maintenance, careful histological analyses revealed that gastric gland epithelial cells fail to undergo normal differentiation in 3-6 week-old occludin$^{-/-}$ mice and, additionally, become hyperplastic in 40 week-old mutant animals. This susceptibility of occludin$^{-/-}$ mice for developing hyperplasia but not cancer is not surprising considering the ability of these animals to form apparently normal TJs. In future studies, however, it would also be interesting to determine whether occludin$^{-/-}$ mice have a higher predisposition for tumor development following treatment with carcinogens.

In one study, it was found that human pancreatic cancer-derived cell lines expressing high or low levels of claudin-4 exhibited weak or strong metastatic potential, respectively, following tail vein injection into nude mice.[87] Providing direct evidence that claudin-4 upregulation decreases metastatic potential, it was shown that over-expressing claudin-4 in a highly metastatic pancreatic cancer line that expressed low endogenous levels of this protein significantly decreases its capacity to form pulmonary metastases.[87] This finding, however, contradicts the results of a recent study that showed upregulated claudin-4 expression in both primary and metastatic pancreatic cancers.[88] Clearly, this discrepancy needs clarification.

TJ Cytoplasmic Proteins

The Ras-effector protein AF-6 is encoded by a gene that localizes on human chromosome 6q27 in a region commonly deleted in ovarian cancer,[89] and was identified as an ALL-1 fusion partner in acute myeloid leukemias.[90] Whereas AF-6[+/-] mice exhibit no recognizable defects, AF-6[-/-] mice die during early embryogenesis.[91] Supporting the notion that the TJ protein AF-6 functions to suppress inappropriate proliferation in at least some cell types, columnar epithelial cells in the neuroectoderm of AF-6[-/-] embryos display both morphological and polarity defects, which result in a hyperplasic-like phenotype.

The TJ-associated MAGUK proteins MAGI-1, MAGI-2 and MAGI-3 have been reported to interact with the important tumor suppressor protein PTEN,[40,92] which functions to negatively regulate PI3K signaling.[93] Evidence obtained in cell lines suggests that these MAGUK proteins augment the activity of PTEN, possibly by inducing its localization to the membrane and thereby increasing access to phosphatidylinositol 3,4,5 phosphate lipid substrates. Considering that PTEN[+/-] mice are acutely susceptible to a variety of malignancies including breast and prostate cancers and lymphomas,[94-96] it will be important in future studies to generate *MAGI-2* and *MAGI-3* knockout mice and to assess whether these mutant animals likewise exhibit a predisposition for tumor development.

Also noteworthy is that studies conducted in *Drosophila* have revealed a role for the TJ-protein ZO-1 in negatively regulating cellular proliferation. Tamou, a *Drosophila* homologue of the MAGUK protein ZO-1, has been implicated in this process through its impact on cell differentiation.[97] This *Drosophila* protein is required for expression of the transcriptional repressor extramacrochaetae (emc), which inhibits formation of sensory organ precursor cells that differentiate into various sensory organs. Consequently, the absence of tamou decreases *emc* transcription, resulting both in abnormally high numbers of sensory-organ precursor cells and in supernumerary sensory organs. Furthermore, flies hypomorphic for both *tamou* and *emc* develop multiple extra wing veins, a phenotype possibly caused by cellular overproliferation.

In Vitro Studies That Reveal Potential Roles for TJ Proteins in Cancer

As the molecular definition of TJs represents a recent undertaking, animal studies investigating the functions of TJ proteins are still somewhat limited. As might be expected, reports utilizing in vitro systems are more numerous, and select findings having relevance to cancer are summarized below.

TJ Transmembrane Proteins

Evidence has been reported showing that occludin downregulation is responsible in part for oncogene-mediated transformation of a rat salivary gland epithelial cell line. Expression of oncogenic Raf-1 in these cells leads to the functional loss of TJs and the acquisition of a stratified phenotype due to the lack of cell-cell contact growth control,[98] abnormalities that correlated with downregulation of occludin and claudin-1, as well as with altered distribution patterns of ZO-1 and E-cadherin. Importantly, reconstitution of occludin expression in the Raf-1-transformed cells promotes reversal of their abnormal growth properties and reacquisition of normal claudin-1 protein levels, functional TJs, and properly localized E-cadherin at lateral membranes.

Several studies indicate that claudin downregulation also contributes to the transformed properties of tumor cell lines. In breast cancer cell lines, claudin-1 expression is absent or greatly diminished[99] and, significantly, over-expression of claudin-1 not only reduces paracellular flux[100] and increases TER in monolayers[101] but also promotes apoptosis of three-dimensional culture tumor spheroids.[102] Moreover, expression of claudin-1 and several other TJ proteins are repressed by over-expression of Snail, a transcription factor involved in the epithelial-mesenchymal transition seen in many cancers.[103] Also notable is that over-expression of claudin-4 in pancreatic cancer cells is linked to a reduced invasive potential in vitro and to an inability to form colonies in soft agar.[87]

TJ Cytoplasmic Proteins

Studies of the macromolecular TJ complex Par3/Par6/aPKC reveal that dysregulated cell-polarity signaling can also cause cell transformation. This complex is crucial for TJ formation and epithelial cell polarity,[48,104] and its activity depends on an association with the Rho-family GTPases Rac1 and Cdc42, which bind to Par6.[44,105] Consistent with a complex whose function relies on proper subunit stoichiometry, over-expression of any individual component of the Par3/Par6/aPKC complex acts in a dominant-negative fashion to block TJ formation.[44,53,61] With respect to a possible link to cancer, it has been reported that this complex can also potentiate transformation independent of Raf[106] in fibroblasts.

Studies conducted with cell lines have likewise revealed that the suspected tumor suppressor proteins ZO-1 and ZO-2 possess several growth-suppressive activities that could be envisioned as becoming inactivated by TJ disruption in cancer cells. ZO-1 is known to translocate from the TJ into the nucleus of either subconfluent or migrating epithelial cells.[71] In confluent epithelial cells, however, ZO-1 binds to the nuclear transcription factor ZONAB and sequesters it to TJs, an interaction that blocks ZONAB's growth-stimulatory activity, possibly by inhibiting cyclin-dependent kinase 4.[107] Paralleling these observations with ZO-1, ZO-2 similarly interacts with the growth-stimulatory transcription factors Jun, Fos, and C/EBP[108] and, in confluent cells, co-localizes with these transcription factors at TJs. Moreover, ZO-2 over-expression inhibits AP-1-dependent transcription, presumably by functionally inactivating Jun and Fos by sequestering them at TJs. Also noteworthy is that ZO-2 binds to the nuclear scaffold attachment factor B (SAFB),[109] a suspected tumor suppressor whose expression is lost in approximately 20% of breast cancers,[110,111] and that ZO-2 over-expression can inhibit the in vitro transforming activities of several different oncoproteins, including activated Ras.[112]

In addition to having the ability to bind PTEN and augment its activity in cells, the TJ-associated MAGUK protein MAGI-1 also interacts with the guanine nucleotide exchange factors (GEFs) PDZ-GEF[113] and NET1,[114] which function to stimulate the activity of Rap1[115] and RhoA,[116] respectively. In addition, a NET1 truncation mutant was shown to represent a novel oncogene expressed in a human neuroepithelioma cell line.[117] Thus, it is tempting to speculate that MAGI-1 sequesters PDZ-GEF and NET1 in an inactive state at TJs and that, in cancer cells, disruption of TJs may lead to oncogenic, constitutive activation of these GEFs.

The TJ multi-PDZ protein MUPP1 has been shown to interact with claudin-8,[118] the proteoglycan NG2,[119] and the tyrosine kinase receptor c-kit[120] and, additionally, MUPP1 overexpression in MDCK epithelial cells can reduce paracellular conductance.[118] Considering the known important involvement of claudins in TJ formation, NG2 in cell motility and melanoma cell spreading,[121] and the c-kit proto-oncogene product in promoting cellular proliferation and survival,[122] it is possible that dysregulation of MUPP1 promotes TJ disruption and oncogenesis, although direct evidence supporting this hypothesis is presently lacking.

Not surprisingly, several different growth-stimulatory signaling pathways have also been shown to interfere with TJ integrity in epithelial cells. Reports using a variety of different cell lines demonstrate that stimulation of EGF receptors or the Ras/Raf/ERK pathway, but not the PI3K pathway, reduces TJ barrier function.[123,124] It was also recently shown that activation of PKC-alpha and its translocation to the membrane increases TJ permeability in adenocarcinomas derived from either rat or human colon.[62]

While to date most in vitro studies have relied primarily on either tumor lines having multiple mutations or cell lines over-expressing individual TJ proteins, the advent of siRNA technology is expected to play an increasingly prominent role in the design of future in vitro studies seeking to delineate roles for TJ proteins in cancer and metastasis.

The TJ and Viral Oncoproteins

Because small DNA tumor viruses typically initiate infections in quiescent cells and replication of their genomes depends heavily on host enzymes specifically expressed during S-phase of the cell cycle, these agents have devised a multitude of strategies to force quiescent cells to re-enter the cell cycle. In general, the oncoproteins encoded by these viruses serve this function by targeting cellular factors having key roles in regulating cell proliferation. Consequently, studies of DNA tumor virus oncoproteins have led to the discovery of several extremely important cellular regulatory proteins. Indeed, the well-known tumor suppressor p53 was first discovered through its interaction with the SV40 large T antigen (LTag) oncoprotein,[125,126] and has subsequently been identified as a target of multiple viral oncoproteins.[127-129] Other examples include the discoveries of PI3K[130] and p300/CBP[131] through their interactions with the middle T-antigen of mouse polyomavirus and the E1A protein of human adenovirus, respectively. In addition, much of what we know about the function of the pRb tumor suppressor was gleaned from studies of its interactions with SV40 LTag and adenovirus E1A.[132,133] The more recent finding that TJ-associated PDZ proteins and PP2A are targets for several different viral oncoproteins provides additional evidence that functional perturbations of TJs can contribute directly to the development of cancers (Fig. 2).

Human Papillomavirus (HPV)

Numerous epidemiological studies have identified the sexually-transmitted HPV as the primary etiologic agent of greater than 90% of cervical cancers.[134] There are over 75 different types of HPV, and these are divided into two groups termed "high-risk" and "low-risk" based on the frequency of their association with cervical cancers. The E6 and E7 proteins represent the major oncogenic determinants of high-risk HPVs, and the oncogenic potential of HPV E6 depends not only on its ability to target p53 for degradation[129] but also on its interaction with a variety of other cellular factors.[135-137]

E6 proteins derived from high-risk but not low-risk HPVs possess a functional PDZ domain-binding motif that mediates interactions with several different TJ-associated PDZ proteins, including MUPP1,[138] MAGI-1,[139,140] MAGI-2 and MAGI–3.[141] Such interactions promote degradation of these PDZ proteins in cells, and E6 mutants specifically unable to bind PDZ proteins exhibit substantially reduced oncogenic potential.[135,142,143] Expression of high-risk HPV-16 E6 in epithelial cells destabilizes TJs, as evidenced by ZO-1 mislocalization,[144] although whether this effect on TJs is due to interactions of E6 with one or more TJ-associated PDZ protein is not currently known. Furthermore, there is evidence that the MUPP1-related protein PATJ represents an additional TJ-associated PDZ protein target of HPV E6 (Latorre and Javier, manuscript in preparation). It is also conceivable that E6-mediated degradation of the adherens junction-localized human homologues of the *Drosophila* tumor suppressors Dlg and Scribble contribute to TJ destabilization, considering that disruption of adherens junctions also promotes carcinogenesis[145] and could presumably indirectly disrupt TJs. Moreover, chronic downregulation of Scribble has been linked to progression of normal uterine cervical tissue to invasive cervical cancer, suggesting that Scribble degradation by E6 is a contributing factor to cervical cancer progression.[146]

Simian Virus 40 (SV40)

SV40 infects non-human primates, but many people were exposed to this virus as a contaminant of the polio vaccines administered in the 1950's.[147] SV40 can cause a variety of tumors including lung and kidney adenocarcinomas in rodent models[148] and, in addition,

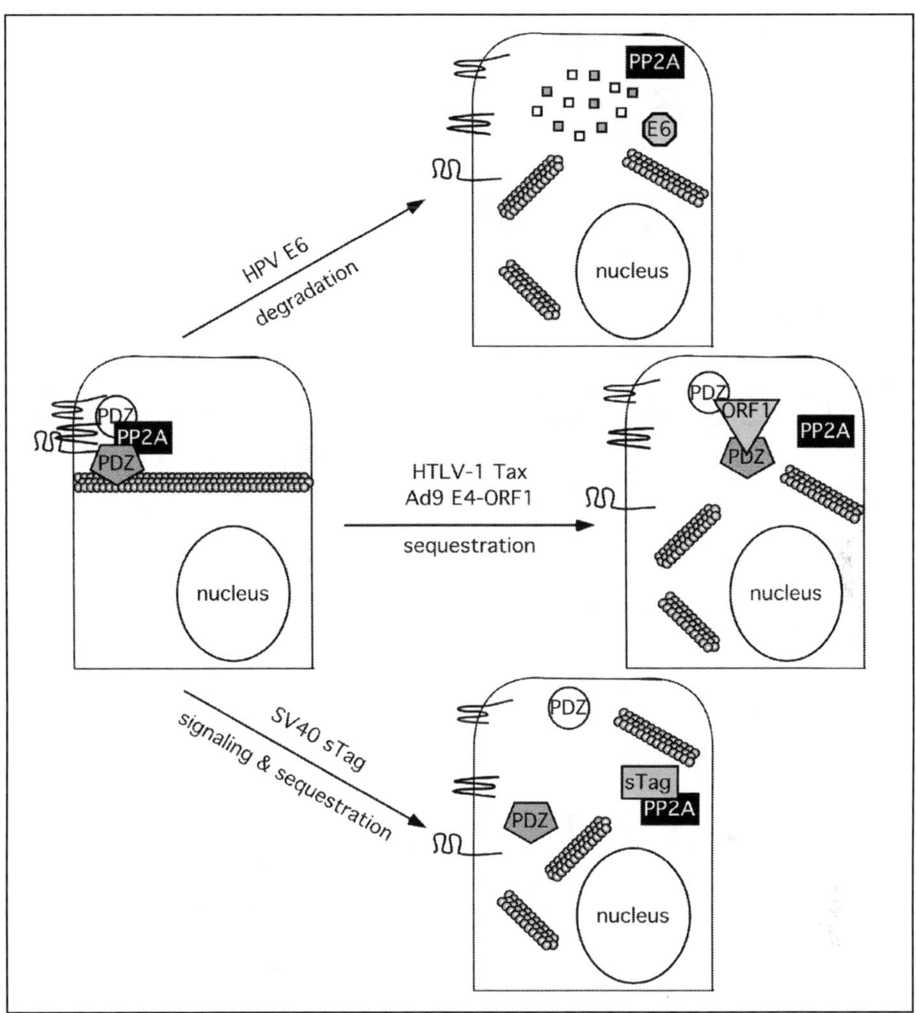

Figure 2. Disruption of TJs by viral oncoproteins. In normal epithelial cells, TJs are ordered macromolecular complexes comprised mainly of occludin, claudins, JAMs, PDZ proteins, and other signaling proteins. In cells transformed by certain viruses, the viral oncoproteins either aberrantly sequester (Ad9 E4-ORF1, HTLV-1 TAX, SV40 sTag) or promote degradation (HPV E6) of various TJ components. Ad9 E4-ORF1, HTLV-1 Tax, and HPV E6 directly target PDZ proteins, whereas SV40 sTag inactivates PP2A. With the exception of SV40 sTag, the disruption of TJ-associated non-PDZ proteins is hypothetical.

recent epidemiological and molecular analyses have revealed increasing evidence that SV40 is likewise associated with human malignancies.[149,150] Although LTag is the major oncogenic determinant of SV40, under certain conditions, SV40-induced tumorigenesis also requires SV40 small t antigen (sTag),[148] which binds to and inactivates PP2A in cells.[151-153] Significantly, sTag was also recently found to disrupt TJs in MDCK cells in a PP2A-dependent manner, as determined by mislocalization of PP2A and other TJ-associated proteins, multilayering and indistinguishable cell borders, decreased TER, and increased paracellular permeability of cells.[154] These effects were specific to TJs because SV40 sTag had no effect on adherens junctions. These findings with SV40 sTag imply a key role for PP2A dysregulation in TJ disruption and cellular transformation.

Human Adenovirus (Ad)

Adenoviruses naturally infect epithelial cells and, as is the case for many small DNA tumor viruses, have evolved several mechanisms to provide a cellular milieu amenable to viral replication. The E1A protein is a major oncogenic determinant for most human adenoviruses, and multiple functions contribute to E1A's transforming activity.[155] Whereas the amino-terminal region of E1A is responsible for its oncogenic potential, the carboxyl-terminus negatively regulates this activity.[156] Indeed, E1A mutants with carboxyl-terminal deletions display enhanced transforming, tumorigenic, and metastatic activities.[156] Among the phenotypes acquired by deletion of the E1A carboxyl-terminal region is the ability to disrupt cell-cell adhesion complexes, including TJs.[157] For example, these E1A mutants mislocalize ZO-1 and increase paracellular permeability in epithelial cells. Two cellular proteins that interact with the E1A carboxyl-terminus are CtBP,[158] a transcriptional repressor of LEF/TCF, and the dual-specificity kinases Dyrk1A/B.[159] While it is not yet clear whether the failure of E1A to bind one or both of these cellular factors is required for TJ disruption, it has been shown that expression of CtBP is lost in malignant melanomas[160] and Dyrks are likely involved in regulating cell-cycle progression and cell differentiation.[161]

Whereas the primary oncogenic determinants of most adenoviruses are E1A and E1B, human adenovirus type 9 is unique in that the E4 region-encoded open reading frame 1 (E4-ORF1) oncoprotein is its major oncogenic determinant.[162] Like high-risk HPV E6 oncoproteins, the E4-ORF1 oncoprotein has a PDZ domain-binding motif that mediates interactions with several TJ-associated PDZ proteins, including ZO-2,[112] MAGI-1,[139] MUPP1,[138] and PATJ (Latorre and Javier, manuscript in preparation). In addition, disruption of this motif abolishes the tumorigenic potential of E4-ORF1,[163,164] and wild-type E4-ORF1 aberrantly sequesters these PDZ proteins in the cytoplasm of cells.[112,138,139] Importantly, similar to high-risk HPV E6, E4-ORF1 also disrupts TJs in epithelial cells (Latorre and Javier, manuscript in preparation).

Human T-Cell Leukemia Virus-1 (HTLV-1)

HTLV-1, the etiologic agent of adult T-cell leukemia, is the only known oncogenic human retrovirus. The viral Tax protein is believed to account for the tumorigenic potential of HTLV-1.[165] Similar to a difference between the E6 proteins from high-risk and low-risk HPV, the Tax protein of tumorigenic HTLV-1 but not non-tumorigenic HTLV-2 interacts with cellular PDZ proteins, and these interactions are linked to its transforming potential in fibroblasts.[166] Tax was recently shown to interact with the TJ-associated MAGUK protein MAGI-3 and alter its subcellular localization,[167] hinting that the ability of Tax to perturb TJ-related complexes in T-cells may contribute to the development of adult T-cell leukemia.

Concluding Remarks

Although the molecular analysis of the TJ is still in its infancy, multiple lines of evidence implicate these membrane structures in the control of normal cell proliferation and polarity. Furthermore, recent studies have shown that TJ disruption is a hallmark of malignant and metastatic cancers, and evidence that some TJ-associated proteins are targeted by viral oncoproteins underscores their possible functions as tumor suppressors. While studies thus far have mostly been limited to investigating the effects of TJ protein over-expression in established cell lines, the use of siRNA technology is expected to aid in validating these in vitro findings and in extending our knowledge about how TJ proteins can regulate cellular proliferation. The generation of tissue-specific gene knockout mice and multiple gene-knockout mice are additional approaches that promise to broaden our understanding of the important roles TJ proteins play in cancer.

References

1. Mori M, Sawada N, Kokai Y et al. Role of tight junctions in the occurrence of cancer invasion and metastasis. Med Electron Microsc 1999; 32(4):193-198.
2. Birchmeier C, Birchmeier W, Brand-Saberi B. Epithelial-mesenchymal transitions in cancer progression. Acta Anat (Basel) 1996; 156(3):217-226.
3. Gumbiner BM. Cell adhesion: the molecular basis of tissue architecture and morphogenesis. Cell 1996; 84(3):345-357.
4. Farquhar MG, Palade GE. Junctional complexes in various epithelia. J Cell Biol 1963; 17:375-412.
5. Lallemand D, Curto M, Saotome I et al. NF2 deficiency promotes tumorigenesis and metastasis by destabilizing adherens junctions. Genes Dev 2003; 17(9):1090-1100.
6. Yajnik V, Paulding C, Sordella R et al. DOCK4, a GTPase activator, is disrupted during tumorigenesis. Cell 2003; 112(5):673-684.
7. Furuse M, Hirase T, Itoh M et al. Occludin: a novel integral membrane protein localizing at tight junctions. J Cell Biol 1993; 123(6 Pt 2):1777-1788.
8. Martin-Padura I, Lostaglio S, Schneemann M et al. Junctional adhesion molecule, a novel member of the immunoglobulin superfamily that distributes at intercellular junctions and modulates monocyte transmigration. J Cell Biol 1998; 142(1):117-127.
9. Aurrand-Lions M, Johnson-Leger C, Wong C et al. Heterogeneity of endothelial junctions is reflected by differential expression and specific subcellular localization of the three JAM family members. Blood 2001; 98(13):3699-3707.
10. Furuse M, Fujita K, Hiiragi T et al. Claudin-1 and -2: novel integral membrane proteins localizing at tight junctions with no sequence similarity to occludin. J Cell Biol 1998; 141(7):1539-1550.
11. Morita K, Furuse M, Fujimoto K et al. Claudin multigene family encoding four-transmembrane domain protein components of tight junction strands. Proc Natl Acad Sci USA 1999; 96(2):511-516.
12. Fujimoto K. Freeze-fracture replica electron microscopy combined with SDS digestion for cytochemical labeling of integral membrane proteins. Application to the immunogold labeling of intercellular junctional complexes. J Cell Sci 1995; 108 (Pt 11):3443-3449.
13. Saitou M, Ando-Akatsuka Y, Itoh M et al. Mammalian occludin in epithelial cells: its expression and subcellular distribution. Eur J Cell Biol 1997; 73(3):222-231.
14. Furuse M, Sasaki H, Fujimoto K et al. A single gene product, claudin-1 or -2, reconstitutes tight junction strands and recruits occludin in fibroblasts. J Cell Biol 1998; 143(2):391-401.
15. Tsukita S, Furuse M. Occludin and claudins in tight-junction strands: leading or supporting players? Trends Cell Biol 1999; 9(7):268-273.
16. Furuse M, Itoh M, Hirase T et al. Direct association of occludin with ZO-1 and its possible involvement in the localization of occludin at tight junctions. J Cell Biol 1994; 127(6 Pt 1):1617-1626.
17. Itoh M, Morita K, Tsukita S. Characterization of ZO-2 as a MAGUK family member associated with tight as well as adherens junctions with a binding affinity to occludin and alpha catenin. J Biol Chem 1999; 274(9):5981-5986.
18. Haskins J, Gu L, Wittchen ES et al. ZO-3, a novel member of the MAGUK protein family found at the tight junction, interacts with ZO-1 and occludin. J Cell Biol 1998; 141(1):199-208.
19. Wittchen ES, Haskins J, Stevenson BR. Protein interactions at the tight junction. Actin has multiple binding partners, and ZO-1 forms independent complexes with ZO-2 and ZO-3. J Biol Chem 1999; 274(49):35179-35185.
20. Sheth P, Basuroy S, Li C et al. Role of phosphatidylinositol 3-kinase in oxidative stress-induced disruption of tight junctions. J Biol Chem 2003; 278(49):49239-49245.
21. Nusrat A, Chen JA, Foley CS et al. The coiled-coil domain of occludin can act to organize structural and functional elements of the epithelial tight junction. J Biol Chem 2000; 275(38):29816-29822.
22. Balda MS, Whitney JA, Flores C et al. Functional dissociation of paracellular permeability and transepithelial electrical resistance and disruption of the apical-basolateral intramembrane diffusion barrier by expression of a mutant tight junction membrane protein. J Cell Biol 1996; 134(4):1031-1049.
23. Aurrand-Lions M, Duncan L, Ballestrem C et al. JAM-2, a novel immunoglobulin superfamily molecule, expressed by endothelial and lymphatic cells. J Biol Chem 2001; 276(4):2733-2741.
24. Bazzoni G. The JAM family of junctional adhesion molecules. Curr Opin Cell Biol 2003; 15(5):525-530.
25. Liang TW, DeMarco RA, Mrsny RJ et al. Characterization of huJAM: evidence for involvement in cell-cell contact and tight junction regulation. Am J Physiol Cell Physiol 2000; 279(6):C1733-1743.

26. Itoh M, Sasaki H, Furuse M et al. Junctional adhesion molecule (JAM) binds to PAR-3: a possible mechanism for the recruitment of PAR-3 to tight junctions. J Cell Biol 2001; 154(3):491-497.

27. Liu Y, Nusrat A, Schnell FJ et al. Human junction adhesion molecule regulates tight junction resealing in epithelia. J Cell Sci 2000; 113 (Pt 13):2363-2374.

28. Hamazaki Y, Itoh M, Sasaki H et al. Multi-PDZ domain protein 1 (MUPP1) is concentrated at tight junctions through its possible interaction with claudin-1 and junctional adhesion molecule. J Biol Chem. 2002; 277(1):455-461.

29. Bazzoni G, Martinez-Estrada OM, Orsenigo F et al. Interaction of junctional adhesion molecule with the tight junction components ZO-1, cingulin, and occludin. J Biol Chem. 2000; 275(27):20520-20526.

30. Hirabayashi S, Tajima M, Yao I et al. JAM4, a junctional cell adhesion molecule interacting with a tight junction protein, MAGI-1. Mol Cell Biol. 2003; 23(12):4267-4282.

31. Yamamoto T, Harada N, Kano K et al. The Ras target AF-6 interacts with ZO-1 and serves as a peripheral component of tight junctions in epithelial cells. J Cell Biol 1997; 139(3):785-795.

32. Ebnet K, Schulz CU, Meyer Zu Brickwedde MK et al. Junctional adhesion molecule interacts with the PDZ domain-containing proteins AF-6 and ZO-1. J Biol Chem 2000; 275(36):27979-27988.

33. Heiskala M, Peterson PA, Yang Y. The roles of claudin superfamily proteins in paracellular transport. Traffic 2001; 2(2):93-98.

34. Morita K, Sasaki H, Fujimoto K et al. Claudin-11/OSP-based tight junctions of myelin sheaths in brain and Sertoli cells in testis. J Cell Biol 1999; 145(3):579-588.

35. Roh MH, Liu CJ, Laurinec S et al. The carboxyl terminus of zona occludens-3 binds and recruits a mammalian homologue of discs lost to tight junctions. J Biol Chem 2002; 277(30):27501-27509.

36. Itoh M, Furuse M, Morita K et al. Direct binding of three tight junction-associated MAGUKs, ZO-1, ZO-2, and ZO-3, with the COOH termini of claudins. J Cell Biol 1999; 147(6):1351-1363.

37. Stevenson BR, Siliciano JD, Mooseker MS et al. Identification of ZO-1: a high molecular weight polypeptide associated with the tight junction (zonula occludens) in a variety of epithelia. J Cell Biol 1986; 103(3):755-766.

38. Jesaitis LA, Goodenough DA. Molecular characterization and tissue distribution of ZO-2, a tight junction protein homologous to ZO-1 and the Drosophila discs-large tumor suppressor protein. J Cell Biol 1994; 124(6):949-961.

39. Roh MH, Makarova O, Liu CJ et al. The Maguk protein, Pals1, functions as an adapter, linking mammalian homologues of Crumbs and Discs Lost. J Cell Biol 2002; 157(1):161-172.

40. Wu Y, Dowbenko D, Spencer S et al. Interaction of the tumor suppressor PTEN/MMAC with a PDZ domain of MAGI3, a novel membrane-associated guanylate kinase. J Biol Chem 2000; 275(28):21477-21485.

41. Ide N, Hata Y, Nishioka H et al. Localization of membrane-associated guanylate kinase (MAGI)-1/ BAI-associated protein (BAP) 1 at tight junctions of epithelial cells. Oncogene 1999; 18(54):7810-7815.

42. Laura RP, Ross S, Koeppen H et al. MAGI-1: a widely expressed, alternatively spliced tight junction protein. Exp Cell Res 2002; 275(2):155-170.

43. Izumi Y, Hirose T, Tamai Y et al. An atypical PKC directly associates and colocalizes at the epithelial tight junction with ASIP, a mammalian homologue of Caenorhabditis elegans polarity protein PAR-3. J Cell Biol 1998; 143(1):95-106.

44. Joberty G, Petersen C, Gao L et al. The cell-polarity protein Par6 links Par3 and atypical protein kinase C to Cdc42. Nat Cell Biol 2000; 2(8):531-539.

45. Gonzalez-Mariscal L, Betanzos A, Avila-Flores A. MAGUK proteins: structure and role in the tight junction. Semin Cell Dev Biol 2000; 11(4):315-324.

46. Sheng M, Sala C. PDZ domains and the organization of supramolecular complexes. Annu Rev Neurosci 2001; 24:1-29.

47. Roh MH, Margolis B. Composition and function of PDZ protein complexes during cell polarization. Am J Physiol Renal Physiol 2003; 285(3):F377-387.

48. Hurd TW, Gao L, Roh MH et al. Direct interaction of two polarity complexes implicated in epithelial tight junction assembly. Nat Cell Biol 2003; 5(2):137-142.

49. Straight SW, Shin K, Fogg VC et al. Loss of PALS1 Expression Leads to Tight Junction and Polarity Defects. Mol Biol Cell 2004; 15(4):1981-1990.

50. Reichert M, Muller T, Hunziker W. The PDZ domains of zonula occludens-1 induce an epithelial to mesenchymal transition of Madin-Darby canine kidney I cells. Evidence for a role of beta-catenin/ Tcf/Lef signaling. J Biol Chem 2000; 275(13):9492-9500.

51. Wittchen ES, Haskins J, Stevenson BR. Exogenous expression of the amino-terminal half of the tight junction protein ZO-3 perturbs junctional complex assembly. J Cell Biol 2000; 151(4):825-836.

52. Wittchen ES, Haskins J, Stevenson BR. NZO-3 expression causes global changes to actin cytoskeleton in Madin-Darby canine kidney cells: linking a tight junction protein to Rho GTPases. Mol Biol Cell 2003; 14(5):1757-1768.
53. Gao L, Joberty G, Macara IG. Assembly of epithelial tight junctions is negatively regulated by Par6. Curr Biol 2002; 12(3):221-225.
54. Ryeom SW, Paul D, Goodenough DA. Truncation mutants of the tight junction protein ZO-1 disrupt corneal epithelial cell morphology. Mol Biol Cell 2000; 11(5):1687-1696.
55. Citi S, Sabanay H, Jakes R et al. Cingulin, a new peripheral component of tight junctions. Nature 1988; 333(6170):272-276.
56. Zhong Y, Saitoh T, Minase T et al. Monoclonal antibody 7H6 reacts with a novel tight junction-associated protein distinct from ZO-1, cingulin and ZO-2. J Cell Biol 1993; 120(2):477-483.
57. Saha C, Nigam SK, Denker BM. Involvement of Galphai2 in the maintenance and biogenesis of epithelial cell tight junctions. J Biol Chem 1998; 273(34):21629-21633.
58. Denker BM, Saha C, Khawaja S et al. Involvement of a heterotrimeric G protein alpha subunit in tight junction biogenesis. J Biol Chem 1996; 271(42):25750-25753.
59. Nunbhakdi-Craig V, Machleidt T, Ogris E et al. Protein phosphatase 2A associates with and regulates atypical PKC and the epithelial tight junction complex. J Cell Biol 2002; 158(5):967-978.
60. Avila-Flores A, Rendon-Huerta E, Moreno J et al. Tight-junction protein zonula occludens 2 is a target of phosphorylation by protein kinase C. Biochem J 2001; 360(Pt 2):295-304.
61. Nagai-Tamai Y, Mizuno K, Hirose T et al. Regulated protein-protein interaction between aPKC and PAR-3 plays an essential role in the polarization of epithelial cells. Genes Cells 2002; 7(11):1161-1171.
62. Mullin JM, Laughlin KV, Ginanni N et al. Increased tight junction permeability can result from protein kinase C activation/translocation and act as a tumor promotional event in epithelial cancers. Ann N Y Acad Sci 2000; 915:231-236.
63. Yoo J, Nichols A, Mammen J et al. Bryostatin-1 enhances barrier function in T84 epithelia through PKC-dependent regulation of tight junction proteins. Am J Physiol Cell Physiol 2003; 285(2):C300-309.
64. Jacinto A, Martinez-Arias A, Martin P. Mechanisms of epithelial fusion and repair. Nat Cell Biol 2001; 3(5):E117-123.
65. Vermeer PD, Einwalter LA, Moninger TO et al. Segregation of receptor and ligand regulates activation of epithelial growth factor receptor. Nature 2003; 422(6929):322-326.
66. Swift JG, Mukherjee TM, Rowland R. Intercellular junctions in hepatocellular carcinoma. J Submicrosc Cytol 1983; 15(3):799-810.
67. Cochand-Priollet B, Raison D, Molinie V et al. Altered gap and tight junctions in human thyroid oncocytic tumors: a study of 8 cases by freeze-fracture. Ultrastruct Pathol 1998; 22(6):413-420.
68. Soler AP, Miller RD, Laughlin KV et al. Increased tight junctional permeability is associated with the development of colon cancer. Carcinogenesis 1999; 20(8):1425-1431.
69. Woods DF, Bryant PJ. Molecular cloning of the lethal(1)discs large-1 oncogene of Drosophila. Dev Biol 1989; 134(1):222-235.
70. Willott E, Balda MS, Fanning AS et al. The tight junction protein ZO-1 is homologous to the Drosophila discs-large tumor suppressor protein of septate junctions. Proc Natl Acad Sci USA 1993; 90(16):7834-7838.
71. Gottardi CJ, Arpin M, Fanning AS et al. The junction-associated protein, zonula occludens-1, localizes to the nucleus before the maturation and during the remodeling of cell-cell contacts. Proc Natl Acad Sci USA 1996; 93(20):10779-10784.
72. Islas S, Vega J, Ponce L et al. Nuclear localization of the tight junction protein ZO-2 in epithelial cells. Exp Cell Res 2002; 274(1):138-148.
73. Morin PJ, Sparks AB, Korinek V et al. Activation of beta-catenin-Tcf signaling in colon cancer by mutations in beta-catenin or APC. Science 1997; 275(5307):1787-1790.
74. He TC, Sparks AB, Rago C et al. Identification of c-MYC as a target of the APC pathway. Science 1998; 281(5382):1509-1512.
75. Tetsu O, McCormick F. Beta-catenin regulates expression of cyclin D1 in colon carcinoma cells. Nature 1999; 398(6726):422-426.
76. Shtutman M, Zhurinsky J, Simcha I et al. The cyclin D1 gene is a target of the beta-catenin/LEF-1 pathway. Proc Natl Acad Sci USA 1999; 96(10):5522-5527.
77. Mann B, Gelos M, Siedow A et al. Target genes of beta-catenin-T cell-factor/lymphoid-enhancer-factor signaling in human colorectal carcinomas. Proc Natl Acad Sci USA 1999; 96(4):1603-1608.
78. Martin TA, Jiang WG. Tight junctions and their role in cancer metastasis. Histol Histopathol 2001; 16(4):1183-1195.

79. Papadopoulos MC, Saadoun S, Woodrow CJ et al. Occludin expression in microvessels of neoplastic and non-neoplastic human brain. Neuropathol Appl Neurobiol 2001; 27(5):384-395.

80. Liebner S, Fischmann A, Rascher G et al. Claudin-1 and claudin-5 expression and tight junction morphology are altered in blood vessels of human glioblastoma multiforme. Acta Neuropathol (Berl) 2000; 100(3):323-331.

81. Groothuis DR, Vriesendorp FJ, Kupfer B et al. Quantitative measurements of capillary transport in human brain tumors by computed tomography. Ann Neurol 1991; 30(4):581-588.

82. Quan C, Lu SJ. Identification of genes preferentially expressed in mammary epithelial cells of Copenhagen rat using subtractive hybridization and microarrays. Carcinogenesis 2003; 24(10):1593-1599.

83. Zhong Y, Enomoto K, Tobioka H et al. Sequential decrease in tight junctions as revealed by 7H6 tight junction-associated protein during rat hepatocarcinogenesis. Jpn J Cancer Res 1994; 85(4):351-356.

84. Hoover KB, Liao SY, Bryant PJ. Loss of the tight junction MAGUK ZO-1 in breast cancer: relationship to glandular differentiation and loss of heterozygosity. Am J Pathol 1998; 153(6):1767-1773.

85. Rangel LB, Agarwal R, D'Souza T et al. Tight junction proteins claudin-3 and claudin-4 are frequently overexpressed in ovarian cancer but not in ovarian cystadenomas. Clin Cancer Res 2003; 9(7):2567-2575.

86. Saitou M, Furuse M, Sasaki H et al. Complex phenotype of mice lacking occludin, a component of tight junction strands. Mol Biol Cell 2000; 11(12):4131-4142.

87. Michl P, Barth C, Buchholz M et al. Claudin-4 expression decreases invasiveness and metastatic potential of pancreatic cancer. Cancer Res 2003; 63(19):6265-6271.

88. Nichols LS, Ashfaq R, Iacobuzio-Donahue CA. Claudin 4 protein expression in primary and metastatic pancreatic cancer: support for use as a therapeutic target. Am J Clin Pathol 2004; 121(2):226-230.

89. Saito S, Sirahama S, Matsushima M et al. Definition of a commonly deleted region in ovarian cancers to a 300-kb segment of chromosome 6q27. Cancer Res 1996; 56(24):5586-5589.

90. Prasad R, Gu Y, Alder H et al. Cloning of the ALL-1 fusion partner, the AF-6 gene, involved in acute myeloid leukemias with the t(6;11) chromosome translocation. Cancer Res 1993; 53(23):5624-5628.

91. Zhadanov AB, Provance DW, Jr., Speer CA et al. Absence of the tight junctional protein AF-6 disrupts epithelial cell-cell junctions and cell polarity during mouse development. Curr Biol 1999; 9(16):880-888.

92. Wu X, Hepner K, Castelino-Prabhu S et al. Evidence for regulation of the PTEN tumor suppressor by a membrane-localized multi-PDZ domain containing scaffold protein MAGI-2. Proc Natl Acad Sci USA 2000; 97(8):4233-4238.

93. Wu X, Senechal K, Neshat MS et al. The PTEN/MMAC1 tumor suppressor phosphatase functions as a negative regulator of the phosphoinositide 3-kinase/Akt pathway. Proc Natl Acad Sci USA 1998; 95(26):15587-15591.

94. Stambolic V, Tsao MS, Macpherson D et al. High incidence of breast and endometrial neoplasia resembling human Cowden syndrome in pten+/- mice. Cancer Res 2000; 60(13):3605-3611.

95. Podsypanina K, Ellenson LH, Nemes A et al. Mutation of Pten/Mmac1 in mice causes neoplasia in multiple organ systems. Proc Natl Acad Sci USA 1999; 96(4):1563-1568.

96. Di Cristofano A, Pesce B, Cordon-Cardo C et al. Pten is essential for embryonic development and tumour suppression. Nat Genet 1998; 19(4):348-355.

97. Takahisa M, Togashi S, Suzuki T et al. The Drosophila tamou gene, a component of the activating pathway of extramacrochaetae expression, encodes a protein homologous to mammalian cell-cell junction-associated protein ZO-1. Genes Dev 1996; 10(14):1783-1795.

98. Li D, Mrsny RJ. Oncogenic Raf-1 disrupts epithelial tight junctions via downregulation of occludin. J Cell Biol 2000; 148(4):791-800.

99. Swisshelm K, Machl A, Planitzer S et al. SEMP1, a senescence-associated cDNA isolated from human mammary epithelial cells, is a member of an epithelial membrane protein superfamily. Gene 1999; 226(2):285-295.

100. Hoevel T, Macek R, Mundigl O et al. Expression and targeting of the tight junction protein CLDN1 in CLDN1-negative human breast tumor cells. J Cell Physiol 2002; 191(1):60-68.

101. Inai T, Kobayashi J, Shibata Y. Claudin-1 contributes to the epithelial barrier function in MDCK cells. Eur J Cell Biol 1999; 78(12):849-855.

102. Hoevel T, Macek R, Swisshelm K et al. Reexpression of the TJ protein CLDN1 induces apoptosis in breast tumor spheroids. Int J Cancer 2004; 108(3):374-383.

103. Ikenouchi J, Matsuda M, Furuse M et al. Regulation of tight junctions during the epithelium-mesenchyme transition: direct repression of the gene expression of claudins/occludin by Snail. J Cell Sci 2003; 116(Pt 10):1959-1967.

104. Kim SK. Cell polarity: new PARtners for Cdc42 and Rac. Nat Cell Biol 2000; 2(8):E143-145.

105. Lin D, Edwards AS, Fawcett JP et al. A mammalian PAR-3-PAR-6 complex implicated in Cdc42/ Rac1 and aPKC signalling and cell polarity. Nat Cell Biol 2000; 2(8):540-547.

106. Qiu RG, Abo A, Steven Martin G. A human homolog of the C. elegans polarity determinant Par-6 links Rac and Cdc42 to PKCzeta signaling and cell transformation. Curr Biol 2000; 10(12):697-707.

107. Balda MS, Garrett MD, Matter K. The ZO-1-associated Y-box factor ZONAB regulates epithelial cell proliferation and cell density. J Cell Biol 2003; 160(3):423-432.

108. Betanzos A, Huerta M, Lopez-Bayghen E et al. The tight junction protein ZO-2 associates with Jun, Fos and C/EBP transcription factors in epithelial cells. Exp Cell Res 2004; 292(1):51-66.

109. Traweger A, Fuchs R, Krizbai IA et al. The tight junction protein ZO-2 localizes to the nucleus and interacts with the heterogeneous nuclear ribonucleoprotein scaffold attachment factor-B. J Biol Chem 2003; 278(4):2692-2700.

110. Oesterreich S, Allredl DC, Mohsin SK et al. High rates of loss of heterozygosity on chromosome 19p13 in human breast cancer. Br J Cancer 2001; 84(4):493-498.

111. Townson SM, Sullivan T, Zhang Q et al. HET/SAF-B overexpression causes growth arrest and multinuclearity and is associated with aneuploidy in human breast cancer. Clin Cancer Res 2000; 6(9):3788-3796.

112. Glaunsinger BA, Weiss RS, Lee SS et al. Link of the unique oncogenic properties of adenovirus type 9 E4-ORF1 to a select interaction with the candidate tumor suppressor protein ZO-2. EMBO J 2001; 20(20):5578-5586.

113. Mino A, Ohtsuka T, Inoue E et al. Membrane-associated guanylate kinase with inverted orientation (MAGI)-1/brain angiogenesis inhibitor 1-associated protein (BAP1) as a scaffolding molecule for Rap small G protein GDP/GTP exchange protein at tight junctions. Genes Cells 2000; 5(12):1009-1016.

114. Dobrosotskaya IY. Identification of mNET1 as a candidate ligand for the first PDZ domain of MAGI-1. Biochem Biophys Res Commun 2001; 283(4):969-975.

115. Liao Y, Satoh T, Gao X et al. RA-GEF-1, a guanine nucleotide exchange factor for Rap1, is activated by translocation induced by association with Rap1*GTP and enhances Rap1-dependent B-Raf activation. J Biol Chem 2001; 276(30):28478-28483.

116. Alberts AS, Treisman R. Activation of RhoA and SAPK/JNK signalling pathways by the RhoA-specific exchange factor mNET1. EMBO J 1998; 17(14):4075-4085.

117. Chan AM, Takai S, Yamada K et al. Isolation of a novel oncogene, NET1, from neuroepithelioma cells by expression cDNA cloning. Oncogene 1996; 12(6):1259-1266.

118. Jeansonne B, Lu Q, Goodenough DA et al. Claudin-8 interacts with multi-PDZ domain protein 1 (MUPP1) and reduces paracellular conductance in epithelial cells. Cell Mol Biol (Noisy-le-grand) 2003; 49(1):13-21.

119. Barritt DS, Pearn MT, Zisch AH et al. The multi-PDZ domain protein MUPP1 is a cytoplasmic ligand for the membrane-spanning proteoglycan NG2. J Cell Biochem 2000; 79(2):213-224.

120. Mancini A, Koch A, Stefan M et al. The direct association of the multiple PDZ domain containing proteins (MUPP-1) with the human c-Kit C-terminus is regulated by tyrosine kinase activity. FEBS Lett 2000; 482(1-2):54-58.

121. Burg MA, Grako KA, Stallcup WB. Expression of the NG2 proteoglycan enhances the growth and metastatic properties of melanoma cells. J Cell Physiol 1998; 177(2):299-312.

122. Wehrle-Haller B. The role of Kit-ligand in melanocyte development and epidermal homeostasis. Pigment Cell Res 2003; 16(3):287-296.

123. Chen Y, Lu Q, Schneeberger EE et al. Restoration of tight junction structure and barrier function by down-regulation of the mitogen-activated protein kinase pathway in ras-transformed Madin-Darby canine kidney cells. Mol Biol Cell 2000; 11(3):849-862.

124. Grande M, Franzen A, Karlsson JO et al. Transforming growth factor-beta and epidermal growth factor synergistically stimulate epithelial to mesenchymal transition (EMT) through a MEK-dependent mechanism in primary cultured pig thyrocytes. J Cell Sci 2002; 115(Pt 22):4227-4236.

125. Lane DP, Crawford LV. T antigen is bound to a host protein in SV40-transformed cells. Nature 1979; 278(5701):261-263.

126. Linzer DI, Levine AJ. Characterization of a 54K dalton cellular SV40 tumor antigen present in SV40-transformed cells and uninfected embryonal carcinoma cells. Cell 1979; 17(1):43-52.

127. Querido E, Marcellus RC, Lai A et al. Regulation of p53 levels by the E1B 55-kilodalton protein and E4orf6 in adenovirus-infected cells. J Virol 1997; 71(5):3788-3798.
128. Sarnow P, Ho YS, Williams J et al. Adenovirus E1b-58kd tumor antigen and SV40 large tumor antigen are physically associated with the same 54 kd cellular protein in transformed cells. Cell 1982; 28(2):387-394.
129. Scheffner M, Werness BA, Huibregtse JM et al. The E6 oncoprotein encoded by human papillomavirus types 16 and 18 promotes the degradation of p53. Cell 1990; 63(6):1129-1136.
130. Whitman M, Kaplan DR, Schaffhausen B et al. Association of phosphatidylinositol kinase activity with polyoma middle-T competent for transformation. Nature 1985; 315(6016):239-242.
131. Arany Z, Sellers WR, Livingston DM et al. E1A-associated p300 and CREB-associated CBP belong to a conserved family of coactivators. Cell 1994; 77(6):799-800.
132. DeCaprio JA, Ludlow JW, Figge J et al. SV40 large tumor antigen forms a specific complex with the product of the retinoblastoma susceptibility gene. Cell 1988; 54(2):275-283.
133. Whyte P, Buchkovich KJ, Horowitz JM et al. Association between an oncogene and an anti-oncogene: the adenovirus E1A proteins bind to the retinoblastoma gene product. Nature 1988; 334(6178):124-129.
134. Munoz N, Bosch FX, de Sanjose S et al. The role of HPV in the etiology of cervical cancer. Mutat Res 1994; 305(2):293-301.
135. Nguyen ML, Nguyen MM, Lee D et al. The PDZ ligand domain of the human papillomavirus type 16 E6 protein is required for E6's induction of epithelial hyperplasia in vivo. J Virol 2003; 77(12):6957-6964.
136. Elbel M, Carl S, Spaderna S et al. A comparative analysis of the interactions of the E6 proteins from cutaneous and genital papillomaviruses with p53 and E6AP in correlation to their transforming potential. Virology 1997; 239(1):132-149.
137. Kiyono T, Foster SA, Koop JI et al. Both Rb/p16INK4a inactivation and telomerase activity are required to immortalize human epithelial cells. Nature 1998; 396(6706):84-88.
138. Lee SS, Glaunsinger B, Mantovani F et al. Multi-PDZ domain protein MUPP1 is a cellular target for both adenovirus E4-ORF1 and high-risk papillomavirus type 18 E6 oncoproteins. J Virol 2000; 74(20):9680-9693.
139. Glaunsinger BA, Lee SS, Thomas M et al. Interactions of the PDZ-protein MAGI-1 with adenovirus E4-ORF1 and high-risk papillomavirus E6 oncoproteins. Oncogene 2000; 19(46):5270-5280.
140. Thomas M, Glaunsinger B, Pim D et al. HPV E6 and MAGUK protein interactions: determination of the molecular basis for specific protein recognition and degradation. Oncogene 2001; 20(39):5431-5439.
141. Thomas M, Laura R, Hepner K et al. Oncogenic human papillomavirus E6 proteins target the MAGI-2 and MAGI-3 proteins for degradation. Oncogene 2002; 21(33):5088-5096.
142. Kiyono T, Hiraiwa A, Fujita M et al. Binding of high-risk human papillomavirus E6 oncoproteins to the human homologue of the Drosophila discs large tumor suppressor protein. Proc Natl Acad Sci USA 1997; 94(21):11612-11616.
143. Nguyen MM, Nguyen ML, Caruana G et al. Requirement of PDZ-containing proteins for cell cycle regulation and differentiation in the mouse lens epithelium. Mol Cell Biol 2003; 23(24):8970-8981.
144. Nakagawa S, Huibregtse JM. Human scribble (Vartul) is targeted for ubiquitin-mediated degradation by the high-risk papillomavirus E6 proteins and the E6AP ubiquitin-protein ligase. Mol Cell Biol 2000; 20(21):8244-8253.
145. Cavallaro U, Christofori G. Cell adhesion in tumor invasion and metastasis: loss of the glue is not enough. Biochim Biophys Acta 2001; 1552(1):39-45.
146. Nakagawa S, Yano T, Nakagawa K et al. Analysis of the expression and localisation of a LAP protein, human scribble, in the normal and neoplastic epithelium of uterine cervix. Br J Cancer 2004; 90(1):194-199.
147. Shah K, Nathanson N. Human exposure to SV40: review and comment. Am J Epidemiol 1976; 103(1):1-12.
148. Choi YW, Lee IC, Ross SR. Requirement for the simian virus 40 small tumor antigen in tumorigenesis in transgenic mice. Mol Cell Biol 1988; 8(8):3382-3390.
149. Carbone M, Pass HI, Miele L et al. New developments about the association of SV40 with human mesothelioma. Oncogene 2003; 22(33):5173-5180.
150. Vilchez RA, Kozinetz CA, Arrington AS et al. Simian virus 40 in human cancers. Am J Med 2003; 114(8):675-684.
151. Hahn WC, Dessain SK, Brooks MW et al. Enumeration of the simian virus 40 early region elements necessary for human cell transformation. Mol Cell Biol 2002; 22(7):2111-2123.

152. Pallas DC, Shahrik LK, Martin BL et al. Polyoma small and middle T antigens and SV40 small t antigen form stable complexes with protein phosphatase 2A. Cell 1990; 60(1):167-176.
153. Scheidtmann KH, Mumby MC, Rundell K et al. Dephosphorylation of simian virus 40 large-T antigen and p53 protein by protein phosphatase 2A: inhibition by small-t antigen. Mol Cell Biol 1991; 11(4):1996-2003.
154. Nunbhakdi-Craig V, Craig L, Machleidt T et al. Simian virus 40 small tumor antigen induces deregulation of the actin cytoskeleton and tight junctions in kidney epithelial cells. J Virol 2003; 77(5):2807-2818.
155. Braithwaite AW, Cheetham BF, Li P et al. Adenovirus-induced alterations of the cell growth cycle: a requirement for expression of E1A but not of E1B. J Virol 1983; 45(1):192-199.
156. Chinnadurai G. Modulation of oncogenic transformation by the human adenovirus E1A C-terminal region. Curr Top Microbiol Immunol 2004; 273:139-161.
157. Fischer RS, Quinlan MP. The C terminus of E1A regulates tumor progression and epithelial cell differentiation. Virology 1998; 249(2):427-439.
158. Schaeper U, Boyd JM, Verma S et al. Molecular cloning and characterization of a cellular phosphoprotein that interacts with a conserved C-terminal domain of adenovirus E1A involved in negative modulation of oncogenic transformation. Proc Natl Acad Sci USA 1995; 92(23):10467-10471.
159. Zhang Z, Smith MM, Mymryk JS. Interaction of the E1A oncoprotein with Yak1p, a novel regulator of yeast pseudohyphal differentiation, and related mammalian kinases. Mol Biol Cell 2001; 12(3):699-710.
160. Poser I, Golob M, Weidner M et al. Down-regulation of COOH-terminal binding protein expression in malignant melanomas leads to induction of MIA expression. Cancer Res 2002; 62(20):5962-5966.
161. Becker W, Joost HG. Structural and functional characteristics of Dyrk, a novel subfamily of protein kinases with dual specificity. Prog Nucleic Acid Res Mol Biol 1999; 62:1-17.
162. Thomas DL, Shin S, Jiang BH et al. Early region 1 transforming functions are dispensable for mammary tumorigenesis by human adenovirus type 9. J Virol 1999; 73(4):3071-3079.
163. Weiss RS, Gold MO, Vogel H et al. Mutant adenovirus type 9 E4 ORF1 genes define three protein regions required for transformation of CREF cells. J Virol 1997; 71(6):4385-4394.
164. Frese KK, Lee SS, Thomas DL et al. Selective PDZ protein-dependent stimulation of phosphatidylinositol 3-kinase by the adenovirus E4-ORF1 oncoprotein. Oncogene 2003; 22(5):710-721.
165. Gatza ML, Watt JC, Marriott SJ. Cellular transformation by the HTLV-I Tax protein, a jack-of-all-trades. Oncogene 2003; 22(33):5141-5149.
166. Hirata A, Higuchi M, Niinuma A et al. PDZ domain-binding motif of human T-cell leukemia virus type 1 Tax oncoprotein augments the transforming activity in a rat fibroblast cell line. Virology 2004; 318(1):327-336.
167. Ohashi M, Sakurai M, Higuchi M et al. Human T-cell leukemia virus type 1 Tax oncoprotein induces and interacts with a multi-PDZ domain protein, MAGI-3. Virology 2004; 320(1):52-62.
168. Kimura Y, Shiozaki H, Hirao M et al. Expression of occludin, tight-junction-associated protein, in human digestive tract. Am J Pathol 1997; 151(1):45-54.
169. Yin F, Qiao T, Shi Y et al. [In situ hybridization of tight junction molecule occludin mRNA in gastric cancer]. Zhonghua Zhong Liu Za Zhi 2002; 24(6):557-560.
170. Busch C, Hanssen TA, Wagener C et al. Down-regulation of CEACAM1 in human prostate cancer: correlation with loss of cell polarity, increased proliferation rate, and Gleason grade 3 to 4 transition. Hum Pathol 2002; 33(3):290-298.
171. Tobioka H, Isomura H, Kokai Y et al. Polarized distribution of carcinoembryonic antigen is associated with a tight junction molecule in human colorectal adenocarcinoma. J Pathol 2002; 198(2):207-212.
172. Tobioka H, Isomura H, Kokai Y et al. Occludin expression decreases with the progression of human endometrial carcinoma. Hum Pathol 2004; 35(2):159-164.
173. Miwa N, Furuse M, Tsukita S et al. Involvement of claudin-1 in the beta-catenin/Tcf signaling pathway and its frequent upregulation in human colorectal cancers. Oncol Res 2000; 12(11-12):469-476.
174. Long H, Crean CD, Lee WH et al. Expression of Clostridium perfringens enterotoxin receptors claudin-3 and claudin-4 in prostate cancer epithelium. Cancer Res 2001; 61(21):7878-7881.
175. Hough CD, Sherman-Baust CA, Pizer ES et al. Large-scale serial analysis of gene expression reveals genes differentially expressed in ovarian cancer. Cancer Res 2000; 60(22):6281-6287.
176. Terris B, Blaveri E, Crnogorac-Jurcevic T et al. Characterization of gene expression profiles in intraductal papillary-mucinous tumors of the pancreas. Am J Pathol 2002; 160(5):1745-1754.

177. Kominsky SL, Argani P, Korz D et al. Loss of the tight junction protein claudin-7 correlates with histological grade in both ductal carcinoma in situ and invasive ductal carcinoma of the breast. Oncogene 2003; 22(13):2021-2033.
178. Al Moustafa AE, Alaoui-Jamali MA, Batist G et al. Identification of genes associated with head and neck carcinogenesis by cDNA microarray comparison between matched primary normal epithelial and squamous carcinoma cells. Oncogene 2002; 21(17):2634-2640.
179. Rangel LB, Sherman-Baust CA, Wernyj RP et al. Characterization of novel human ovarian cancer-specific transcripts (HOSTs) identified by serial analysis of gene expression. Oncogene 2003; 22(46):7225-7232.
180. Katoh M. CLDN23 gene, frequently down-regulated in intestinal-type gastric cancer, is a novel member of CLAUDIN gene family. Int J Mol Med 2003; 11(6):683-689.
181. Kaihara T, Kusaka T, Nishi M et al. Dedifferentiation and decreased expression of adhesion molecules, E-cadherin and ZO-1, in colorectal cancer are closely related to liver metastasis. J Exp Clin Cancer Res 2003; 22(1):117-123.
182. Kaihara T, Kawamata H, Imura J et al. Redifferentiation and ZO-1 reexpression in liver-metastasized colorectal cancer: possible association with epidermal growth factor receptor-induced tyrosine phosphorylation of ZO-1. Cancer Sci 2003; 94(2):166-172.
183. Kleeff J, Shi X, Bode HP et al. Altered expression and localization of the tight junction protein ZO-1 in primary and metastatic pancreatic cancer. Pancreas 2001; 23(3):259-265.
184. Chlenski A, Ketels KV, Tsao MS et al. Tight junction protein ZO-2 is differentially expressed in normal pancreatic ducts compared to human pancreatic adenocarcinoma. Int J Cancer 1999; 82(1):137-144.
185. Chlenski A, Ketels KV, Korovaitseva GI et al. Organization and expression of the human zo-2 gene (tjp-2) in normal and neoplastic tissues. Biochim Biophys Acta 2000; 1493(3):319-324.
186. Fang CM, Xu YH. Down-regulated expression of atypical PKC-binding domain deleted asip isoforms in human hepatocellular carcinomas. Cell Res 2001; 11(3):223-229.

CHAPTER 10

Regulation of Paracellular Transport across Tight Junctions by the Actin Cytoskeleton

Matthias Bruewer and Asma Nusrat

Abstract

The epithelial lining of luminal organs such as the gastrointestinal and respiratory tract forms a regulated, selectively permeable barrier between luminal contents and the underlying tissue compartments. Paracellular permeability across epithelial and endothelial cells is in large part regulated by an apical intercellular junction also referred to as the tight junction (TJ). The tight junction and its subjacent adherens junction (AJ) constitute the apical junctional complex (AJC). The AJC is composed of a multiprotein complex, which affiliates with the underlying apical perijunctional F-actin ring. Such AJC association with the perijunctional F-actin ring is vital for maintaining its structure and function in health. Stimuli such as nutrients, internal signaling molecules and cytokines influence the apical F-actin organization and also modulate the AJC structure and paracellular permeability. Here we review some of the key stimuli that influence F-actin organization, AJC structure and paracellular permeability.

Introduction

The tight junction (TJ) is the apical most intercellular junction in epithelial and endothelial cells and in association with the subjacent adherens junction (AJ) it constitutes the apical junctional complex (AJC). TJs form regulated, selectively permeable barriers between two distinct compartments.[1,2] Thus, for example in the intestinal and respiratory tract, TJs interface luminal contents and underlying tissue compartments. TJs do not just represent static structural elements but they are dynamically regulated to control paracellular solute and ion transport in diverse physiologic states. In part such regulation in health and dysregulation in disease occurs secondary to signaling events influencing the underlying perijunctional actin cytoskeleton.[3,4]

Columnar epithelial cells have a prominent perijunctional actin-myosin II ring that encircles the apical pole of polarized cells. This perijunctional filamentous actin ring is readily visualized by fluorescence labeling of filamentous actin and by electron microscopy. In fact the major interface of this ring with the lateral membrane occurs just below the TJ in the AJ. However, actin filaments project from this ring and interface with specific sites of membrane kisses in TJs that represent regions where membranes from apposing cells come into close apposition (Fig. 1).[5] Thus, it is logical to envision regulation of TJ structure and paracellular solute transport by factors that influence lateral tension within the perijunctional actin-myosin II ring.[6-8]

Tight Junctions, edited by Lorenza Gonzalez-Mariscal. ©2006 Landes Bioscience and Springer Science+Business Media.

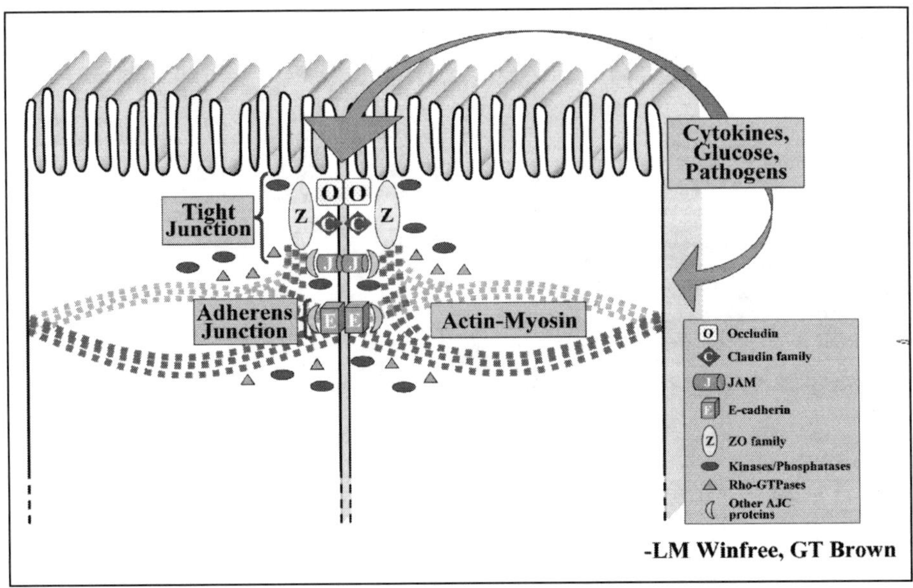

Figure 1. Protein composition and regulation of the apical junctional complex (AJC). Transmembrane proteins in the apical junctional complex affiliate with an underlying perijunctional actin myosin ring. Signaling molecules and extracellular stimuli such as cytokines, nutrients and pathogens regulate the AJC.

Our knowledge of TJ protein composition and regulation of paracellular transport is rapidly expanding. It is clearly evident that transmembrane proteins in TJs such as occludin, claudin(s) and junction adhesion molecule (JAM)-1 affiliate with cytoplasmic plaque proteins that in turn have been implicated in the association of the TJ protein complex with the actin cytoskeleton.[9-12] Prototypes of such actin associating proteins are the zonula occludens proteins (ZO-1, ZO-2 and ZO-3). Cingulin also interacts with the zonula occludens proteins and enterocyte myosin heavy chain.[13] The overall organization of TJ associated proteins is illustrated in Figure 1. Structure and functional properties of these proteins are detailed in other chapters of this issue.

Current knowledge of TJs is consistent with a view that they are specialized membrane microdomains[14-16] that might function as molecular platforms involved in actin organization (Rho-GTPases, EFA6), cell signaling (c-src and c-yes), membrane trafficking (VAP-33, Rab3b, Rab13, Rab 8, Sec6, and Sec8), and cell polarity (Par3 and Par6).[17-20] Thus TJs represent very dynamic platforms that regulate paracellular movement of ions and solutes in many physiologic and pathologic states. Diverse stimuli that influence TJ function and its subjacent actin cytoskeleton are discussed below.

Modulation of Myosin Light Chain-Phosphorylation in the Perijunctional F-Actin Ring Influences TJ Function

Circumferential contraction and therefore tension in the apical perijunctional actin-myosin ring regulates solute transport across the paracellular space.[21-25] Such modifications in the perijunctional actin-myosin ring are achieved by the phosphorylation of the myosin light chain (MLC) of myosin II by MLC kinase (MLCK), which in turn acts on bipolar F-actin fibers in the perijunctional F-actin ring. Increased epithelial permeability in MLCK-dependent fashion has been demonstrated in model intestinal epithelial cell lines exposed to *enteropathogenic Escherichia coli*, transmigrating polymorphonuclear leukocytes or cytokines such as interferon gamma (IFN-γ) and tumor necrosis factor alpha (TNF-α).[26-28] In addition, physiologic agonists have also been demonstrated to influence paracellular transport by modifications in the perijunctional actin-myosin ring. A classic scenario involves the uptake of glucose in the intestinal tract.[29] Early ultrastructural examination of the intestinal mucosa has demonstrated that intrajunctional dilatations and condensation of the perijunctional cytoskeleton occur with Na[+]-glucose cotransport induced increase in permeability.[22,30,31] Such modifications in the actin cytoskeleton support the concept of cytoskeletal regulation of paracellular permeability. In fact, subsequent studies linked the activation of enterocyte Na[+]-glucose cotransporter with phosphorylation of MLC.[32] Na[+]-glucose cotransport induces cytoplasmic alkalinization that is dependent on the activation of the brush border Na[+]/H[+] exchanger isoform NHE3.[33] Inhibition of the NHE3 exchanger reduces MLC phosphorylation that is associated with an increase in transepithelial resistance (TER) to passive ion flow.[33] It has therefore been proposed that NHE3 activation may be a critical component of the signaling pathway for Na[+]-glucose cotransport-dependent TJ regulation. Further evidence supporting the role of MLC phosphorylation in TJ regulation comes from studies using epithelial cells transfected with truncated MLCK gene construct lacking the inhibitory domain required for kinase regulation.[34] Expression of this construct in model epithelial cell lines induced an increase in myosin regulatory light chain phosphorylation and an increase in paracellular permeability.[27,34] Upstream regulation of the MLCK by protein kinase C has been proposed as a mechanism regulating the TJ permeability.[35]

Other physiological agonists influencing the cytoskeleton and junctional transport through modulation of the actin cytoskeleton include responses to histamine[36] and lysophosphatidic acid[37] that induce phosphorylation of MLCs. Ethanol and low concentration of extracellular calcium increase activity of MLCK and disrupt the TJ protein complex by influencing ZO-1 and occludin organization in the AJC.[38,39] Agonists that influence TJ permeability in rat hepatocytes include angiotensin II, vasopressin and epinephrine.[40]

Modulation of Barrier Function by Rho GTPases

The Rho family of small GTPases, comprising Rho, Rac and Cdc42, are believed to play an important role in regulating and maintaining the perijunctional actin ring, TJ structure/function, and assembly of polarized epithelial cells.[41-46] Rho function is modulated by a set of regulatory proteins and is activated through GDP-GTP exchange, which is promoted by guanine nucleotide exchange factors (GEF) and is inactivated through GTPase-activating proteins.[47,48] Rho guanine nucleotide dissociation inhibitors mediate stabilization of inactive GDP-bound form of Rho.[49] Conformational changes then allow the GTPases to interact with multiple effector molecules involved in actin cytoskeletal control[50,51] Rho activity cycles are rapidly reversible, and are terminated upon hydrolysis of GTP by GTPase-activating proteins.

Several insights have been gained, based on the use of diverse pharmacological and molecular tools that interfere with function of the Rho family of GTPases. We have previously shown that inactivation of Rho GTPases (with *Clostridium botulinum* C3 transferase and *Clostridium difficile* toxins A and B) is associated with a compromised barrier function in T84 intestinal epithelial cells manifested both as functional decrease in TER, increase in paracellular flux of labeled dextran (3kDa) and a structural redistribution of ZO-1 and occludin away from the lateral plasma membrane. This effect was associated with reorganization of perijunctional F-actin.[45] Moreover, we demonstrated, that disassembly of TJs induced by *Clostridium difficile* toxins A and B reduced the hyperphosphorylated occludin species and ZO-1 in "raft-like" membrane microdomains.[14] The downstream effector of Rho referred to as Rho kinase (ROCK) has also been documented to regulate TJ function.[52] However, ROCK inhibition induces profound reorganization of the apical F-actin cytoskeleton without influencing TJ protein distribution in the lateral membrane. These findings imply that ROCK mediated effects on TJ function are primarily due to its influence on the apical actin cytoskeleton in epithelial cells. These observations were further supported by transfection studies, in which a dominant negative mutant of ROCK induced loss of the apical F-actin–rich brush border and a reduction in the apical perijunctional F-actin ring without influencing occludin localization. Studies using transfected epithelial cell lines expressing dominant negative mutants of Rho GTPases demonstrated increase in paracellular permeability without influencing the TJ protein organization.[53] In addition to the above downstream effector, an upstream GEF of the Dbl family of proto-oncogenes that activates Rho has been shown to associate with TJs.[54] A link between the TJ cytoplasmic plaque protein ZO-3 and RhoA related signaling has been proposed.[17] These studies reported that transfection of the amino terminal half of ZO-3 (NZO-3) in MDCK cells resulted in decreased RhoA GTPase activity and a change in cellular F-actin organization. The authors proposed a model whereby altered interactions between ZO-3 and an AJ protein, p120 catenin in NZO-3 expressing cells influences RhoA GTPase activity.

Recent studies reported, that not only inactivation, but also activation of the Rho family of proteins enhances paracellular permeability.[43,53,55] Using an elegant inducible transfection system in MDCK cells, convincing evidence for an involvement of RhoA, Rac1 and Cdc42 in regulation of epithelial barrier function has been obtained.[53,55,56] In this system, induction of dominant-active RhoA, Rac1 and Cdc42 activity was correlated with inability of epithelial cells to develop high TER. Moreover, increased paracellular permeability to molecules of different sizes was accompanied by redistribution of occludin and ZO-1 from the lateral membrane as well as modifications in junctional associated F-actin cytoskeleton. Using the same system, we found that activation as well as inactivation of RhoA, Rac1 or Cdc42 induced time-dependent disruptions in epithelial gate function and distinct morphological alterations in apical and basal F-actin pools. Constitutive activation of Rho A and Cdc42 induced redistribution of occludin, ZO-1, claudin-1, claudin-2 and JAM-1 from the lateral membrane. Constitutively active Rac1 on the other hand primarily influenced claudin-1 and -2 organization in TJs. These structural alterations were accompanied by changes in the biochemical properties of the TJ proteins.[97] Interestingly, an increased activation of RhoA has been described in biopsies of patients with Crohn's disease indicating that RhoA may be involved in the cascade that leads to impaired barrier function in these patients.[57] Reported effects of the activation of Rho GTPases by *Escherichia coli* cytotoxic necrotizing factor (CNF)-1 on epithelial TER have to date been diverse. While one report suggests a lack of CNF-1 effect on TER,[58] two other reports using intestinal epithelial cell lines T84 and Caco2 document CNF-1 induced an increase in paracellular permeability.[43,59] In the latter study, increased paracellular permeability was associated with significant redistribution of the TJ proteins occludin, ZO-1, claudin-1 and JAM-1 following basolateral exposure of epithelial cells to CNF-1. In parallel, CNF-1 incubation resulted in decreased apical F-actin that was accompanied by formation of prominent basal F-actin cables.[43] Thus, increased activation of Rho appears to disrupt the continuity between adjacent F-actin pools in microvilli, perijunctional ring and the terminal web that in turn could destabilize the "scaffold" of the TJ protein complex.[43]

Interestingly, CNF-1 treatment induced internalization of TJ proteins into endosomal/caveolar-like membranous structures, evidenced by colocalization of TJ proteins with caveolin-1 by immunogold electron-microscopy.[43] By immunolabeling and confocal microscopy TJ proteins were observed to colocalize with internalized early and recycling endosomal markers (EEA-1, Rab-11). This provides novel evidence that increased activation of Rho-GTPases induces internalization of TJ proteins into endosomal structures. Interestingly, dominant-active Rac1 and Cdc42 have been shown to affect endocytic trafficking in epithelial cells.[55,60,61] The colocalization of TJ proteins with markers of recycling endosomes also suggests that recycling of TJ proteins back to the lateral membrane could occur, thereby providing a route for the rapid reestablishment of barrier function during the recovery phase following injury and internalization of TJ proteins. Given that inactivation as well as activation of Rho-GTPases adversely affects epithelial barrier function, it is likely that a delicate balance of Rho activity/quiescence is required for the maintenance of the optimal epithelial/endothelial barrier function.

Modulation of Barrier Function by Cytokines

Many cytokines have been shown to influence epithelial TJ function and the actin cytoskeleton both in vivo and in vitro. The cytokines IL-1, IL-4, IL-10, IL-13, TNF-α, and IFN-γ have all been shown to regulate TJs of both epithelia and endothelia.[62-64] In addition, IL-1β influences TJ permeability through an effect on the claudin family of transmembrane proteins thought to be important in maintaining junctional integrity in astrocytes.[65] A complete review of all the cytokines shown to modulate epithelial barrier function is beyond the scope of this article. Our review therefore has focused on the influence of few select cytokines on TJ structure/function and its adjoining actin cytoskeleton.

Interferon-γ

Interferon-γ (IFN-γ) is a 20- to 25-kDa glycoprotein released by activated T cells and natural killer cells in inflammatory states. In vitro models have been extensively used to examine the influence of this pro-inflammatory cytokine on intercellular junctions of epithelial cells. The initial studies addressing the influence of this cytokine on TJs utilized model epithelial cell line, T84.[63,66,67] These studies demonstrated that IFN-γ induced a time-dependent increase in paracellular permeability that was accompanied by disorganization of apical F-actin and loss of ZO-1 from TJs.[63,68] These morphological effects were associated with a change in the differential detergent solubility profiles of ZO-1 and ZO-2. The investigators did not observe IFN-γ induced change of phosphorylation status of these proteins. This was unexpected as phosphorylation status of TJ proteins is considered to modulate assembly of the TJ protein complex.[69] Using the model T84 intestinal epithelial cell line, we have recently reported a IFN-γ-induced disruption of epithelial gate and fence function that was associated with differential internalization of TJ transmembrane proteins occludin, JAM-1, Claudin-1 and −4.[70] We have also observed a concomitant reorganization of apical F-actin (our unpublished results). In contrast ZO-1 maintained its localization in the TJs. It is intriguing to speculate why ZO-1 localization and expression levels in our study were only slightly affected by IFN-γ. ZO-1 is a key TJ cytoplasmic plaque protein that provides a scaffold upon which other proteins can be assembled.[71,72] We hypothesized that ZO-1 maintains its localization to provide this scaffold for efficient reassembly of TJ proteins upon cytokine withdrawal, which would be required for rapid and critical reestablishment of epithelial barrier function. Recent results from our laboratory support a IFN-γ induced internalization of TJ proteins into endosomal structures and this event is in part mediated by restructuring of the apical actin cytoskeleton (unpublished observations) (Fig. 2). A similar mechanism of TJ protein endocytosis has been observed following depletion of extracellular calcium and disassembly of TJs.[73] Thus, the actin cytoskeleton appears to be essential in not only maintaining a functioning TJ but is also required for the regulated disassembly and reassembly of the TJ.

An inflammatory response is regulated by a complex array of inhibitory and stimulatory cytokines, and thus it is likely that effects produced by IFN-γ are modulated by other cytokines. Several studies using different epithelial cell lines have shown, that TNF-α, another pro-inflammatory cytokine, can act synergistically with IFN-γ to increase paracellular permeability,[27,70,74,75] most likely due to TNF-α-induced up-regulation of the IFN-γ receptor. Coyne et al[74] demonstrated in human epithelial airway cells, that combined treatment of TNF-α and IFN-γ induced profound effects on TJ barrier function, which could be blocked by inhibitors of protein kinase C. These studies emphasized the importance of the link between the actin cytoskeleton and TJs in regulation of barrier function both in the baseline state and following exposure to pro-inflammatory cytokines such as IFN-γ and TNF-α. In contrast to TNF-α, TGFβ or IL-10 have a negative influence on the IFN-γ induced changes in paracellular permeability.[76,77]

The above-described effects of IFN-γ and TNF-α on TJs, although complex, might have pathophysiological relevance because an increase in paracellular permeability across intestinal epithelial cells has been observed in patients with inflammatory bowel diseases

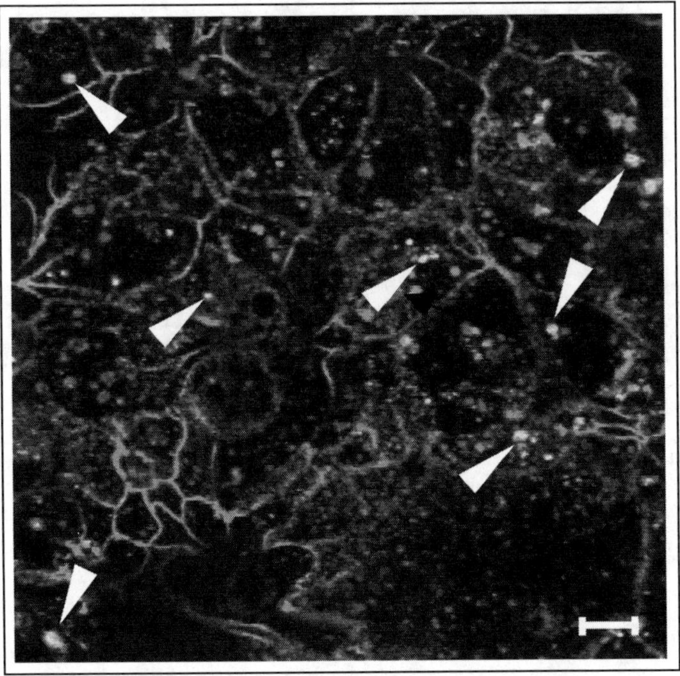

Figure 2. Colocalization of internalized occludin with the early endosomal marker EEA-1. T84 cell mono-layers incubated for 48 h with IFN-γ were double-stained for occludin (green) and the early endosomal marker, EEA-1 (red) and analyzed by confocal microscopy. Internalized occludin colocalizes with EEA-1 (yellow, arrowheads) (scale bar 10 μm). A color version of this figure is available at http://www.Eurekah.com.

(IBD).[78] Since enhanced paracellular permeability across intestinal epithelium also occurs in first-degree relatives of patients with Crohn's disease, altered TJ permeability may be a contributing factor in this process,[79,80] whereas disturbed barrier function in patients with ulcerative colitis is more likely secondary to the array of inflammatory signals that characterize this state.[79,80] In this regard it is important, that redistribution of AJC proteins has been observed in tissues from patients with active IBD.[81,82]

Tumor Necrosis Factor-α

Tumor necrosis factor-α (TNF-α) is a 17-kDa proinflammatory cytokine produced mainly by mononuclear cells, and it influences barrier function of some epithelial cells. A biphasic response of TNF-α on TER has been reported in a porcine renal epithelial cell line, LLC-PK1.[83] In this study, an initial fall in TER and increased paracellular permeability was followed by an increase in TER. The latter phase correlated with decreased relative anion selectivity of TJs. A role of tyrosine kinase and protein kinase A in mediating such effects of TNF-α on this cell type were proposed. In contrast to the findings above, TNF-α induced a fall in TER without the subsequent rebound in the intestinal epithelial cell lines HT-29 and Caco2. The decrease in TER was associated with increased paracellular permeability to mannitol in HT-29 mono-layers but not in the Caco2 cells. In the latter, fall in TER was associated with increased paracellular permeability to Na^+ and Cl^-, implying altered charge but not size selectivity in TJs. The only morphological correlate in the above studies was a decrease in the number of TJ strand complexity by freeze-fracture EM in HT29 cells.[84]

TNF-α has however been implicated in modulating claudin-1 expression and ZO-1 organization in TJs.[85] Studies in T84 cells did not report an effect of TNF-α alone on barrier function,[66,70,86] indicating that TNF-α can exert different effects on barrier function depending on the target epithelium. In contrast to epithelial cells in pulmonary endothelial monolayers, TNF-α can induce an increase in permeability by influencing the actin cytoskeleton. In such endothelial cells, TNF-α significantly increased MLC phosphorylation, formation of prominent stress fiber and paracellular gaps.[87]

Hepatocyte Growth Factor

Hepatocyte growth factor (HGF) HGF is a 103 kDa disulfide-linked, heterodimeric protein, that is produced chiefly by mesenchyme-derived cells and influences epithelial permeability in a paracrine fashion via ligation with its receptor, c-met. It was previously documented that analogous to IFN-γ, HGF induced a delayed decrease in TER of T84 epithelial monolayers over a period of 48 h.[88] Structural studies to analyze the influence of HGF on intercellular junctions have not yielded unifying results. Depending on the origin of epithelial cells, variable effects of HGF on protein organization in the TJ versus its subjacent AJ have been proposed. Our studies in T84 epithelial cells have suggested an initial effect of HGF on apical F-actin organization that is accompanied by alterations in epithelial paracellular permeability (our unpublished observation). Such observations further emphasize the importance of apical F-actin structures in regulating paracellular permeability. In renal MDCK cells, HGF has been documented to influence AJC assembly.[89] HGF-induced inhibition of junction assembly was associated with an increase in the Triton X-100 insoluble pool of E-cadherin[90] and plakoglobin[89] without influencing their total concentration thereby suggesting a cytokine induced change in their cytoskeletal association.

Other Cytokines

TGFβ1 has been reported to enhance barrier function in human enterocytes and to promote intestinal epithelial restitution.[91] Moreover, TGFβ1 curtails the effects of barrier reducing cytokines such as IFN-γ, Il-4 and −10.[76,92] In contrast, TGFβ1 has been shown to prevent glucocorticoid stimulated TJ formation and to reduce TER in 31EG4 polarized murine mammary epithelial cells.[93,94] Such effects were accompanied by redistribution of ZO-1 from the lateral membrane and restructuring of perijunctional F-actin.[93] Subsequently, it has been demonstrated that glucocorticoid induced downregulation of RhoA was required for its regulation of TJ and actin cytoskeletal organization.[95]

Exposure of the epidermal A431 cell line to epidermal growth factor (EGF) promotes TJ assembly. EGF facilitates restructuring of apical F-actin and induces phosphorylation of TJ cytoplasmic plaque proteins ZO-1 and ZO-2.[96] Other cytokines such as interleukin (IL)-1, IL-4, IL-13, TGF-alpha, insulin-like growth factor (IGF)-I and -II, PDGF and vascular endothelial growth factor (VEGF) have been documented to decrease the barrier properties of epithelial cells. Mechanisms ranging from redistribution of TJ proteins to alterations in the actin cytoskeletal organization have been proposed to mediate the cytokine-induced effects on epithelial barrier function.

References

1. Diamond JM. Twenty-first Bowditch lecture. The epithelial junction: Bridge, gate, and fence. Physiologist 1977; 20(1):10-18.
2. van Meer G, Simons K. The function of tight junctions in maintaining differences in lipid composition between the apical and the basolateral cell surface domains of MDCK cells. EMBO J 1986; 5(7):1455-1464.
3. Dejana E, Corada M, Lampugnani MG. Endothelial cell-to-cell junctions. FASEB J 1995; 9(10):910-918.
4. Gumbiner BM. Cell adhesion: The molecular basis of tissue architecture and morphogenesis. Cell 1996; 84(3):345-357.

5. Staehelin LA. Structure and function of intercellular junctions. Int Rev Cytol 1974; 39:191-283.
6. Madara JL, Moore R, Carlson S. Alteration of intestinal tight junction structure and permeability by cytoskeletal contraction. Am J Physiol 1987; 253(6 Pt 1):C854-861.
7. Madara JL, Parkos C, Colgan S et al. The movement of solutes and cells across tight junctions. Ann NY Acad Sci 1992; 664:47-60.
8. Pitelka DR, Taggart BN. Mechanical tension induces lateral movement of intramembrane components of the tight junction: Studies on mouse mammary cells in culture. J Cell Biol 1983; 96(3):606-612.
9. Furuse M, Hirase T, Itoh M et al. Occludin: A novel integral membrane protein localizing at tight junctions. J Cell Biol 1993; 123(6 Pt 2):1777-1788.
10. Furuse M, Itoh M, Hirase T et al. Direct association of occludin with ZO-1 and its possible involvement in the localization of occludin at tight junctions. J Cell Biol 1994; 127(6 Pt 1):1617-1626.
11. Fanning AS, Jameson BJ, Jesaitis LA et al. The tight junction protein ZO-1 establishes a link between the transmembrane protein occludin and the actin cytoskeleton. J Biol Chem 1998; 273(45):29745-29753.
12. Liu Y, Nusrat A, Schnell FJ et al. Human junction adhesion molecule regulates tight junction resealing in epithelia. J Cell Sci 2000; 113(Pt 13):2363-2374.
13. Citi S, Cordenonsi M. Tight junction proteins. Biochim Biophys Acta 1998; 1448(1):1-11.
14. Nusrat A, von Eichel-Streiber C, Turner JR et al. Clostridium difficile toxins disrupt epithelial barrier function by altering membrane microdomain localization of tight junction proteins. Infect Immun 2001; 69(3):1329-1336.
15. Nusrat A, Parkos CA, Verkade P et al. Tight junctions are membrane microdomains. J Cell Sci 2000; 113(Pt 10):1771-1781.
16. Harhaj NS, Barber AJ, Antonetti DA. Platelet-derived growth factor mediates tight junction redistribution and increases permeability in MDCK cells. J Cell Physiol 2002; 193(3):349-364.
17. Wittchen ES, Haskins J, Stevenson BR. NZO-3 expression causes global changes to actin cytoskeleton in Madin-Darby canine kidney cells: Linking a tight junction protein to Rho GTPases. Mol Biol Cell 2003; 14(5):1757-1768.
18. van der Wouden JM, Maier O, van ISC et al. Membrane dynamics and the regulation of epithelial cell polarity. Int Rev Cytol 2003; 226:127-164.
19. Meyer TN, Hunt J, Schwesinger C et al. Galpha12 regulates epithelial cell junctions through Src tyrosine kinases. Am J Physiol Cell Physiol 2003; 285(5):C1281-1293.
20. Luton F, Klein S, Chauvin JP et al. EFA6, exchange factor for ARF6, regulates the actin cytoskeleton and associated tight junction in response to E-cadherin engagement. Mol Biol Cell 2003.
21. Madara JL. Tight junction dynamics: Is paracellular transport regulated? Cell 1988; 53(4):497-498.
22. Madara JL, Pappenheimer JR. Structural basis for physiological regulation of paracellular pathways in intestinal epithelia. J Membr Biol 1987; 100(2):149-164.
23. Madara JL, Stafford J, Dharmsathaphorn K et al. Structural analysis of a human intestinal epithelial cell line. Gastroenterology 1987; 92(5 Pt 1):1133-1145.
24. Turner JR, Madara JL. Physiological regulation of intestinal epithelial tight junctions as a consequence of Na(+)-coupled nutrient transport. Gastroenterology 1995; 109(4):1391-1396.
25. Hecht G, Pothoulakis C, LaMont JT et al. Clostridium difficile toxin A perturbs cytoskeletal structure and tight junction permeability of cultured human intestinal epithelial monolayers. J Clin Invest 1988; 82(5):1516-1524.
26. Yuhan R, Koutsouris A, Savkovic SD et al. Enteropathogenic Escherichia coli-induced myosin light chain phosphorylation alters intestinal epithelial permeability. Gastroenterology 1997; 113(6):1873-1882.
27. Zolotarevsky Y, Hecht G, Koutsouris A et al. A membrane-permeant peptide that inhibits MLC kinase restores barrier function in vitro models of intestinal disease. Gastroenterology 2002; 123(1):163-172.
28. Edens HA, Levi BP, Jaye DL et al. Neutrophil transepithelial migration: Evidence for sequential, contact-dependent signaling events and enhanced paracellular permeability independent of transjunctional migration. J Immunol 2002; 169(1):476-486.
29. Madara JL. Sodium-glucose cotransport and epithelial permeability. Gastroenterology 1994; 107(1):319-320.
30. Pappenheimer JR. Physiological regulation of transepithelial impedance in the intestinal mucosa of rats and hamsters. J Membr Biol 1987; 100(2):137-148.
31. Pappenheimer JR, Reiss KZ. Contribution of solvent drag through intercellular junctions to absorption of nutrients by the small intestine of the rat. J Membr Biol 1987; 100(2):123-136.

32. Turner JR, Rill BK, Carlson SL et al. Physiological regulation of epithelial tight junctions is associated with myosin light-chain phosphorylation. Am J Physiol 1997; 273(4 Pt 1):C1378-1385.

33. Turner JR, Black ED, Ward J et al. Transepithelial resistance can be regulated by the intestinal brush-border Na(+)/H(+) exchanger NHE3. Am J Physiol Cell Physiol 2000; 279(6):C1918-1924.

34. Hecht G, Pestic L, Nikcevic G et al. Expression of the catalytic domain of myosin light chain kinase increases paracellular permeability. Am J Physiol 1996; 271(5 Pt 1):C1678-1684.

35. Turner JR, Angle JM, Black ED et al. PKC-dependent regulation of transepithelial resistance: Roles of MLC and MLC kinase. Am J Physiol 1999; 277(3 Pt 1):C554-562.

36. Lum H, Malik AB. Regulation of vascular endothelial barrier function. Am J Physiol 1994; 267(3 Pt 1):L223-241.

37. van Nieuw Amerongen GP, Vermeer MA, van Hinsbergh VW. Role of RhoA and Rho kinase in lysophosphatidic acid-induced endothelial barrier dysfunction. Arterioscler Thromb Vasc Biol 2000; 20(12):E127-133.

38. Ma TY, Nguyen D, Bui V et al. Ethanol modulation of intestinal epithelial tight junction barrier. Am J Physiol 1999; 276(4 Pt 1):G965-974.

39. Ma TY, Tran D, Hoa N et al. Mechanism of extracellular calcium regulation of intestinal epithelial tight junction permeability: Role of cytoskeletal involvement. Microsc Res Tech 2000; 51(2):156-168.

40. Lowe PJ, Miyai K, Steinbach JH et al. Hormonal regulation of hepatocyte tight junctional permeability. Am J Physiol 1988; 255(4 Pt 1):G454-461.

41. Braga VM, Machesky LM, Hall A et al. The small GTPases Rho and Rac are required for the establishment of cadherin-dependent cell-cell contacts. J Cell Biol 1997; 137(6):1421-1431.

42. Hall A. Rho GTPases and the actin cytoskeleton. Science 1998; 279(5350):509-514.

43. Hopkins AM, Walsh SV, Verkade P et al. Constitutive activation of Rho proteins by CNF-1 influences tight junction structure and epithelial barrier function. J Cell Sci 2003; 116(Pt 4):725-742.

44. Nobes CD, Hall A. Rho, rac and cdc42 GTPases: Regulators of actin structures, cell adhesion and motility. Biochem Soc Trans 1995; 23(3):456-459.

45. Nusrat A, Giry M, Turner JR et al. Rho protein regulates tight junctions and perijunctional actin organization in polarized epithelia. Proc Natl Acad Sci USA 1995; 92(23):10629-10633.

46. Ridley AJ, Hall A. Distinct patterns of actin organization regulated by the small GTP-binding proteins Rac and Rho. Cold Spring Harb Symp Quant Biol 1992; 57:661-671.

47. Chuang TH, Xu X, Knaus UG et al. GDP dissociation inhibitor prevents intrinsic and GTPase activating protein-stimulated GTP hydrolysis by the Rac GTP-binding protein. J Biol Chem 1993; 268(2):775-778.

48. Fukumoto Y, Kaibuchi K, Hori Y et al. Molecular cloning and characterization of a novel type of regulatory protein (GDI) for the rho proteins, ras p21-like small GTP-binding proteins. Oncogene 1990; 5(9):1321-1328.

49. Hall A. G proteins and small GTPases: Distant relatives keep in touch. Science 1998; 280(5372):2074-2075.

50. Aspenstrom P. The Rho GTPases have multiple effects on the actin cytoskeleton. Exp Cell Res 1999; 246(1):20-25.

51. Hall A. The cellular functions of small GTP-binding proteins. Science 1990; 249(4969):635-640.

52. Walsh SV, Hopkins AM, Chen J et al. Rho kinase regulates tight junction function and is necessary for tight junction assembly in polarized intestinal epithelia. Gastroenterology 2001; 121(3):566-579.

53. Jou TS, Schneeberger EE, Nelson WJ. Structural and functional regulation of tight junctions by RhoA and Rac1 small GTPases. J Cell Biol 1998; 142(1):101-115.

54. Benais-Pont G, Punn A, Flores-Maldonado C et al. Identification of a tight junction-associated guanine nucleotide exchange factor that activates Rho and regulates paracellular permeability. J Cell Biol 2003; 160(5):729-740.

55. Rojas R, Ruiz WG, Leung SM et al. Cdc42-dependent modulation of tight junctions and membrane protein traffic in polarized Madin-Darby canine kidney cells. Mol Biol Cell 2001; 12(8):2257-2274.

56. Wojciak-Stothard B, Potempa S, Eichholtz T et al. Rho and Rac but not Cdc42 regulate endothelial cell permeability. J Cell Sci 2001; 114(Pt 7):1343-1355.

57. Segain JP, Raingeard de la Bletiere D, Sauzeau V et al. Rho kinase blockade prevents inflammation via nuclear factor kappa B inhibition: Evidence in Crohn's disease and experimental colitis. Gastroenterology 2003; 124(5):1180-1187.

58. Hofman P, Flatau G, Selva E et al. Escherichia coli cytotoxic necrotizing factor 1 effaces microvilli and decreases transmigration of polymorphonuclear leukocytes in intestinal T84 epithelial cell monolayers. Infect Immun 1998; 66(6):2494-2500.

59. Gerhard R, Schmidt G, Hofmann F et al. Activation of Rho GTPases by Escherichia coli cytotoxic necrotizing factor 1 increases intestinal permeability in Caco2 cells. Infect Immun 1998; 66(11):5125-5131.

60. Jou TS, Leung SM, Fung LM et al. Selective alterations in biosynthetic and endocytic protein traffic in Madin-Darby canine kidney epithelial cells expressing mutants of the small GTPase Rac1. Mol Biol Cell 2000; 11(1):287-304.

61. Leung SM, Rojas R, Maples C et al. Modulation of endocytic traffic in polarized Madin-Darby canine kidney cells by the small GTPase RhoA. Mol Biol Cell 1999; 10(12):4369-4384.

62. Oshima T, Laroux FS, Coe LL et al. Interferon-gamma and interleukin-10 reciprocally regulate endothelial junction integrity and barrier function. Microvasc Res 2001; 61(1):130-143.

63. Youakim A, Ahdieh M. Interferon-gamma decreases barrier function in T84 cells by reducing ZO-1 levels and disrupting apical actin. Am J Physiol 1999; 276(5 Pt 1):G1279-1288.

64. Ahdieh M, Vandenbos T, Youakim A. Lung epithelial barrier function and wound healing are decreased by IL-4 and IL-13 and enhanced by IFN-gamma. Am J Physiol Cell Physiol 2001; 281(6):C2029-2038.

65. Duffy HS, John GR, Lee SC et al. Reciprocal regulation of the junctional proteins claudin-1 and connexin43 by interleukin-1beta in primary human fetal astrocytes. J Neurosci 2000; 20(23):RC114.

66. Madara JL, Stafford J. Interferon-gamma directly affects barrier function of cultured intestinal epithelial monolayers. J Clin Invest 1989; 83(2):724-727.

67. Adams RB, Planchon SM, Roche JK. IFN-gamma modulation of epithelial barrier function. Time course, reversibility, and site of cytokine binding. J Immunol 1993; 150(6):2356-2363.

68. Sugi K, Musch MW, Field M et al. Inhibition of Na^+,K^+-ATPase by interferon gamma down-regulates intestinal epithelial transport and barrier function. Gastroenterology 2001; 120(6):1393-1403.

69. Tsukamoto T, Nigam SK. Role of tyrosine phosphorylation in the reassembly of occludin and other tight junction proteins. Am J Physiol 1999; 276(5 Pt 2):F737-750.

70. Bruewer M, Luegering A, Kucharzik T et al. Proinflammatory cytokines disrupt epithelial barrier function by apoptosis-independent mechanisms. J Immunol 2003; 171(11):6164-6172.

71. Itoh M, Furuse M, Morita K et al. Direct binding of three tight junction-associated MAGUKs, ZO-1, ZO-2, and ZO-3, with the COOH termini of claudins. J Cell Biol 1999; 147(6):1351-1363.

72. Wittchen ES, Haskins J, Stevenson BR. Protein interactions at the tight junction. Actin has multiple binding partners, and ZO-1 forms independent complexes with ZO-2 and ZO-3. J Biol Chem 1999; 274(49):35179-35185.

73. Ivanov AI, Nusrat A, Parkos CA. Endocytosis of epithelial apical junctional proteins by a clathrin-mediated pathway into a unique storage compartment. Mol Biol Cell 2004; 15(1):176-188.

74. Coyne CB, Vanhook MK, Gambling TM et al. Regulation of airway tight junctions by proinflammatory cytokines. Mol Biol Cell 2002; 13(9):3218-3234.

75. Fish SM, Proujansky R, Reenstra WW. Synergistic effects of interferon gamma and tumour necrosis factor alpha on T84 cell function. Gut 1999; 45(2):191-198.

76. Planchon SM, Martins CA, Guerrant RL et al. Regulation of intestinal epithelial barrier function by TGF-beta 1. Evidence for its role in abrogating the effect of a T cell cytokine. J Immunol 1994; 153(12):5730-5739.

77. Madsen KL, Lewis SA, Tavernini MM et al. Interleukin 10 prevents cytokine-induced disruption of T84 monolayer barrier integrity and limits chloride secretion. Gastroenterology 1997; 113(1):151-159.

78. Tsukada Y, Nakamura T, Iimura M et al. Cytokine profile in colonic mucosa of ulcerative colitis correlates with disease activity and response to granulocytapheresis. Am J Gastroenterol 2002; 97(11):2820-2828.

79. Irvine EJ, Marshall JK. Increased intestinal permeability precedes the onset of Crohn's disease in a subject with familial risk. Gastroenterology 2000; 119(6):1740-1744.

80. Soderholm JD, Olaison G, Lindberg E et al. Different intestinal permeability patterns in relatives and spouses of patients with Crohn's disease: An inherited defect in mucosal defence? Gut 1999; 44(1):96-100.

81. Kucharzik T, Walsh SV, Chen J et al. Neutrophil transmigration in inflammatory bowel disease is associated with differential expression of epithelial intercellular junction proteins. Am J Pathol 2001; 159(6):2001-2009.

82. Gassler N, Rohr C, Schneider A et al. Inflammatory bowel disease is associated with changes of enterocytic junctions. Am J Physiol Gastrointest Liver Physiol 2001; 281(1):G216-228.

83. Peralta Soler A, Mullin JM, Knudsen KA et al. Tissue remodeling during tumor necrosis factor-induced apoptosis in LLC-PK1 renal epithelial cells. Am J Physiol 1996; 270(5 Pt 2):F869-879.

84. Schmitz H, Fromm M, Bentzel CJ et al. Tumor necrosis factor-alpha (TNFalpha) regulates the epithelial barrier in the human intestinal cell line HT-29/B6. J Cell Sci 1999; 112(Pt 1):137-146.
85. Poritz LS, Garver KI, Tilberg AF et al. Tumor necrosis factor alpha disrupts tight junction assembly(1,2). J Surg Res 2004; 116(1):14-18.
86. Colgan SP, Parkos CA, Matthews JB et al. Interferon-gamma induces a cell surface phenotype switch on T84 intestinal epithelial cells. Am J Physiol 1994; 267(2 Pt 1):C402-410.
87. Petrache I, Verin AD, Crow MT et al. Differential effect of MLC kinase in TNF-alpha-induced endothelial cell apoptosis and barrier dysfunction. Am J Physiol Lung Cell Mol Physiol 2001; 280(6):L1168-1178.
88. Nusrat A, Parkos CA, Bacarra AE et al. Hepatocyte growth factor/scatter factor effects on epithelia. Regulation of intercellular junctions in transformed and nontransformed cell lines, basolateral polarization of c-met receptor in transformed and natural intestinal epithelia, and induction of rapid wound repair in a transformed model epithelium. J Clin Invest 1994; 93(5):2056-2065.
89. Pasdar M, Li Z, Marreli M et al. Inhibition of junction assembly in cultured epithelial cells by hepatocyte growth factor/scatter factor is concomitant with increased stability and altered phosphorylation of the soluble junctional molecules. Cell Growth Differ 1997; 8(4):451-462.
90. Balkovetz DF, Sambandam V. Dynamics of E-cadherin and gamma-catenin complexes during dedifferentiation of polarized MDCK cells. Kidney Int 1999; 56(3):910-921.
91. Howe K, Gauldie J, McKay DM. TGF-beta effects on epithelial ion transport and barrier: Reduced Cl- secretion blocked by a p38 MAPK inhibitor. Am J Physiol Cell Physiol 2002; 283(6):C1667-1674.
92. Di Leo V, Yang PC, Berin MC et al. Factors regulating the effect of IL-4 on intestinal epithelial barrier function. Int Arch Allergy Immunol 2002; 129(3):219-227.
93. Woo PL, Cha HH, Singer KL et al. Antagonistic regulation of tight junction dynamics by glucocorticoids and transforming growth factor-beta in mouse mammary epithelial cells. J Biol Chem 1996; 271(1):404-412.
94. Guan Y, Woo PL, Rubenstein NM et al. Transforming growth factor-alpha abrogates the glucocorticoid stimulation of tight junction formation and reverses the steroid-induced down-regulation of fascin in rat mammary epithelial tumor cells by a Ras-dependent pathway. Exp Cell Res 2002; 273(1):1-11.
95. Rubenstein NM, Guan Y, Woo PL et al. Glucocorticoid down-regulation of RhoA is required for the steroid-induced organization of the junctional complex and tight junction formation in rat mammary epithelial tumor cells. J Biol Chem 2003; 278(12):10353-10360.
96. Van Itallie CM, Balda MS, Anderson JM. Epidermal growth factor induces tyrosine phosphorylation and reorganization of the tight junction protein ZO-1 in A431 cells. J Cell Sci 1995; 108(Pt 4):1735-1742.
97. Bruewer M, Hopkins AM, Hobert ME et al. RhoA, Rac1, and Cdc42 exert distinct effects on epithelial barrier via selective structural and biochemical modulation of junctional proteins and F-actin. Am J Physiol Cell Physiol 2004; 287(2):C327-35.

CHAPTER 11

Regulation of Tight Junctions' Functional Integrity:
Role of a Urinary Factor, Lipids and Ouabain

Liora Shoshani, David Flores-Benítez, Lorenza Gonzalez-Mariscal and Rubén Gerardo Contreras

Abstract

The growing molecular complexity of the tight junction (TJ), its ability to modulate the degree of sealing according to physiological requirements, and the fact that its transepithelial electrical resistance (TER) ranges over several orders of magnitude, indicate that there must be a number of agents modulating its permeability. The interest to find these agents stems from an urgency to make the TJ tighter (e.g., to prevent antigen absorption in autoimmune diseases), or to make it leakier (e.g., to allow the absorption of orally administered pharmaceutical drugs). In the present chapter we discuss three forms of modifying the degree of sealing of the TJ: (1) A peptidic factor extracted from urine that makes the TJ tighter, decreases the cellular content of claudin-2, and prompts the relocalization of claudin-4; (2) An experimentally induced modification of the lipidic composition of the plasma membrane that changes some basic attributes of the TJ; and (3) ouabain, a substance that specifically inhibits Na^+,K^+-ATPase both as an enzyme and as a ion pump, and induces an inotropic activity in heart muscle fibers. Yet we discuss here a newly found property of ouabain: the triggering of a cascade of phosphorylations that results in the opening of the TJ, as part of an overall cell detachment. Interestingly, at concentrations of ouabain within physiological ranges that do not fully detach the cell, the release of the grip induced by this substance sends specific nuclear addressed molecules (NACos) from the diverse sites of attachment to the nucleus, where they modulate the expression of genes that influence proliferation and differentiation.

Introduction

Tight junctions (TJs) are located at the most apical region of the lateral membrane of vertebrate epithelia and endothelia. They were first described as a seal of intercellular spaces.[1] The scope has since drastically changed due to a series of observations. First: TJs contribute to the selective permeability barrier across epithelia.[2] Second: their degree of sealing, usually gauged through the transepithelial electrical resistance (TER),[3] varies over several orders of magnitude in close correlation with the difference in composition between the two fluid compartments separated by the epithelium (see Table 1). Third: the TJs consist of -so far- 40 different protein species, arranged in a complex cluster that links the plasma membrane to the cytoskeleton.[1,4,5] Fourth: some of these molecules have high homology with tumor-suppressor proteins, carry nuclear localization signals and PDZ sequences for the assembly of protein scaffolds at different sub-cellular locations, and their interactions are accompanied by specific

Tight Junctions, edited by Lorenza Gonzalez-Mariscal. ©2006 Landes Bioscience and Springer Science+Business Media.

Table 1. Transepithelial electrical resistance (TER) of different mammalian epithelia

Tissue	Species	TER ($\Omega \cdot cm^2$)	Refs.
Proximal renal tubule	Dog, rabbit	6-7	130,131
Descending thin limbs of Henle's loop	Rabbit	700	132
Distal convoluted tubule	Rat	350	31
Cortical collecting tubules	Rabbit	867	33
Inner medullary collecting duct	Hamster	1000	34
Gallbladder	Rabbit	20	133
Small intestine	Rat (duodenum)	98	134
	Rat (jejunum)	51	135
	Rat (ileum)	88	134
	Rabbit (ileum)	100	
Colon	Rabbit	286	136
	Rabbit	385	137
Urinary bladder	Rabbit	5000-10000	35

changes of phosphorylation. These data suggests that, besides of acting as a diffusion barrier, the TJ participates in many other important biological functions (for recent reviews see refs. 6,7). Fifth: there is a relatively new group of substances named NACos, pertaining to the diverse attaching structures (TJs, adherent junctions, desmosomes, etc.) that respond to a lowering of the degree of attachment with a progressive shuttling to the nucleus (see Chapter 7). Their importance stems from the fact that, once in the nucleus, NACos modulate the expression of genes involved in proliferation and differentiation.[6,7]

Obviously, a structure with a molecular machinery so complex, and adapting to so many physiological situations, must be regulated by a number of natural signals. It must be also involved in a multitude of diseases caused by the failure of any of its multiple components. This explains and justifies the present effort to find natural signals and pharmacological factors and procedures to control the hermeticity of the TJ. The present chapter summarizes recent studies performed in our laboratory to analyze three regulatory factors that modulate TJ permeability: urinary factors, lipids and ouabain.

Regulatory Factors of the Tight Junctions

Given the complexity of the TJ structure it is not surprising to find that its function is susceptible to be modified by a broad range of extra- and intra-cellular physiological substances (see Table 2). Thus, the degree of sealing is modulated in response to physiological requirements and pharmacological challenge, through modifications in the molecular composition of the TJ. Some of the reported extracellular factors that modify the function of the TJs include: hormones such as the parathyroid hormone-related peptide[8] and progastrin;[9] cytokines_like interferon-γ and tumor necrosis factor[10-12] and growth factors as hepatocyte growth factor (HGF), TGF-β2 and TGF-β3,[13] vascular endothelial growth factor (VEGF)[14] and epidermal growth factor (EGF).[15]

There are also numerous reports in which the regulating factor is not completely identified. Thus, Marmorstein et al[16] found an epithelial permeability factor (EPF) in human serum that lowers the TER and opens the TJs of kidney epithelial cells. EPF is a protein that forms multimers that interact with the basolateral surface of the epithelium and causes constriction of the cytoskeleton, and an increase in permeability at specific sites along the TJ. Jaeger et al[17] reported evidence of a factor secreted at the apical side of epithelial monolayers, which modulates TJ structure and permeability. While the factors reported by Marmorstein et al and Jaeger et al, are of proteic nature, Gorodeski and Goldfarb[18] studied a lipidic factor in seminal fluid that increases TER acting through a Protein Kinase C (PKC) dependent mechanism.

Table 2. Factors that modify the functional integrity of tight junctions

Factor	Tissue	Effects	Refs.
		Factors that increase TER	
Conditioned epithelial media	A6, Caco-2, MDCK	Decreases TJs' depth	17
Conditioned neural retinal media	Chicken RPE	Increases continuity of TJs strands observed by freeze fracture	138
Urine extract (DLU)	MDCK	Decreases claudin-2 expression. Promotes lateral membrane localization of claudin-4	67, *
EGF	MDCK	Decreases claudin-2 expression. Increases claudin-1, -3 and -4 expression and promotes their localization to the lateral membrane	15
HGF	Human pulmonary artery endothelium	Increases cortical F-actin and β-catenin at cell-cell contacts	139
Lipid of Seminal fluid	CaSki		18
TGF-β1	T84	Inhibits interleukins effects on TJs	140,141
		Factors that decrease TER	
Complement factor C5a	MDCK		142
HGF	Bovine RPE	Induces loss of fence function, decreases ZO-1 and occludin content. De-localizes TJs' proteins	143
IL-1β	ARPE-19	Upregulates claudin-1 and downregulates occludin genes	144
IFN-γ, TNF-α	Mouse cholangiocytes	Increases ^{3}H-inulin transepithelila flux when applied to the basolateral domain	145
Serum factor of 67 kDa,	MDCK, PBCEC	Condensates cortical actin ring. In PBCEC redistributes ZO-1, occludin and claudin-5 from the cell periphery to the perinuclear cytoplasm.	16,23,146
VEGF	BMEC	Increases sucrose permeability, decreases occludin and ZO-1 TJs content	20
PDGF	MDCK	Increases paracellular dextran flux, redistributes occludin, ZO-1, claudin-1 and E-cadherin	147

* Contreras and Flores-Benitez, unpublished results

Several intracellular factors have been reported that regulate TJ's functional integrity. Most of them are members of known signaling pathways that regulate either the cytoskeleton dynamics[19] or the expression of proteins involved in the junctional complex.[20] It seems that, the final targets of these modifications are proteins that constitute the TJs. Hence, it has been shown that changes in TJs permeability are usually accompanied by variations in the expression and/or cellular distribution of occludin,[21] claudins,[22-24] ZO-1[25,26] and JAM.[27]

Urinary Factors

The mammalian kidney clears the whole volume of plasma several times a day and eliminates waste products in the urine. In an adult man, approximately 170 liters of glomerular filtrate produced each day are extensively modified as they proceed along the nephron and reduced to a mere 1-2 liters of urine. The tubular structural/functional adaptation necessary to perform this function is the progressive tightening of the TJs of the segments that the fluid content meets in its way to the urinary bladder. This tightening is accounted for by the permeability reduction of the paracellular route.[28,29] Thus, the TER measured in different mammalian nephrons (see Table 1), increases from a mere 5-8 $\Omega.cm^2$ at the level of the proximal tubule,[30] to 150-600 $\Omega.cm^2$ at the distal one,[31,32] 860-2,000 at the collecting duct[33,34] and a further increase to 10,000 or even 135,000 $\Omega.cm^2$ at the level of the urinary bladder.[35-37] This functional changes on TJs permeability is accompanied by an increase in the number of junctional strands, as observed in freeze fracture replicas[38] as well as by the differential expression of specific TJ's proteins. While these structural and physiological modifications of the TJs makes sense from a teleologic point of view, as the wall of the tube withstands a progressively steeper gradient between the tubular content and the surrounding interstice, we still ignore the factors and mechanisms responsible for these adaptations.

Differential Distribution of TJ Proteins among the Various Segments of the Mammalian Nephron

Studies with frozen sections and in manually micro-dissected tubules of the kidney have shown that the integral TJ protein occludin displays a clear cell border pattern in the distal segments of the nephron, whereas in the proximal ones only discrete punctua are detected at the cellular boundaries. The adaptor TJ proteins ZO-1 and ZO-2 are instead present at the TJ region of leaky as well as tight segments. However the amount of such proteins present at the distal portions is significantly higher than in the proximal segment.[39]

As has been stated in Chapter 2, the expression of occludin is closely related to the degree of junctional sealing.[40] Therefore it is not surprising that this protein is conspicuously present at the junctions of the tighter segments of the nephron. However, the role of occludin remains somewhat obscure, due to the fact that many epithelia seem unaffected in occludin knock out mice.[41] Claudins integrate a family of at least 20 homologous proteins, in which each member exhibits a tissue specific distribution. In the mammalian nephron, claudins that decrease paracellular permeability for cations tend to concentrate at the tighter renal segments, while the opposite is true for those that function as cationic pores (Fig. 1). Thus, the proximal tubule of rabbit and mouse, conspicuously expresses claudin-2.[42-44] This claudin markedly decreases TER in epithelial cultured cells[45] by selectively increasing the permeability to cations[46] suggesting that claudin-2 is a paracellular cation channel. In contrast, claudin-8, which is characterized for being impermeable to Na^+ and other cations,[47] is expressed in Henle's loop, distal and collecting renal tubules.[44,48] In the distal segment, the paracellular barrier is specially tight to monovalent cations, resisting transtubular gradients up to 1000:1 for H^+,[49] 20:1 for K^+[50] and 1:30 for Na^+.[51] Therefore the expression of claudin-8 and the absence of claudin-2 in the distal segment is in agreement with the paracellular permeability properties of this tight renal segment. Furthermore, the induction of claudin-8 expression in a renal cell line (MDCK), is associated with the down regulation of endogenous claudin-2.[47] Since in this experiment, the number and complexity of TJ strands was not altered, it has been proposed that claudin-8 replaces endogenous claudin-2, inserting into already existing TJ strands.

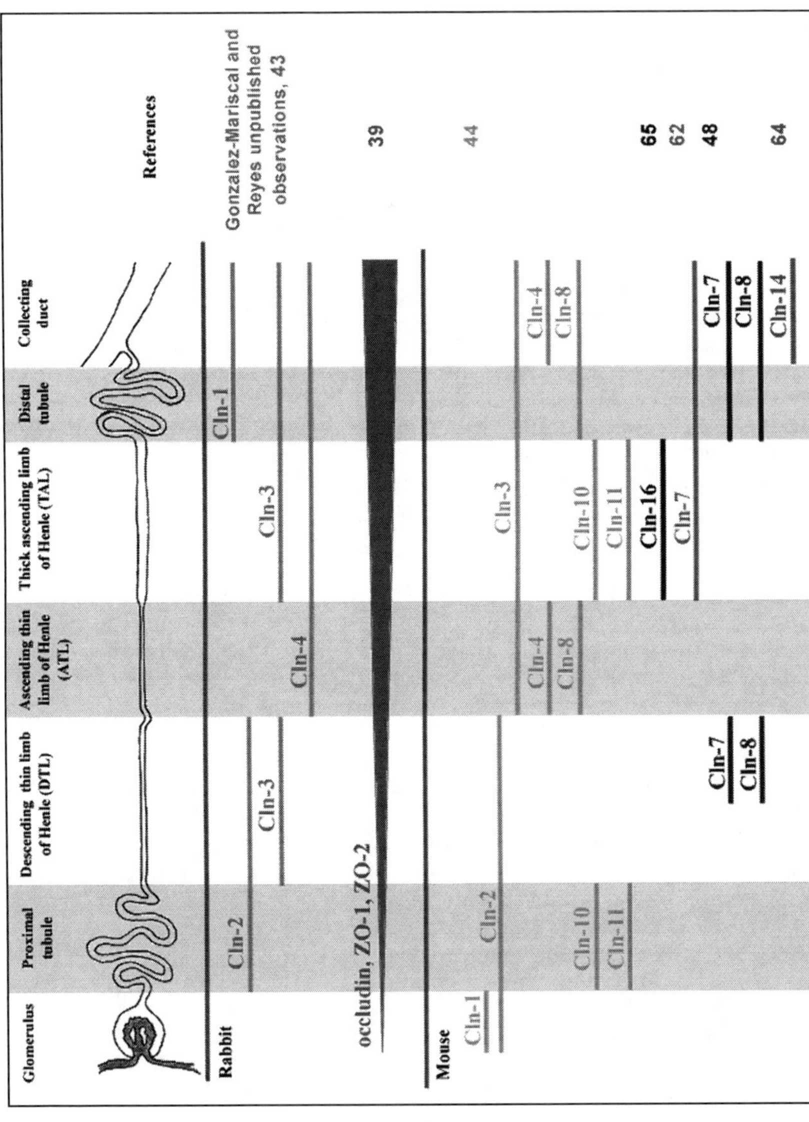

Figure 1. Tight junction's proteins distribution along the rabbit and mouse nephron. The upper part is a schematic representation of different segments of the nephron. The purple bar in crescendo represent the distribution of occludin, ZO-1 and ZO-2 proteins in rabbit nephron.[39] The bars represent the distribution of distinct claudins along the nephron as reported by several authors.

Claudins-3 and -4 are expressed at the TJ region of relatively tighter renal segments from rabbit since the newborn stage. Nevertheless, certain differences in the distribution of claudins-3 and –4 exist between mouse and rabbit. For example, claudin-3 is present in the descending thin limb (DTL) and thick ascending limb (TAL) of Henle, but is absent in Henle's ascending thin limb (ATL) of rabbit (Gonzalez-Mariscal and Reyes, unpublished observations). While in mouse claudin-3 is not found at the DTL, it is present at the ATL segment. Claudin-4 is present in the distal tubule of rabbit (Gonzalez-Mariscal and Reyes, unpublished observations) but not of mouse[44] (Fig. 1). These results might reflect differences that exist between these mammals, but could be extensive to other species as well. As expected for its location, claudin-4 reduces paracellular monovalent cation permeability in cultured epithelial monolayers. This effect can be abolished by altering the net charge of the first extracellular loop,[52,53] thus reinforcing the hypothesis that the charged residues present in this loop are responsible for the paracellular ionic selectivity exerted by claudins.

WNK4 is a protein kinase homologous expressed on TJs of the distal nefhron.[54] Pathogenic mutation of WNK4 causes a rare human hypertension linked to hiperkalemia called pseudohypoaldosteronism type II (PHAII). The mutation induces the kinase activity of WNK4 and a subsequent increase in paracellular Cl- permeability. Mutant WNK is able to bind and phosphorylate claudin-1 and 4 in cell cultures, indicating that claudins are important molecular targets of WNK4 kinase, and that claudin phosphorylation may be the pathogenic mechanism involved in PHAII.[55]

Claudin-11, like claudin-4, has a positively charged residue on the second half of its first extracellular loop, responsible for the discrimination against cations exerted by this claudin upon transfection into epithelial cells.[56] In the kidney, claudin-11 like claudin-4 is expressed in TAL, however, it is also present in the leakier nephron segment: the proximal tubule. This was unexpected, since the proximal tubule is enriched with claudin-2, the paracellular cationic channel. On the other hand, freeze-fracture studies of the proximal tubule have revealed TJ composed of one or two strands arranged in a parallel manner and mostly discontinuous with intervening spaces of various widths.[38,57,57-59] It is notorious, that a similar pattern is obtained upon transfection of L-fibroblasts with claudin-11.[60] Furthermore, it has been proposed that because of branching, TJ networks are capable of generating high TER with only a few strands.[61] Therefore the presence of claudin-11 at the proximal tubule might contribute to the high permeability of this segment by generating a pattern of unbranched TJ strands.

Claudin-7 is expressed in the same tight nephron segments as claudin-8 (distal and collecting tubules),[62] although minor differences have arisen regarding its distribution along Henle's loop (Fig 1). Upon transfection into MDCK cells, claudin-7 generates a decrease of the paracellular conductance and mannitol flux. Interestingly, over-expression of claudin-7 is accompanied by an up-regulation of claudin-4, another protein that reduces paracellular permeability. In kidney, claudin-7 in contrast to claudin-8, localizes primarily to the basolateral membrane. The significance of such distribution, also encountered in rat uterus (Gonzalez-Mariscal and Mendoza, unpublished observations), remains unknown, and might be speculatively related to an additional role of certain claudins in cell-substrate attachment.

Claudin-14, initially described for its expression in the sensory epithelium of the organ of Corti in the ear,[63] is present in the kidney at the collecting tubules. This distribution coincides with the fact that in MDCK cells, claudin-14 behaves as a cation restrictive barrier.[64]

Claudin-16, also called paracellin-1, is specifically expressed in the thick ascending limb of Henle's loop where it mediates the specific paracellular conductance of magnesium and calcium.[65] Mutations that interfere with normal claudin-16 expression produce an autosomal recessive disease characterized by hypomagnesemia with hypercalciuria and nephrocalcinosis.[65] Mutations that inactivate the carboxyl terminal PDZ binding motif of claudin-16 disable its association with ZO-1 and lead to its lysosomal mistargeting. Under this condition the patients display a serious but self-limiting hypercalciuria.[66]

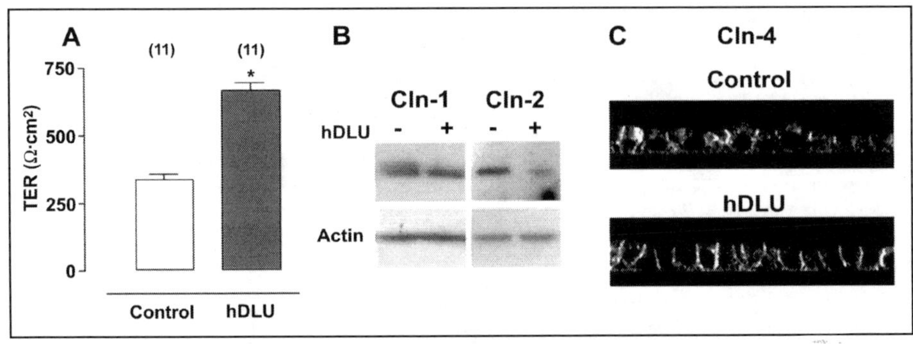

Figure 2. Human urine contains at least one factor that increases the degree of sealing of Tight Junctions (TJs) by changing claudin composition and localization. A) Transepithelial Electrical Resistance (TER) of MDCK cells plated on nitrocelluloce filters after 24 h of incubations with an human urine extensively dialyzed, lyophilized and reconstituted in complete culture medium (hDLU, 10 µg of protein/ml). B) Immunoblot of MDCK cell extracts treated as indicated above. C) Fluorescence confocal lateral views of MDCK cells incubated with hDLU and stained with antibodies against claudin-4.

Urinary Factors Capable to Modulate TJ Sealing

Gallardo et al[67] have found a peptidic factor in dialyzed and lyophilized urine (DLU) of several mammalian species (Fig. 2A), that enhances the TER of monolayers of MDCK cells in a dose dependent manner, and may therefore provide a clue to the gradually progressive sealing of TJs along the mammalian nephron. Because the volume of intratubular fluid is drastically reduced and concentrated in its way along the nephron, this factor would make the TJs of distal segments more hermetic. It consists of a thermo-resistant peptide of 30-50 kDa that bears negative charge. This factor that, as mentioned above is detectable in the urine from different animal species, including human, is able to increase the TER of monolayers prepared with established cell lines derived from different animal species. The factor seems to be continuously required, as its effect is reversed upon removal.

Although the intrinsic mechanism of action of the urinary factor is yet under study, it has been found that the freeze fracture pattern of TJ as well as the distribution of ZO-1 are not modified by DLU.[67] Nevertheless, DLU induces changes in the composition and localization of claudins. Figure 2 shows that MDCK cells, often taken as a culture model system for the distal nephron segments,[68] decrease their cellular content of claudin-2 (Fig. 2B) and localize claudin-4 in the lateral membrane domain (Fig. 2C) after incubation in media containing DLU. Of course the extract could also act by changing other aspects of TJ, such as the phosphorylation state of some junctional proteins.

Jeager et al[17] have found that MDCK cells secrete a substance that enhances TER when applied to a second MDCK monolayer. This observation supports the idea that the substance found in DLU may in fact originate from kidney tubules. Epithelial Growth Factor (EGF) is abundantly expressed in urine[69] and in MDCK cells it binds to its receptor (EGFR) and increases TER through a mechanism resembling the one used by hDLU. Thus, it inhibits the expression and junctional localization of claudin-2 and increases claudin-1, 3, and 4 expression and cellular redistribution to the TJs.[15]

Lipids

The Lipid Model versus the Protein Model of TJ

TJs appear in freeze-fracture replicas as continuous anastomosing strands. These strands have been proposed to be composed of either proteins[70] or lipids.[71,72] The "protein model" depicts the strands as polymers of protein molecules in the plane of the plasma membrane that join, through their external domains, to the corresponding polymer of the neighboring cell. This protein model is strongly supported by the demonstration that the transmembrane proteins occludin[73] and claudins[74] are main constituents of the strands. While fibroblasts transfected with claudins form abundant strands resembling those of epithelial TJs,[75,76] those transfected with occludin express a scarce number of small strands.[75] Co-transfection of both proteins induces the formation of abundant long strands where claudin and occludin copolymerize.[75] However, transfected fibroblasts never form a ring of strands circumventing the cell, indicating that transmembrane proteins are not sufficient for the formation of complete anastomosing strands and that some other factor, perhaps the plaque formed by intracellular TJ's proteins and the cytosqueleton, are required.

The "lipid model" depicts the strands as cylindrical inverted micelles where the polar groups of lipids are directed toward the axis, and the hydrophobic tails are immersed in the lipid matrix of the plasma membrane of both neighboring cells.[71] The lipid model is in keeping with the fact that the lipid-soluble probe dipicrylamine diffuse, with the aid of an applied voltage, from the cell membrane of a previously loaded cell to the plasma membrane of the neighboring one.[77] It is also consistent with photobleaching experiments of large areas of the cell membrane of MDCK cells, that were previously loaded with the lipid fluorescent probe, to measure then the fluorescence recovery or the ability of the probe to diffuse to neighboring cells. Grebenkamper and Galla[78] found with this system that the probes diffuse in the plane of the membrane of the same cell, as well as towards neighboring ones, provided temperature is kept above the melting point of the hydrophobic chains. This result indicates that the lipid probe is able to diffuse through the TJ from one cell to the neighbor, as expected for a TJ of lipidic nature.[78] In agreement with these evidences, phospholipids at the TJ are cytochemically detected with the aid of PLA2 complexed with gold.[79]

However, several classical studies have failed to support or discard the lipid model. Dragsten et al[80,81] studied the diffusion of fluorescent lipid probes added to the outer leaflet of the apical membrane of MDCK cells. These probes were unable to diffuse into the basolateral membrane, presumably due to the barrier formed by the polymerized proteins of the TJ. Only lipid probes with the ability to flip-flop from the outer into the cytoplasmic leaflet were able to circumvent the tight junction.[81,82] Although the results appear inconsistent with the lipid model, it was argued that the first type of lipid probes do not diffuse across strands made of lipids, for the same reason that they are not able to flip-flop between the two leaflets of the membrane, i.e., their hydrophilic heads are too large.[78] In another classical study, van Meer et al[83] co-cultured at confluence different clones of MDCK cells, one of them expressing the glycosphingolipid Forssman antigen (GFA). It was expected that if junction strands were composed of lipids, the outer leaflet of neighboring cells would constitute a sort of lipidic continuum, and co-culture should allow diffusion of GFA from a cell that expresses this molecule towards the neighbors that do not. However, this was not the case. Although this result appears inconsistent with the lipid model, it may not disprove the model because, as pointed out by Grebenkamper and Galla,[78] GFA may be too large to diffuse across the cylindrical lipid micelle. Taken together, the available evidences to support the lipid model are rather conflictive.

Lipids as Membrane Components That Regulates the TJs

Lipid and protein models seem to be the extreme idealizations of an ample range of possibilities. Therefore, strands might not be made exclusively of proteins or lipids; on the other hand, lipids may participate indirectly as a source of second messengers that may in turn regulate TJs' integrity. In agreement with both possibilities, it was observed that changes in the total composition of phospholipids, sphingolipids, cholesterol and fatty acids, do not alter either TER or the structure of the strands. Nevertheless, enrichment with linoleic acid does increase the paracellular flux of dextran without modifying either the TER or the TJs' strands.[84] This apparently contradictory effect of linoleic acid on TER and permeability is similar to the observed by Balda et al[40] in MDCK cells transfected with chicken occludin tagged with an epitope. In this cells, TER and paracellular permeability to dextran increase, demonstrating that, contrary to the prospect, electrical resistance and junctional permeability can be dissociated. The mechanism of action of linoleic acid on TJs' permeability remains unclear, but a possible explanation may be that alterations in the viscosity of the lipidic component may modify the functional properties of the paracellular channels.

Leung et al[85] have investigated the participation of glycosphingolipids (GSLs) in the TJ of cell strains with different degree of sealing. MDCK I and Fisher rat thyroid cells spontaneously express high TER in normal culture conditions, yet when the synthesis of GSLs is inhibited by the fungal metabolites PPMP, ISP1 or FB1, TER markedly decreases to values similar to the low TER, MDCK type II cells. Interestingly, this was not accompanied by structural changes in TJs.[85] ISP1 and FB1 act at early steps in the synthesis of GSLs, hence inhibit the production of ceramide, a GSLs precursor known to act as a lipidic second messenger,[86] whereas PPMP functions at a later stage therefore causing ceramide to accumulate. These results suggest that GSLs and not ceramide regulate the TER due to changes in the lipidic environment of the TJs.

In some cases, lipid modifications change TJ structure. This is illustrated by the increased TER, accompanied by changes in detergent solubility of occludin and ZO-1, when alkylphospholipids are incorporated to the plasma membrane.[87]

Lipids as Sources of Second Messengers That Regulate TJs' Integrity

On the other hand, there are several evidences that support the role of lipids, as sources of second messengers that may in turn regulate TJs' integrity. In this regard, diacylglycerol plays a role in TJs' formation[88,89] and increments TER through a mechanism that involves protein kinase C.[18,90] Interestingly, the lipid phosphatase PTEN (phosphatase and tensin homologue), originally described as tumor suppressor, is associated with the TJ,[91,92] suggesting the involvement of 3-phosphoinositide lipid second messengers in tight junction regulation. Moreover, MDCK cells depleted of cholesterol by incubation with methyl beta-cyclodextrin, an agent that promotes cholesterol efflux from the plasma membrane, accelerate Ca^{2+}-induced TJ formation[93] and transiently increase the TER of mature monolayers,[94] both phenomena attributable to alterations of signaling pathways by lipid second messengers. Although TJs' strands are mainly formed by transmembrane proteins, TJ comprises also a lipidic compartment enriched in GSLs that is able, together with the lipids of the plasma membrane, to activate signaling pathways with strong functional consequences and/or structural changes of TJ.

The TJs Are Lipidic Microdomains

Membrane fractions resistant to cold solubilization with detergents and enriched in cholesterol, GSLs and proteins like caveolin-1, are known as lipid "rafts" or triton-insoluble floating fractions (TIFF). Nusrat et al[95] have shown that hyper-phosphorylated TJs' proteins, occludin and ZO-1, co-precipitate from TIFF and partially co-localize with caveolin-1, indicating that TJs' proteins are included in a raft-like lipidic compartment. Occludin and ZO-1 are displaced from these rafts after TJ disassembly by calcium chelating, leading to the conclusion that TJs are specialized lipid microdomains. The TJ microdomain also includes the exocyst, a site of active vesicle fusion.[96,97] The exocyst seems to interact with other raft like compartments located in its vicinity. Interestingly, apical GPI-anchored proteins are delivered from the trans-Golgi network to a place near to TJs at the basolateral membrane, in which they are transiently retained. Subsequently, the GPI-anchored proteins are rapidly internalized by a non-clathrin dependent pathway to be finally transcytosed to the apical plasma membrane.[98]

Lipidic modifications of TJs' may play a role in some pathological situations. Patients with Chrons' disease express high levels of PLA2 at intestinal epithelia. PLA2 modifies lipid composition in TJs and decreases the degree of sealing. Under conditions where intestinal TJs' seal is compromised, bacterial translocation through weakened TJs may be favored. That is the case in epithelial cells culture models, in which TJs weakness, induced by excessive PLA2 treatment, results in increased bacterial translocation through the paracellular route.[99] Alkylphospholipids, that modify TJs' lipid composition and function, are anti-tumor lipidic drugs. They modulate cancer cell invasion through changes in TJ and adherens junctions.[87] Nevertheless, there is still a lot of research to do for better understanding the role of lipids as TJs regulators.

Ouabain

Ouabain is a steroid of the digitalis family identified in plant-extracts, that has been used for centuries to treat congestive heart failure because of its inotropic effect. Ouabain specifically inhibits the ubiquitous Na^+,K^+-ATPase, producing an increase in cytosolic Na^+ and, through the involvement of the Na^+/Ca^{2+} exchanger, causes a rise in the intracellular concentration of Ca^{2+} ($[Ca^{+2}]_i$) which in turn results in the positive inotropic reaction of the cardiac muscle.[100,101] Recent works suggest that the action of ouabain on the sarcoplasmic reticulum membrane system,[102,103] or on a membrane receptor different from the Na-pump,[104] may also play a significant role. The high specificity and affinity of ouabain for the Na^+,K^+-ATPase led to suspect that it would mimic an endogenous molecule. In keeping with this expectation, a ouabain-like compound was purified from human plasma, adrenals and hypothalamus.[105,106] Immunoassays indicate that ouabain concentration in plasma is around 1 nM,[107,108] although elevated concentrations of endogenous ouabain have been found in numerous physiological and pathological conditions such as physical exercise, hypertension,[109] chronic renal failure, hyper-aldosteronism, congestive heart failure, and preeclampsia (For review see[100]). A plasma concentration as high as 0.7 μM was measured in a patient with an adrenal tumor.[108] Ouabain release to the bloodstream is regulated by the sympathetic nervous system.[110]

The discovery of endogenous ouabain(s) prompted an effort to find its physiological roles. In this respect, we have found a mechanism, termed P→A, that transduces the occupancy of the Na^+,K^+-ATPase (P) by ouabain, into an activation of intracellular signaling pathways that lead to cell detachment by altering molecules involved in cell-cell and cell-substrate attachment (A) (Fig. 3).[111] The reason to include a brief description of P→A in this chapter, stems from the fact that one of the structures that is early and deeply affected by this mechanism is the TJ. In other words, the hormone ouabain, or some other molecule closely involved in its physiological synthesis and degradation, appears to play a paramount role in the control of the TJ.

The P→A Mechanism

To grasp the intrinsic mechanism of P→A we shall now make a digression to describe succinctly the interrelationship between TJs, adherent junctions and signaling. Recent works from different laboratories[20,112,113] have shown that MAPK pathway modulates the barrier function of tight junctions. Thus, in Ras-transformed MDCK cells, occludin, claudin-1, and ZO-1 are absent from cell-cell contacts. Treatment with PD98059, which blocks the activation of MAPK, recruites these proteins to the cell membrane. Moreover, inhibition of MAPK activity stabilizes occludin and ZO-1 by differentially increasing their half-lives. Electron microscopy and TER analysis, show formation of functional tight junctions after MEK1 inhibition.[112] During this process tyrosine phosphorylation of occludin and ZO-1 increases significantly. With regards to claudin-1, while some have observed no change in its tyrosine phosphorylation, others have identified a putative phosphorylation site for MAP kinase in Thr-203, that is required to promote the barrier function of TJs.[113] Interestingly, in human corneal epithelial cells studied by Wang et al,[20] activation of ERK1/2 MAP kinase pathway also induces TJ disruption. Nevertheless, Macek et al[114] found that in human breast tumor cells, the Ras-MEK1 pathway does not affect the expression and function of TJ-associated molecules.

In MDCK cells, ouabain increases the amount of both active MAPK and p190$^{Rho-GAP}$. The latter is a Rho GTPase activating protein. Ouabain also causes redistribution and degradation of the TJ components occludin and ZO-1, and of other cell-cell junctions proteins (desmoplakin, cytokeratin, β-actinin, vinculin and actin). Genistein, an inhibitor of protein tyrosine kinases, and UO126 that inhibits the kinase of ERK1/2 (MEK) block the cell detachment produced by

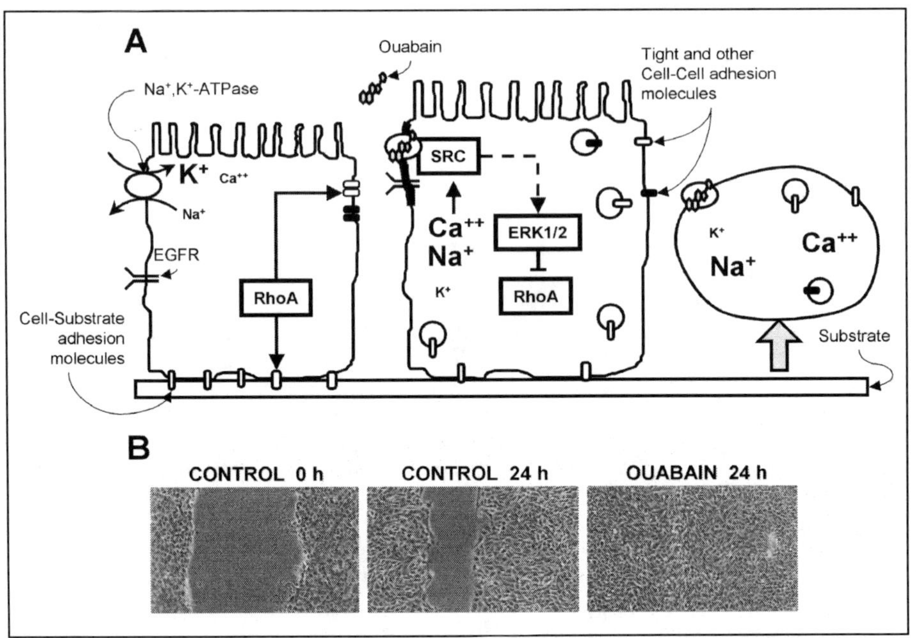

Figure 3. Ouabain induced detachment may participate in cell migration. A, left) An epithelial cell is attached through cell-cell and cell-substrate adhesion molecules and RhoA helps to maintain these molecules functional. Center) Ouabain binding to the Na⁺,K⁺-ATPase inhibitis the transport of K⁺ and Na⁺, produces ion imbalance, cell swelling and the translocation of the enzyme to a caveolae like domain, where it associates with c-src. C-src is then able to transactivate the Epithelial Growth Factor Receptor (EGFR) and, consequently, the ERK1/2 pathway. Activated ERKs inhibit RhoA provoking the internalization (small circles) and degradation of TJ's and other adhesion proteins. Right) The cells detach from each other and from the substrate. B) Ouabain facilitates wound healing. Mature and confluent MDCK monolayers were scratched at cero time (CONTROL 0h). Wound healing was observed 24 h latter in control conditions (CONTROL 24h) or in the presence of 10 nM ouabain (OUABAIN 24h). In this condition ouabain seems to increase cell migration and accelerate the wound healing.

ouabain. The content of p190$^{Rho-GAP}$ is increased by ouabain, suggesting that both the ERK1/2 and Rho pathways are involved.

Another group of observations that may shed light on the possible role of Na⁺,K⁺-ATPase in regulating TJ assembly comes from the group of Rajasekaran,[115-117] which found that in MDCK and human RPE cells, inhibition of Na⁺,K⁺-ATPase either by ouabain or by K⁺ depletion, prevents the formation of TJs and desmosomes and maintains the cells in a non-polarized state. Their findings also confirm a decrease in RhoA activity, that correlates with a failure to form actin stress fibers, previously demonstrated by Contreras et al.[111]

A role for GTPases in TJ junction assembly, was first suggested by Balda et al,[88] using a Ca²⁺ switch protocol.[118] Their observation was followed by numerous works linking both the regulation of junction assembly and the regulation of selective paracellular permeability to the Rho-family GTPases.[119-121] Interestingly, one of the main signaling pathways that regulate TJs is the small GTPase Rho pathway. This Rho-dependent pathway involves Rho effectors such as Rho kinases as well as Rho-specific guanine nucleotide exchange factors (GEFs).[122] Changes in paracellular permeability induced by Rho, might also involve modulation of myosin activity[123,124] as well as a modulation of actin cytoskeleton (for review see ref. 7). In summary, ouabain inhibition of the pump might regulate TJs assembly and function, either directly by a yet unknown mechanism, or indirectly via modulation of the cytoskeleton induced by decreased RhoA activity. This implies that the Na^+,K^+-ATPase would function not just as an ion-pump, but as a signal-transducer as well. As recently reviewed by Xie[125,126] binding of ouabain to Na^+,K^+-ATPase changes the interaction of the enzyme with neighboring membrane proteins, and induces the formation of multiple signaling modules, resulting in activation of Src, transactivation of the EGF receptor (EGFR), and increased production of reactive oxygen species (ROS). Interactions between these signaling pathways lead to activation of several other cascades, including ERK1/2 and p38 MAPKs, phospholipase C, and protein kinase C isozymes, in a cell-specific manner. Ouabain also increases $[Ca^{2+}]_i$ and contractility, induces some early-response protooncogenes, and activates transcription factors AP-1 and NF-kappaB. Interestingly, these two transcription factors are also involved in the regulation of TJs functionality.[11,12] Taken together, ouabain activation of growth-related genes may be part of the interplay between TJs' components and the processes of proliferation and differentiation discussed below (for review see refs. 6,7).

The Possible Physiological Role of P→A Mechanism

We may now speculate on the possible overall physiological role of P→A mechanism. Since junctions maintain the architecture of tissues by holding together neighboring cells, they would need to relax the grip whenever a change in the position or shape of the cells is required to heal a wound, or during organogenesis where different cell types adopt the architecture of a given tissue.[127,128] Furthermore, P→A mechanism may speculatively play a role during the development of metastasis, as cells must dislodge themselves from the mass of the main tumor, and home among cells of distant organs. Since ouabain is a hormone normally present in the blood,[100] it is conceivable that endogenous ouabain-like substances may also promote detachment and removal of damaged epithelial cells. The recently described union of ouabain to a receptor different from Na^+,K^+-ATPase, whose function is not yet elucidated,[104] could even allow the P→A mechanism to be triggered by a receptor other than the α-subunit of the Na^+,K^+-ATPase.[129]

The complete detachment of a cell promoted by the P→A mechanism may be an extreme situation resulting from the relatively high concentrations of ouabain (1-10 μM) used in physiological studies to characterize this mechanism. Yet we have recently found that this mechanism may play a subtler physiological role.[129] In this regard, there is a group of proteins (NACos for Nuclear and Adhesion Complex) pertaining to some of the several types of cell-cell and cell-substrate junctions, that shuttle to the nucleus whenever the adhesion of these sites is decreased.[6] One of the best known NACos is β-catenin, a protein belonging to adherens junctions, that also participates in the Wnt signaling pathway. We have found that ouabain causes β-catenin to travel from adherens junctions to the nucleus and accumulate in discrete spots.[129] This effect seems to be a specific one, as another type of NACos, e.g., ZO-1, is not displaced to the nucleus under the effect of ouabain. Remarkably, the effect on β-catenin is observed with ouabain concentrations as low as 300 nM, that does not cause complete detachment of the cells, but whose effect on cell-cell adhesion is revealed by a significant decrease in TER (c.a. 60%). Therefore NACos shuttling induced by ouabain may in principle activate nuclear factors involved in the functional regulation of the TJs. This subtle release and signaling to the nucleus may also play a role in other processes such as wound healing, illustrated in figure 1B, where a confluent monolayer of MDCK cells was scratched (CONTROL 0h) and allowed to recover 24 hours under control conditions (CONTROL24), or in the presence of 10 nM ouabain. In this condition ouabain seems to increase cell migration and accelerate wound healing.

Concluding Remarks

In the present chapter we focused on three distinct elements that seem to play a major role in the assembly and regulation of the TJ: a mammalian factor extracted from urine, lipids and ouabain. The urinary factor may account for the progressively higher hermeticity of the TJs along the renal tubule. Lipids play a double role as components and regulators of junctional function. Finally ouabain, an hormone known to specifically inhibit the Na⁺,K⁺-ATPase, disassembles the TJ and detaches epithelial cells from the substrate through a complex signaling pathway that sends information to intracellular structures including the nucleus.

References

1. Cereijido M, Shoshani L, Contreras RG. Molecular physiology and pathophysiology of tight junctions. I. Biogenesis of tight junctions and epithelial polarity. Am J Physiol Gastrointest Liver Physiol 2000; 279:G477-G482.
2. Cereijido M, Anderson JM. Tight Junctions. 2nd ed. Boca Raton FL: CRC Press; 2001.
3. Cereijido M, Robbins ES, Dolan WJ et al. Polarized monolayers formed by epithelial cells on a permeable and translucent support. J Cell Biol 1978; 77:853-880.
4. Cereijido M, Shoshani L, Contreras RG. Cell adhesion, polarity, and epithelia in the dawn of metazoan. Phy Rev 2004; 84, in press.
5. Gonzalez-Mariscal L, Betanzos A, Nava P et al. Tight junction proteins. Prog Biophys Mol Biol 2003; 81:1-44.
6. Balda MS, Matter K. Epithelial cell adhesion and the regulation of gene expression. Trends Cell Biol 2003; 13:310-318.
7. Matter K, Balda MS. Signalling to and from tight junctions. Nat Rev Mol Cell Biol 2003; 4:225-236.
8. Stelwagen K, Callaghan MR. Regulation of mammary tight junctions through parathyroid hormone-related peptide-induced activation of apical calcium channels. J Endocrinol 2003; 178:257-264.
9. Hollande F, Lee DJ, Choquet A et al. Adherens junctions and tight junctions are regulated via different pathways by progastrin in epithelial cells. J Cell Sci 2003; 116:1187-1197.
10. Blum MS, Toninelli E, Anderson JM et al. Cytoskeletal rearrangement mediates human microvascular endothelial tight junction modulation by cytokines. Am J Physiol 1997; 273:H286-H294.
11. Ma TY, Iwamoto GK, Hoa NT et al. TNF-alpha-induced increase in intestinal epithelial tight junction permeability requires NF-kappa B activation. Am J Physiol Gastrointest Liver Physiol 2004; 286:G367-G376.
12. Zech JC, Pouvreau I, Cotinet A et al. Effect of cytokines and nitric oxide on tight junctions in cultured rat retinal pigment epithelium. Invest Ophthalmol Vis Sci 1998; 39:1600-1608.
13. Wong CH, Mruk DD, Lui WY et al. Regulation of blood-testis barrier dynamics: an in vivo study. J Cell Sci 2004; 117:783-798.
14. Fischer S, Wobben M, Marti HH et al. Hypoxia-induced hyperpermeability in brain microvessel endothelial cells involves VEGF-mediated changes in the expression of zonula occludens-1. Microvasc Res 2002; 63:70-80.
15. Singh AB, Harris RC. Epidermal growth factor receptor activation differentially regulates claudin expression and enhances transepithelial resistance in Madin-Darby canine kidney cells. J Biol Chem 2004; 279:3543-3552.
16. Marmorstein AD, Mortell KH, Ratcliffe DR et al. Epithelial permeability factor: a serum protein that condenses actin and opens tight junctions. Am J Physiol 1992; 262:C1403-C1410.
17. Jaeger MM, Dodane V, Kachar B. Modulation of tight junction morphology and permeability by an epithelial factor. J Membr Biol 1994; 139:41-48.
18. Gorodeski GI, Goldfarb J. Seminal fluid factor increases the resistance of the tight junctional complex of cultured human cervical epithelium CaSki cells. Fertil Steril 1998; 69:309-317.
19. Walsh SV, Hopkins AM, Nusrat A. Modulation of tight junction structure and function by cytokines. Adv Drug Deliv Rev 2000; 41:303-313.
20. Wang Y, Zhang J, Yi XJ et al. Activation of ERK1/2 MAP kinase pathway induces tight junction disruption in human corneal epithelial cells. Exp Eye Res 2004; 78:125-136.
21. Wang W, Dentler WL, Borchardt RT. VEGF increases BMEC monolayer permeability by affecting occludin expression and tight junction assembly. Am J Physiol Heart Circ Physiol 2001; 280:H434-H440.
22. Tedelind S, Ericson LE, Karlsson JO et al. Interferon-gamma down-regulates claudin-1 and impairs the epithelial barrier function in primary cultured human thyrocytes. Eur J Endocrinol 2003; 149:215-221.
23. Nitz T, Eisenblatter T, Psathaki K et al. Serum-derived factors weaken the barrier properties of cultured porcine brain capillary endothelial cells in vitro. Brain Res 2003; 981:30-40.

24. Chiba H, Gotoh T, Kojima T et al. Hepatocyte nuclear factor (HNF)-4alpha triggers formation of functional tight junctions and establishment of polarized epithelial morphology in F9 embryonal carcinoma cells. Exp Cell Res 2003; 286:288-297.
25. Chang C, Wang X, Caldwell RB. Serum opens tight junctions and reduces ZO-1 protein in retinal epithelial cells. J Neurochem 1997; 69:859-867.
26. Siliciano JD, Goodenough DA. Localization of the tight junction protein, ZO-1, is modulated by extracellular calcium and cell-cell contact in Madin-Darby canine kidney epithelial cells. J Cell Biol 1988; 107:2389-2399.
27. Liu Y, Nusrat A, Schnell FJ et al. Human junction adhesion molecule regulates tight junction resealing in epithelia. J Cell Sci 2000; 113 (Pt 13):2363-2374.
28. Reuss L. Tigbt junction permeability to ions and water. In: Cereijido M, Anderson JM, eds. Tight Junctions. 2nd ed. Boca Raton: CRC Press, 2001:61-88.
29. Reuss L, Finn AL. Electrical properties of the cellular transepithelial pathway in Necturus gallbladder. I. Circuit analysis and steady-state effects of mucosal solution ionic substitutions. J Membr Biol 1975; 25:115-139.
30. Hegel U, Fromter E, Wick T. Der elektrische wandwiderstand des proximalen konvolutes der rattenniere. Pflügers Arch 1967; 294:274-290.
31. Malnic G, Giebisch G. Some electrical properties of distal tubular epithelium in the rat. Am J Physiol 1972; 223:797-808.
32. Seely JF, Boulpaep EL. Renal function studies on the isobaric autoperfused dog kidney. Am J Physiol 1971; 221:1075-1083.
33. Helman SI, Grantham JJ, Burg MB. Effect of vasopressin on electrical resistance of renal cortical collecting tubules. Am J Physiol 1971; 220:1825-1832.
34. Rau WS, Fromter E. Electrical properties of the medullary collecting ducts of the golden hamster kidney. II. The transepithelial resistance. Pflugers Arch 1974; 351:113-131.
35. Lewis SA, Eaton DC, Diamond JM. The mechanism of Na+ transport by rabbit urinary bladder. J Membr Biol 1976; 28:41-70.
36. Lavelle JP, Meyers SA, Ruiz WG et al. Urothelial pathophysiological changes in feline interstitial cystitis: a human model. Am J Physiol Renal Physiol 2000; 278:F540-F553.
37. Wang EC, Lee JM, Johnson JP et al. Hydrostatic pressure-regulated ion transport in bladder uroepithelium. Am J Physiol Renal Physiol 2003; 285:F651-F663.
38. Claude P, Goodenough DA. Fracture faces of zonulae occludentes from "tight" and "leaky" epithelia. J Cell Biol 1973; 58:390-400.
39. Gonzalez-Mariscal L, Namorado MC, Martin D et al. Tight junction proteins ZO-1, ZO-2, and occludin along isolated renal tubules. Kidney Int 2000; 57:2386-2402.
40. Balda MS, Whitney JA, Flores C et al. Functional dissociation of paracellular permeability and transepithelial electrical resistance and disruption of the apical-basolateral intramembrane diffusion barrier by expression of a mutant tight junction membrane protein. J Cell Biol 1996; 134:1031-1049.
41. Saitou M, Furuse M, Sasaki H et al. Complex phenotype of mice lacking occludin, a component of tight junction strands. Mol Biol Cell 2000; 11:4131-4142.
42. Enck AH, Berger UV, Yu AS. Claudin-2 is selectively expressed in proximal nephron in mouse kidney. Am J Physiol Renal Physiol 2001; 281:F966-F974.
43. Reyes JL, Lamas M, Martin D et al. The renal segmental distribution of claudins changes with development. Kidney Int 2002; 62:476-487.
44. Kiuchi-Saishin Y, Gotoh S, Furuse M et al. Differential expression patterns of claudins, tight junction membrane proteins, in mouse nephron segments. J Am Soc Nephrol 2002; 13:875-886.
45. Furuse M, Furuse K, Sasaki H et al. Conversion of zonulae occludentes from tight to leaky strand type by introducing claudin-2 into Madin-Darby canine kidney I cells. J Cell Biol 2001; 153:263-272.
46. Amasheh S, Meiri N, Gitter AH et al. Claudin-2 expression induces cation-selective channels in tight junctions of epithelial cells. J Cell Sci 2002; 115:4969-4976.
47. Yu AS, Enck AH, Lencer WI et al. Claudin-8 expression in Madin-Darby canine kidney cells augments the paracellular barrier to cation permeation. J Biol Chem 2003; 278:17350-17359.
48. Li WY, Huey CL, Yu AS. Expression of claudin-7 and -8 along the mouse nephron. Am J Physiol Renal Physiol 2004; 286:F1063-F1071.
49. Hulter HN, Ilnicki LP, Harbottle JA et al. Impaired renal H+ secretion and NH3 production in mineralocorticoid-deficient glucocorticoid-replete dogs. Am J Physiol 1977; 232:F136-F146.
50. Ethier JH, Kamel KS, Magner PO et al. The transtubular potassium concentration in patients with hypokalemia and hyperkalemia. Am J Kidney Dis 1990; 15:309-315.
51. Liddle GW. Aldosterone antagonists. AMA Arch Intern Med 1958; 102:998-1004.

52. Van Itallie C, Rahner C, Anderson JM. Regulated expression of claudin-4 decreases paracellular conductance through a selective decrease in sodium permeability. J Clin Invest 2001; 107:1319-1327.
53. Colegio OR, Van Itallie CM, McCrea HJ et al. Claudins create charge-selective channels in the paracellular pathway between epithelial cells. Am J Physiol Cell Physiol 2002; 283:C142-C147.
54. Wilson FH, Disse-Nicodeme S, Choate KA et al. Human hypertension caused by mutations in WNK kinases. Science 2001; 293:1107-1112.
55. Yamauchi K, Rai T, Kobayashi K et al. Disease-causing mutant WNK4 increases paracellular chloride permeability and phosphorylates claudins. Proc Natl Acad Sci USA 2004; 101:4690-4694.
56. Van Itallie CM, Fanning AS, Anderson JM. Reversal of charge selectivity in cation or anion-selective epithelial lines by expression of different claudins. Am J Physiol Renal Physiol 2003; 285:F1078-F1084.
57. Tisher CC. Morphology of the renal tubular paracellular route. In: Bradley SE, Purcell EF, eds. The Paracellular Route. New York: Josiah Macy, Jr. Foundation, 1982:183-201.
58. Roesinger B, Schiller A, Taugner R. A freeze-fracture study of tight junctions in the pars convoluta and pars recta of the renal proximal tubule. Cell Tissue Res 1978; 186:121-133.
59. Kuhn K, Reale E. Junctional complexes of the tubular cells in the human kidney as revealed with freeze-fracture. Cell Tissue Res 1975; 160:193-205.
60. Morita K, Sasaki H, Fujimoto K et al. Claudin-11/OSP-based tight junctions of myelin sheaths in brain and Sertoli cells in testis. J Cell Biol 1999; 145:579-588.
61. Cereijido M, González-Mariscal L, Contreras G. Tight junction barrier between higher organisms and environment. News Physiol Sci 1989; 4:72-75.
62. Alexandre MD, Chen YH. Claudin-7 overexpression decreases the paracellular conductance in kidney epithelial cells. Molecular Biology of the Cell 14[Supplement], 459a. 2003.
63. Wilcox ER, Burton QL, Naz S et al. Mutations in the gene encoding tight junction claudin-14 cause autosomal recessive deafness DFNB29. Cell 2001; 104:165-172.
64. Ben Yosef T, Belyantseva IA, Saunders TL et al. Claudin 14 knockout mice, a model for autosomal recessive deafness DFNB29, are deaf due to cochlear hair cell degeneration. Hum Mol Genet 2003; 12:2049-2061.
65. Simon DB, Lu Y, Choate KA et al. Paracellin-1, a renal tight junction protein required for paracellular Mg2+ resorption. Science 1999; 285:103-106.
66. Muller D, Kausalya PJ, Claverie-Martin F et al. A novel claudin 16 mutation associated with childhood hypercalciuria abolishes binding to ZO-1 and results in lysosomal mistargeting. Am J Hum Genet 2003; 73:1293-1301.
67. Gallardo JM, Hernandez JM, Contreras RG et al. Tight junctions are sensitive to peptides eliminated in the urine. J Membr Biol 2002; 188:33-42.
68. Herzlinger DA, Easton TG, Ojakian GK. The MDCK epithelial cell line expresses a cell surface antigen of the kidney distal tubule. J Cell Biol 1982; 93:269-277.
69. Harris RC. Response of rat inner medullary collecting duct to epidermal growth factor. Am J Physiol. 1989; 256:F1117-F1124.
70. Staehelin LA. Structure and function of intercellular junctions. Int Rev Cytol. 1974; 39:191-283.
71. Kachar B, Reese TS. Evidence for the lipidic nature of tight junction strands. Nature 1982; 296:464-466.
72. Pinto dS, Kachar B. On tight-junction structure. Cell 1982; 28:441-450.
73. Furuse M, Hirase T, Itoh M et al. Occludin: a novel integral membrane protein localizing at tight junctions. J Cell Biol 1993; 123:1777-1788.
74. Furuse M, Fujita K, Hiiragi T et al. Claudin-1 and -2: novel integral membrane proteins localizing at tight junctions with no sequence similarity to occludin. J Cell Biol 1998; 141:1539-1550.
75. Furuse M, Sasaki H, Fujimoto K et al. A single gene product, claudin-1 or -2, reconstitutes tight junction strands and recruits occludin in fibroblasts. J Cell Biol 1998; 143:391-401.
76. Kubota K, Furuse M, Sasaki H et al. Ca(2+)-independent cell-adhesion activity of claudins, a family of integral membrane proteins localized at tight junctions. Curr Biol 1999; 9:1035-1038.
77. Turin L, Behe P, Plonsky I et al. Hydrophobic ion transfer between membranes of adjacent hepatocytes: a possible probe of tight junction structure. Proc Natl Acad Sci USA 1991; 88:9365-9369.
78. Grebenkamper K, Galla HJ. Translational diffusion measurements of a fluorescent phospholipid between MDCK-I cells support the lipid model of the tight junctions. Chem Phys Lipids 1994; 71:133-143.
79. Kan FW. Cytochemical evidence for the presence of phospholipids in epithelial tight junction strands. J Histochem Cytochem 1993; 41:649-656.
80. Dragsten PR, Handler JS, Blumenthal R. Fluorescent membrane probes and the mechanism of maintenance of cellular asymmetry in epithelia. Fed Proc 1982; 41:48-53.

81. Dragsten PR, Blumenthal R, Handler JS. Membrane asymmetry in epithelia: is the tight junction a barrier to diffusion in the plasma membrane? Nature 1981; 294:718-722.

82. van Meer G, Simons K. The function of tight junctions in maintaining differences in lipid composition between the apical and the basolateral cell surface domains of MDCK cells. EMBO J 1986; 5:1455-1464.

83. van Meer G, Gumbiner B, Simons K. The tight junction does not allow lipid molecules to diffuse from one epithelial cell to the next. Nature 1986; 322:639-641.

84. Calderon V, Lazaro A, Contreras RG et al. Tight junctions and the experimental modifications of lipid content. J Membr Biol 1998; 164:59-69.

85. Leung LW, Contreras RG, Flores-Maldonado C et al. Inhibitors of glycosphingolipid biosynthesis reduce transepithelial electrical resistance in MDCK I and FRT cells. Am J Physiol Cell Physiol 2003; 284:C1021-C1030.

86. Venkataraman K, Futerman AH. Ceramide as a second messenger: sticky solutions to sticky problems. Trends in Cell Biology 2000; 10:408-412.

87. Leroy A, De Bruyne GKP, Oomen LCJM et al. Alkylphospholipids reversibly open epithelial tight junctions. Anticancer Research 2003; 23:27-32.

88. Balda MS, Gonzalez-Mariscal L, Contreras RG et al. Assembly and sealing of tight junctions: possible participation of G- proteins, phospholipase C, protein kinase C and calmodulin. J Membr Biol 1991; 122:193-202.

89. Balda MS, Gonzalez-Mariscal L, Matter K et al. Assembly of the tight junction: the role of diacylglycerol. J Cell Biol 1993; 123:293-302.

90. Avila-Flores A, Rendon-Huerta E, Moreno J et al. Tight-junction protein zonula occludens 2 is a target of phosphorylation by protein kinase C. Biochem J 2001; 360:295-304.

91. Wu X, Hepner K, Castelino-Prabhu S et al. Evidence for regulation of the PTEN tumor suppressor by a membrane-localized multi-PDZ domain containing scaffold protein MAGI-2. Proc Natl Acad Sci USA 2000; 97:4233-4238.

92. Wu Y, Dowbenko D, Spencer S et al. Interaction of the tumor suppressor PTEN/MMAC with a PDZ domain of MAGI3, a novel membrane-associated guanylate kinase. J Biol Chem 2000; 275:21477-21485.

93. Stankewich MC, Francis SA, Vu QU et al. Alterations in cell cholesterol content modulate Ca2+-induced tight junction assembly by MDCK cells. Lipids 1996; 31:817-828.

94. Francis SA, Kelly JM, McCormack J et al. Rapid reduction of MDCK cell cholesterol by methyl-beta-cyclodextrin alters steady state transepithelial electrical resistance. Eur J Cell Biol 1999; 78:473-484.

95. Nusrat A, Parkos CA, Verkade P et al. Tight junctions are membrane microdomains. J Cell Sci 2000; 113 (Pt 10):1771-1781.

96. Grindstaff KK, Yeaman C, Anandasabapathy N et al. Sec6/8 complex is recruited to cell-cell contacts and specifies transport vesicle delivery to the basal-lateral membrane in epithelial cells. Cell 1998; 93:731-740.

97. Yeaman C, Grindstaff KK, Nelson WJ. Mechanism of recruiting Sec6/8 (exocyst) complex to the apical junctional complex during polarization of epithelial cells. J Cell Sci 2004; 117:559-570.

98. Polishchuk R, Di Pentima A, Lippincott-Schwartz J. Delivery of raft-associated, GPI-anchored proteins to the apical surface of polarized MDCK cells by a transcytotic pathway. Nat Cell Biol 2004; 6:297-307.

99. Sawai T, Drongowski RA, Lampman RW et al. The effect of phospholipids and fatty acids on tight-junction permeability and bacterial translocation. Pediatr Surg Int 2001; 17:269-274.

100. Schoner W. Endogenous cardiac glycosides, a new class of steroid hormones. Eur J Biochem 2002; 269:2440-2448.

101. Blaustein MP. Physiological effects of endogenous ouabain: control of intracellular Ca2+ stores and cell responsiveness. Am J Physiol 1993; 264:C1367-C1387.

102. Ruch SR, Nishio M, Wasserstrom JA. Effect of cardiac glycosides on action potential characteristics and contractility in cat ventricular myocytes: role of calcium overload. J Pharmacol Exp Ther 2003; 307:419-428.

103. Nunez-Duran H, Fernandez P. Evidence for an intracellular site of action in the heart for two hydrophobic cardiac steroids. Life Sci 2004; 74:1337-1344.

104. Ward SC, Hamilton BP, Hamlyn JM. Novel receptors for ouabain: studies in adrenocortical cells and membranes. Hypertension 2002; 39:536-542.

105. Hamlyn JM, Blaustein MP, Bova S et al. Identification and characterization of a ouabain-like compound from human plasma. Proc Natl Acad Sci USA 1991; 88:6259-6263.

106. Tymiak AA, Norman JA, Bolgar M et al. Physicochemical characterization of a ouabain isomer isolated from bovine hypothalamus. Proc Natl Acad Sci USA 1993; 90:8189-8193.

107. Gottlieb SS, Rogowski AC, Weinberg M et al. Elevated concentrations of endogenous ouabain in patients with congestive heart failure. Circulation 1992; 86:420-425.
108. Ruegg UT. Ouabain-a link in the genesis of high blood pressure? Experientia 1992; 48:1102-1106.
109. Ferrandi M, Manunta P, Balzan S et al. Ouabain-like factor quantification in mammalian tissues and plasma: comparison of two independent assays. Hypertension 1997; 30:886-896.
110. Laredo J, Shah JR, Hamilton BP, Hamlyn JM. Alpha-1 adrenergic receptors stimulate secretion of endogenous ouabain from human and bovine adrenocortical cells. In: Taniguchi K, Kayas S, eds. Na/K-ATPase and Related ATPases. Amsterdam: Elsvier Science, 2000:671-679.
111. Contreras RG, Shoshani L, Flores-Maldonado C et al. Relationship between Na(+),K(+)-ATPase and cell attachment. J Cell Sci. 1999; 112 (Pt 23):4223-4232.
112. Chen Y, Lu Q, Schneeberger EE et al. Restoration of tight junction structure and barrier function by down-regulation of the mitogen-activated protein kinase pathway in ras-transformed Madin-Darby canine kidney cells. Mol Biol Cell 2000; 11:849-862.
113. Fujibe M, Chiba H, Kojima T et al. Thr203 of claudin-1, a putative phosphorylation site for MAP kinase, is required to promote the barrier function of tight junctions. Exp Cell Res 2004; 295:36-47.
114. Macek R, Swisshelm K, Kubbies M. Expression and function of tight junction associated molecules in human breast tumor cells is not affected by the Ras-MEK1 pathway. Cell Mol Biol (Noisy-le-grand) 2003; 49:1-11.
115. Rajasekaran AK, Rajasekaran SA. Role of Na-K-ATPase in the assembly of tight junctions. Am J Physiol Renal Physiol 2003; 285:F388-F396.
116. Rajasekaran SA, Hu J, Gopal J et al. Na,K-ATPase inhibition alters tight junction structure and permeability in human retinal pigment epithelial cells. Am J Physiol Cell Physiol 2003; 284:C1497-C1507.
117. Rajasekaran SA, Palmer LG, Moon SY et al. Na,K-ATPase activity is required for formation of tight junctions, desmosomes, and induction of polarity in epithelial cells. Mol Biol Cell 2001; 12:3717-3732.
118. Gonzalez-Mariscal L, Chavez DR, Cereijido M. Tight junction formation in cultured epithelial cells (MDCK). J Membr Biol 1985; 86:113-125.
119. Nusrat A, Giry M, Turner JR et al. Rho protein regulates tight junctions and perijunctional actin organization in polarized epithelia. Proc Natl Acad Sci USA 1995; 92:10629-10633.
120. Zhong C, Kinch MS, Burridge K. Rho-stimulated contractility contributes to the fibroblastic phenotype of Ras-transformed epithelial cells. Mol Biol Cell 1997; 8:2329-2344.
121. Takaishi K, Sasaki T, Kotani H et al. Regulation of cell-cell adhesion by rac and rho small G proteins in MDCK cells. J Cell Biol 1997; 139:1047-1059.
122. Benais-Pont G, Punn A, Flores-Maldonado C et al. Identification of a tight junction-associated guanine nucleotide exchange factor that activates Rho and regulates paracellular permeability. J Cell Biol 2003; 160:729-740.
123. Hecht G, Pestic L, Nikcevic G et al. Expression of the catalytic domain of myosin light chain kinase increases paracellular permeability. Am J Physiol 1996; 271:C1678-C1684.
124. Turner JR, Rill BK, Carlson SL et al. Physiological regulation of epithelial tight junctions is associated with myosin light-chain phosphorylation. Am J Physiol 1997; 273:C1378-C1385.
125. Xie Z. Molecular mechanisms of Na/K-ATPase-mediated signal transduction. Ann N Y Acad Sci 2003; 986:497-503.
126. Xie Z, Cai T. Na+-K+-ATPase-mediated signal transduction: from protein interaction to cellular function. Mol Interv 2003; 3:157-168.
127. Blanco G, Mercer RW. Isozymes of the Na-K-ATPase: heterogeneity in structure, diversity in function. Am J Physiol 1998; 275:F633-F650.
128. Jorgensen PL. Aspects of gene structure and functional regulation of the isozymes of Na,K-ATPase. Cell Mol Biol (Noisy -le-grand) 2001; 47:231-238.
129. Contreras RG, Flores-Maldonado C, Lazaro A et al. Ouabain binding to Na+,K+-ATPase relaxes cell attachment and sends specific signal (NACos) to the nucleus. J Membr Biol 2004; 198:149-158.
130. Boulpaep EL, Seely JF. Electrophysiology of proximal and distal tubules in the autoperfused dog kidney. Am J Physiol 1971; 221:1084-1096.
131. Lutz MD, Cardinal J, Burg MB. Electrical resistance of renal proximal tubule perfused in vitro. Am J Physiol 1973; 225:729-734.
132. Abramow M, Orci L. On the "tightness" of the rabbit descending limb of the loop of Henle-physiological and morphological evidence. Int J Biochem 1980; 12:23-27.
133. Henin S, Cremaschi D, Schettino T et al. Electrical parameters in gallbladders of different species. Their contribution to the origin of the transmural potential difference. J Membr Biol 1977; 34:73-91.

134. Okada Y, Irimajiri A, Inouye A. Electrical properties and active solute transport in rat small intestine. II. Conductive properties of transepithelial routes. J Membr Biol 1977; 31:221-232.
135. Munck BG, Schultz SG. Properties of the passive conductance pathway across in vitro rat jejunum. J Membr Biol 1974; 16:163-174.
136. Schultz SG, Frizzell RA, Nellans HN. Active sodium transport and the electrophysiology of rabbit colon. J Membr Biol 1977; 33:351-384.
137. Wills NK, Lewis SA, Eaton DC. Active and passive properties of rabbit descending colon: a microelectrode and nystatin study. J Membr Biol 1979; 45:81-108.
138. Peng S, Rahner C, Rizzolo LJ. Apical and basal regulation of the permeability of the retinal pigment epithelium. Invest Ophthalmol Vis Sci 2003; 44:808-817.
139. Liu F, Schaphorst KL, Verin AD et al. Hepatocyte growth factor enhances endothelial cell barrier function and cortical cytoskeletal rearrangement: potential role of glycogen synthase kinase-3beta. FASEB J 2002; 16:950-962.
140. Planchon S, Fiocchi C, Takafuji V et al. Transforming growth factor-beta1 preserves epithelial barrier function: identification of receptors, biochemical intermediates, and cytokine antagonists. J Cell Physiol 1999; 181:55-66.
141. Planchon SM, Martins CA, Guerrant RL et al. Regulation of intestinal epithelial barrier function by TGF-beta 1. Evidence for its role in abrogating the effect of a T cell cytokine. J Immunol 1994; 153:5730-5739.
142. Conyers G, Milks L, Conklyn M et al. A factor in serum lowers resistance and opens tight junctions of MDCK cells. Am J Physiol 1990; 259:C577-C585.
143. Jin M, Barron E, He S et al. Regulation of RPE intercellular junction integrity and function by hepatocyte growth factor. Invest Ophthalmol Vis Sci 2002; 43:2782-2790.
144. Abe T, Sugano E, Saigo Y et al. Interleukin-1beta and barrier function of retinal pigment epithelial cells (ARPE-19): aberrant expression of junctional complex molecules. Invest Ophthalmol Vis Sci 2003; 4:4097-4104.
145. Hanada S, Harada M, Koga H et al. Tumor necrosis factor-alpha and interferon-gamma directly impair epithelial barrier function in cultured mouse cholangiocytes. Liver Int 2003; 23:3-11.
146. Mortell KH, Marmorstein AD, Cramer EB. Fetal bovine serum and other sera used in tissue culture increase epithelial permeability. In Vitro Cell Dev Biol 1993; 29A:235-238.
147. Harhaj NS, Barber AJ, Antonetti DA. Platelet-derived growth factor mediates tight junction redistribution and increases permeability in MDCK cells. J Cell Physiol 2002; 193:349-364.

CHAPTER 12

Tight Junctions during Development

Bhavwanti Sheth, Judith Eckert, Fay Thomas and Tom P. Fleming

Abstract

During early development, tight junction biogenesis and the differentiation of the first epithelium in the blastocyst is critical for embryonic patterning and organization. Here, we discuss the programme of exactly timed transcription, translation, and post-translational modification of specific junctional proteins that regulates the stepwise membrane assembly of tight junctions during cleavage in the mouse model. Underlying mechanisms that coordinate these processes are discussed along with newly emerging data from other mammalian species. In the mouse embryo, junction assembly follows the establishment of cell polarity at the 8-cell stage and is characterized by sequential membrane delivery of JAM-1, ZO-1α- and Rab13, cingulin and ZO-2 followed by ZO-1α+ and occludin. Tight junction assembly occurs over three developmental stages; compaction, first differentiative division and cavitation. Post-translational modification of occludin, the late expression of ZO-1α+ isoform and their intracellular colocalisation may all contribute to the rapid coordinated delivery of these two proteins to the membrane, resulting in the final sealing of the tight junction followed by blastocoel cavitation. This coordinated delivery of these two tight junction-associated proteins may therefore provide a rate limiting step for the sealing of tight junctions and regulated timing of blastocoel cavitation. Taken together, our studies in mouse, human and bovine embryos suggest that defects in the tightly controlled programming of early development may contribute to reduced embryo viability.

Introduction

The construction of a compartmentalised body plan during early development in many organisms requires the creation of several types of specialised epithelia. These epithelia are important to the embryo not just for their polarized transport function but because they provide two-dimensional tissue layers which can be further moulded into complex three-dimensional shapes. Basic embryological transitions such as blastula formation (or blastocyst in mammals), gastrulation, neurulation and organogenesis, all rely on the shaping of epithelial tissues into the correct form at the correct time in development. Another important step in the making of an organism is the epithelial to mesenchymal interconversion of cells. This process allows for regulated movement of cell layers and changes in cellular organization and morphology. The formation or loss of epithelia contributes to all the processes mentioned above in the embryo and to later organogenesis, for example, the development of the kidney.

Regulation of the epithelial phenotype, either in the shaping of tissues during early development or organogenesis, is primarily coordinated by the intercellular junctions and the associated cytoskeleton. The central role played by the E-cadherin/catenin system of the adherens junction is now well established as the key to regulating epithelial tissue formation and developmental signalling.[1-3] Tight junctions (TJs), in contrast, have historically been regarded as static membrane complexes maintaining the integrity of the transepithelial permeability barrier and the

---

apicobasal polarity inherent within each cell. However the study of epithelia over the last decade has resulted in the view that TJs are highly dynamic, not only during development but also in the normal functioning of an organism (see Chapters 3, 4, 8 and 9 for examples). This view has come about partly due to the characterisation of many novel proteins and their interactions at the TJ, and partly from the expression/localisation profiles of these proteins. One major finding of these studies has been the discovery that, depending on the cellular environment, several TJ-associated proteins are localised in the nucleus where they may play a role in cell signalling (see Chapter 7).

In this chapter, we examine the molecular events underlying TJ biogenesis during the differentiation of the first epithelium in the mammalian embryo. The relationship between TJ assembly and the formation of a blastocyst will be discussed with respect to the pivotal role played by the adherens junction. For a comparison of mammalian early epithelial and TJ development, with that of the amphibian embryo, Xenopus, see Fleming et al.[4]

Epithelial Differentiation in the Mouse Embryo

After fertilization, the mouse zygote reinitiates cell cycling, activates the embryonic genome and undergoes five cell divisions to generate the blastocyst; a spherical cyst consisting of an outer epithelium (trophectoderm), an inner cell mass (ICM) and a fluid-filled cavity called the blastocoel (Fig. 1A,B)

In the three cell divisions leading to the 8-cell embryo, the blastomeres are undifferentiated and spatially nonpolar. It is only towards the end of the 8-cell stage that cell-cell adhesion begins for the first time in development and polarization of the cell surface and cytoplasm is initiated. During the next cleavage (16-cell) a process of differentiative cell division allocates cells to either the outer trophectoderm layer or to the interior of the embryo, constituting the ICM.[5,6] These inner cells are completely surrounded by neighbouring cells and the ICM at this stage of development is nonepithelial in phenotype. Finally, at the 32-cell stage when the trophectoderm layer completes its differentiation, it has an outward-facing apical membrane and initiates vectorial transport of fluid across from the maternal uterine cavity into the embryo interior to form the blastocoel.

This transport activity is dependent on two aspects of epithelial differentiation which occur exclusively in the trophectoderm lineage; (a) development of tight junctions (Fig. 1C) ensure intercellular sealing required for cavitation, and (b) development of a polarized transport pathway operating across the trophectoderm epithelium driven by the Na^+,K^+-ATPase enzyme, localised at the basolateral membranes.[7-12] Hence, at this stage, the ICM is dependent upon the trophectoderm cells for influx and efflux of molecules and ions. Indeed, transport systems between the maternal environment and the blastocoel cavity for amino acids, sugars, ions,

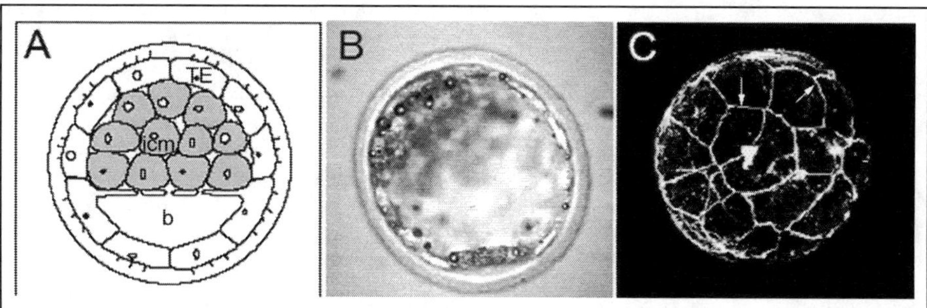

Figure 1. A) Diagrammatic picture of a blastocyst indicating the position of the polar trophectoderm (TE), nonpolar inner cell mass (ICM) and the blastocoel cavity (b). B) Brightfield image of a mouse blastocyst. C) Confocal staining of an early blastocyst showing occludin protein at the TJ sites between all trophectoderm cells.

water (aquaporins), growth factors and other macromolecules have been identified.[7,13-24] The blastocoel therefore, is a controlled microenvironment providing the necessary nutrients and growth factors that may facilitate subsequent differentiation of the ICM. It also provides a fluid environment that the derivatives of the ICM (visceral and parietal endoderm, embryonic ecto-derm) can expand into, immediately after implantation. After full expansion of the blastocoel, it is the trophectoderm that initiates implantation into the uterine wall and continues to differ-entiate into the extraembryonic lineages comprising the trophoblast, ectoplacental cone, and ultimately the chorio-allantoic placenta. The protected ICM gives rise to the entire fetus and the extraembryonic endoderm lineages.[25] Thus, forming the blastocoel is an integral compo-nent of early development. This in turn is dependent upon tight junction biogenesis in the trophectoderm to generate the necessary permeability seal.[26,27]

Compaction: A Prerequisite for Tight Junction Formation

Trophectoderm differentiation in the mouse embryo begins at the 8-cell stage when activa-tion of the E-cadherin/catenin system results in the adherence of cells into a tight ball (called a morula); a process known as compaction.[28,29] This coincides with the redistribution of E-cadherin and catenin proteins from a uniform membrane localisation to cell contact sites and the formation of a nascent apicolateral junctional complex.[26] During compaction, the cells become polarized and exhibit an outer apical membrane rich in microvilli and a basolateral membrane devoid of microvilli.[30-32] The cytoskeleton and specific organelles within the cyto-plasm also polarize along the apicobasal axis.[13,33,34] It is the initiation of the cadherin adhesion that catalyses the onset of polarization and regulates its apicobasal orientation.[35] Similarly, adherens junction formation is also a prerequisite for the membrane assembly of most of the TJ constituents, which begins close to the time of compaction.

Three Phases of Tight Junction Biogenesis

In the mouse embryo, membrane assembly of TJ proteins develops over three developmen-tal stages, with a further maturation of the junction occurring at each subsequent cleavage. For details on the molecular composition of TJs, see the reviews by Matter and Balda,[36] D'Atri and Citi,[37] Gonzalez-Mariscal et al,[38] and Chapters 2-6 in this book.

Immediately following compaction, ZO-1[39-42] is detected for the first time at the apicolateral membranes of 8-cell blastomeres.[43] ZO-1, one of a family of MAGuK (membrane-associated guanylate kinase) proteins (see detailed review by Gonzalez-Mariscal et al,[38,44]) associates cyto-plasmically with the C-terminus of all three transmembrane proteins at the TJ, occludin, claudin and JAM. ZO-1 is expressed as two alternatively spliced isoforms,[45,46] either with or without an 80 amino acid α-domain (ZO-1α+/-). Our studies have shown that it is exclusively the shorter isoform ZO-1α-, which initiates assembly following compaction.[47] This isoform is expressed both from the maternal genome (inherited from the oocyte) and also an early tran-script from the embryonic genome, which in the mouse is activated around the 2-cell stage. ZO-1α- is therefore detectable at all stages of cleavage at mRNA and protein levels. Immunof-luorescence using an isoform-specific antibody, indicated that ZO-1α- first assembles at the apicolateral contact sites as punctuate foci which coalesce several hours later (probably after division into the 16-cell stage) to become a continuous belt-like ring around adjacent cells.[43,47] The polarized assembly of ZO-1α- is dependent upon stable E-cadherin adhesion. Thus, inhi-bition of compaction with an E-cadherin neutralising antibody will randomise the site of ZO-1α- membrane assembly and delay its occurrence.[43]

Rab GTPase proteins are involved in the targeting of proteins to specific destinations dur-ing intracellular transport.[48-50] In fact Rab13 has been reported to associate with the TJ in mature epithelial cells, where it is thought to have a specific role in targeting and docking of TJ proteins[51,52] Rab13 is also expressed throughout mouse cleavage and assembles at the apicolateral sites in precise colocalisation with ZO-1α-.[53] The early assembly of Rab13 in the embryo may suggest that it is involved not only in maintaining the integrity of the junction in mature cells but also in the specification of this domain during differentiation and TJ biogenesis. Rab13 is isoprenylated in vivo, which may facilitate its anchorage to the lipid bilayer during membrane attachment.[54] Regulation of membrane association and disassociation of Rab13 is achieved by the 17kD protein, Δ-phosphodiesterase (δ-PDE), which in turn exhibits C-terminal sequences necessary for interaction with PDZ motifs found on ZO-1 and other members of the MAGuK family.[55] Rab13 is therefore suitably equipped to initiate ZO-1 targeting to the site of junction formation.[50]

In recent studies, we have found that the TJ transmembrane protein JAM-1 (also known as JAM-A,[56]), a member of the immunoglobulin superfamily,[57] also assembles at the cell membranes of the early 8-cell embryo.[57a] Immunolocalisation of JAM-1 during the 8-cell stage was unusual in that its membrane assembly initiated prior to compaction (Fig. 2A), in fact within the first hour of the fourth cell cycle. Thus, unlike any other member of the TJ constituents analysed in our developmental model, JAM-1 assembly at the cell-cell contacts does not require cadherin-mediated adhesion. Also of interest is the fact that JAM-1 colocalises transiently with the apical microvilli pole (Fig. 2B), suggesting an involvement in the establishment of epithelial cell polarity, consistent with experimental studies on mature cell lines.[58,59] The expression pattern observed for JAM-1 protein may indicate that it participates in the recruitment of TJ plaque proteins (ZO-1α-, cingulin) to the membrane contacts, through its binding sites at the C-terminus.[59]

In the second phase of TJ assembly at the 16-cell stage, cingulin, a cytoplasmic plaque protein,[60-62] colocalises with ZO-1α- at the apicolateral contacts.[63] Cingulin protein, like ZO-1α-, is expressed from both the maternal and embryonic genomes of the mouse.[64] The maternal pool localises to the cytocortex of the egg membrane, where it may bind to myosin.[62] This pool of protein is internalized through endocytosis and degraded during cleavage such that it does not appear to participate directly in TJ biogenesis. Embryonically expressed cingulin, once assembled in the 16-cell morula, exhibits a longer half-life compared with that of the maternal pool.[64] Moreover, the turnover time of this assembled protein decreases if E-cadherin adhesion is abolished. These data suggest that E-cadherin adhesion in the early embryo is required not only for spatial organization of the tight junction but also for stabilization of assembled components.

In addition to cingulin, ZO-2 protein also assembles at the TJ in the 16-cell morula where it colocalises with the E-cadherin complex.[65] At the mRNA level, mouse ZO-2 is transcribed in the zygote and throughout cleavage. ZO-2 was first identified as a ZO-1 binding protein in coprecipitation experiments.[66] The two proteins interact through their PDZ2 domains forming a highly stable complex that can be detected even under harsh extraction conditions in cell lines.[67,68] Like other members of the MAGuK family,[44] ZO-2 sequence has recently been shown to possess nuclear localisation motifs which allow it to shuttle between the nucleus and the TJ, depending on the cell proliferation status of the cells.[69,70] During early development, we have also detected ZO-2 in the nuclei of 2, 4 and 8-cell embryos by immunofluorescence. Nuclear staining persists during compaction at the 8-cell stage and gradually diminishes as ZO-2 begins to assemble at cell-cell contacts in the trophectoderm (Fig. 2C,D). Our data indicate that ZO-2 may contribute to both nuclear signalling and structural organization associated with epithelial differentiation (Sheth et al, manuscript in preparation).

The third and final phase of TJ maturation in the early embryo is critical for the timing of blastocyst cavitation. During the late 16- and early 32-cell stages, the second ZO-1 isoform, ZO-1α+, is transcribed de novo, rapidly translated and detected at the protein level by immunoprecipitation and immunofluorescence techniques.[47] Careful examination of the timing of ZO-1α+ membrane assembly showed that assembly first occurs at the early 32-cell stage. Just prior to this, ZO-1α+ is detected at perinuclear sites and its assembly at the membrane is sensitive to brefeldin A, an inhibitor of protein transport from the Golgi to the cell surface. The critical nature of ZO-1α+ delayed expression and membrane assembly in the timing of TJ formation has been revealed by its intracellular colocalisation with the transmembrane proteins

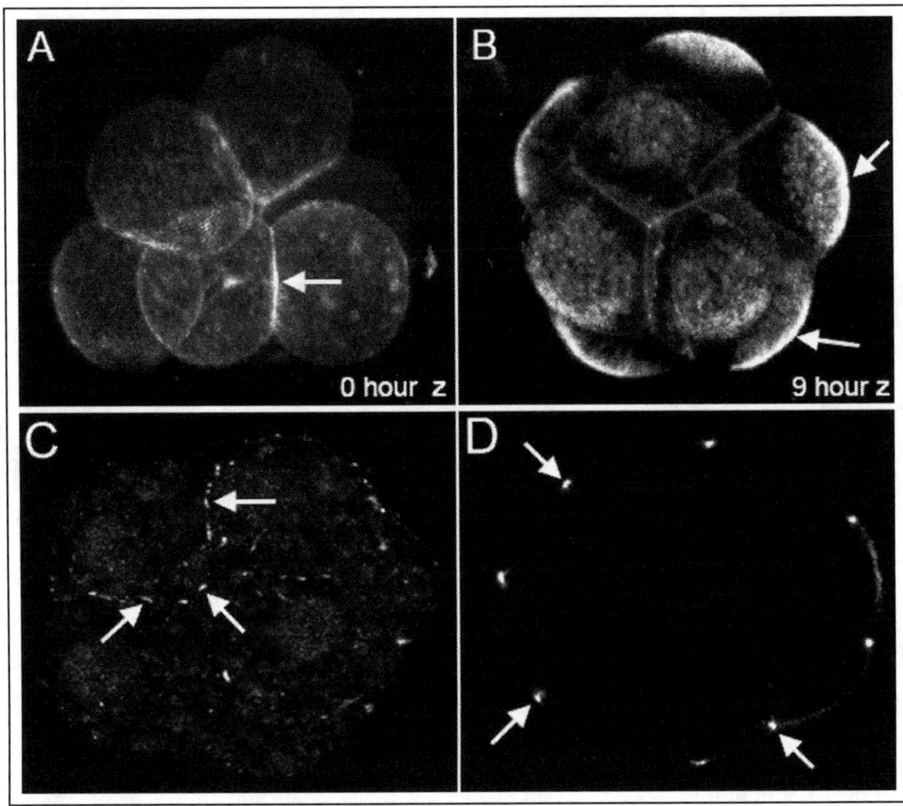

Figure 2. Confocal z-series showing JAM-1 protein assembles at the cell-cell borders (arrow) in 8-cell embryos before compaction (A, 0 hour after division into 8-cell) and at the apical poles in the compact 8-cell embryo (B, 9 hours after division). Immunolabelled 16-cell morula (C) showing initial assembly of ZO-2 at the cell-cell contacts (arrows, mid-plane) and apical localisation at the 32-cell stage (D, arrows).

occludin and claudin 1/3[47,71] (Thomas et al, manuscript in preparation). Occludin[72] and claudin[73] genes encode unrelated proteins, both possesing four transmembrane domains with two extracellular loops and intracellular N-and C-terminal domains, the latter interacting with a series of cytoplasmic plaque proteins which link the TJ to the actin cytoskeleton (see Chapters 2, 3 and 6). Claudins represent a multi-gene family with different isoforms expressed and localised within distinct tissue TJs. Here they constitute the major component of the freeze-fracture strands and the inherent tissue-specific variability in the paracellular barrier function.[38,74]

In contrast to ZO-1α+, occludin is transcribed and translated throughout cleavage but membrane assembly of occludin is delayed until after its colocalisation with ZO-1α+ in the Golgi.[75] Immunoblotting studies revealed that occludin protein exists as four separate bands ranging from 58-72kD and varying in their relative amounts during cleavage. Post-translational modification of occludin has been reported in many cell types.[76-78] In the embryo, one occludin band in particular (65-67kD) becomes prominent and undergoes phosphorylation close to the time of blastocyst formation. It is this form of occludin that exclusively enters a Triton-insoluble pool and appears to be the form involved in TJ formation, since only the assembled occludin is detergent-insoluble.[75]

Occludin assembly at the TJ in the 32-cell embryo is followed very rapidly by polarized fluid transport across the trophectoderm layer and the onset of blastocoel cavitation. Recent studies using a commercial antibody, recognising both claudin 1 and 3 isoforms, have shown that the TJ transmembrane protein claudin-1/3[79] also assembles during this late phase of TJ biogenesis. It is only from the time of cavitation that the embryo becomes impermeable to electron dense tracer and vital dyes.[80] This suggests that the timing of occludin membrane assembly (and possibly also that of caudin-1/3) may act as a limiting factor in TJ sealing and thereby regulate the timing of cavitation in the embryo. In order to achieve this, the delayed transcription and rapid translation of ZO-1α+ may play an important role as the rate-limiting step. Intracellular colocalisation of ZO-1α+ and occludin in the Golgi could aid the rapid delivery to the membrane of both proteins as a complex to be inserted into the TJ site.[47] Indeed inhibition of membrane assembly by transient brefeldin A treatment, results in inhibition of cavitation. The timing of ZO-1α+ transcription may therefore regulate the transition from morula to blastocyst and the timing of completion of TJ biogenesis in the early embryo.

The late timing of occludin membrane assembly during trophectoderm differentiation raises the question of the membrane binding partners of ZO-1 α- and cingulin during earlier cleavage. ZO-1 binds directly to the C-terminus of occludin[81] while interaction between cingulin and occludin, either direct or indirect, has been identified in *Xenopus* epithelial cells.[82] Double-label confocal microscopy has shown that ZO-1α- colocalises with both α- and β-catenins at the 8- and 16-cell stages but segregates to a distinct, more apical site once occludin assembly has occurred.[53] Following occludin assembly, all tight junction proteins studied (ZO-1α-, ZO-1α+, cingulin and ZO-2) are precisely colocalised at a site apical to and separate from proteins of the adherens junction. Thus, we and others,[83] have concluded that during epithelial differentiation, the adherens junction and TJ emerge from a common, single junctional complex, activated from compaction in the case of the embryo, and characterised by colocalisation of TJ plaque proteins (ZO-1α-) with E-cadherin/catenin complexes.

Cell Adhesion and Regulation of Tight Junction Assembly

The trophectoderm model for TJ assembly during development has provided strong evidence for E-cadherin adhesion as a permissive state for subsequent assembly of the TJ. Thus, if E-cadherin adhesion during the 8-cell stage is prevented by neutralising antibodies, membrane assembly of ZO-1α- is delayed and occurs in a nonpolarized, irregular manner.[43] Moreover, in E-cadherin and α-catenin knock-out embryos, normal trophectoderm differentiation is inhibited[84-86] and ZO-1 assembly is disturbed.[87] E-cadherin null embryos were able to compact and undergo delayed cell polarization due to the maternal pool of protein, although subsequent differentiation of the trophectoderm layer was disorganised. These studies clearly demonstrate that E-cadherin adhesion plays a pivotal role in the differentiation of the first epithelium which in turn is essential for further embryonic development. However, our recent demonstration that JAM-1 assembly is independent of E-cadherin adhesion (see above) suggests that not all TJ constituents require cadherin adhesion to be active.

Activation of E-cadherin/catenin adhesion in the embryo, unlike the completion of TJ formation, does not appear to be regulated by the timing of expression of any particular protein constituent. Indeed, E-cadherin, α- and β-catenins are all present on membranes throughout cleavage.[28,29,88] Rather, phosphorylation of one or more components may be involved since protein kinase C (PKC) activation with phorbol esters induces premature compaction at the 4-cell stage[89] and PKCα isoform redistributes to the cell contact sites at the time of compaction.[90]

Many TJ proteins are also phosphorylated in a manner that may affect their binding to each other and their assembly at the junction.[44,76,77] Although a large amount of work in this area by many investigators suggests a critical role of phosphorylation on the function of TJ proteins, the physiological relevance of this phosphorylation remains unclear. Several PKC isoforms have even been localised close to the TJs in cell-lines.[91-93] Indeed, spatial and temporal expression patterns of several PKC isoforms appear to be developmentally regulated[94] in the early embryo.

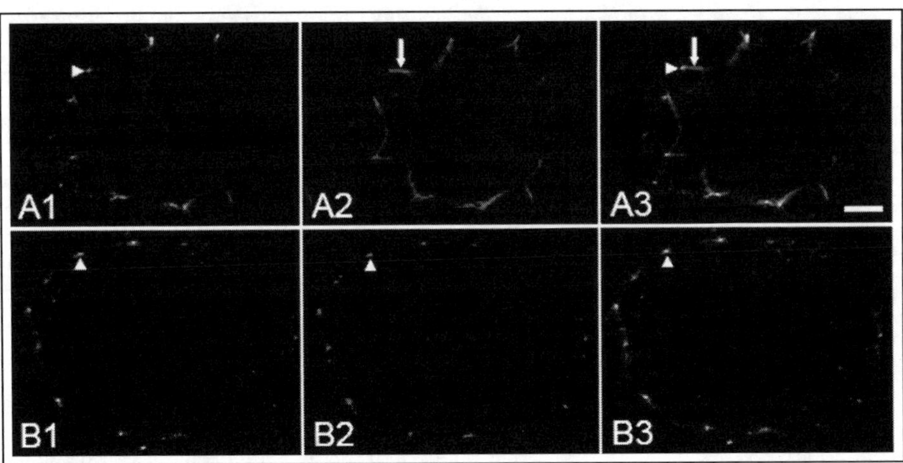

Figure 3. Colocalisation of cGEF-H1 with α-catenin (A1-A3) showing a more apical localisation (arrowhead, A1) compared to the apicolateral catenin staining (arrow, A2). An overlay of the two images shows the segregated positioning of the two proteins (A3). Double-labelling cGEF-H1 (B1) with ZO-1 (B2) shows the precise colocalisation of these two proteins at the apical tight junction sites (arrowhead and yellow overlay, B3). A color version of this figure is available online at www.Eurekah.com.

Applying PKC activators and inhibitors to the embryo model we have recently shown that PKC signalling is involved in regulation of ZO-2 and ZO-1α+ membrane assembly. Moreover, the results indicate a role for PKCδ and ζ isoforms in the synthesis of TJs and an involvement of different PKC isotype networks for each TJ component.[95]

The Rho-family of GTPases is well known for their regulation of the actin cytoskeleton[96] and although a large body of evidence indicates that Rho GTPases are important regulators of epithelial TJs, little is known about the mechanisms that control these GTPases. Activation of Rho GTPases generally requires a guanine nucleotide exchange factor (GEF) and a TJ-associated GEF for Rho has recently been identified and shown to regulate paracellular permeability of small hydrophilic tracers.[97] Colocalisation studies of cGEF-H1 with either ZO-1 or α-catenin in mouse blastocysts revealed its presence apical to the adherens junction in the trophectoderm cells whilst the ICM was devoid of staining for cGEF-H1 (Fig. 3). These experiments revealed the association of a GEF with a mature TJ. Future studies to investigate the spatial and temporal expression of cGEF-H1 during earlier stages of development may shed light on the role of GEFs and therefore Rho-GTPases, in the synthesis of a TJ.

Tight Junctions in Other Mammalian Preimplantation Embryos

All mammalian embryos undergo compaction but at different cell cycles. In the pig, compaction and cell polarization occur more gradually than in the mouse and coincide with relocation of E-cadherin to contact sites.[98] Bovine embryos also compact somewhat later[99] than in the mouse (16-32-cell stage). E-cadherin expression and membrane localisation have also been identified in human embryos[100] and compaction occurs on day 4 when approximately 16-cells are present.[101]

These other models of early embryo TJ biogenesis are broadly comparable with the mouse and illustrate a similar ontogeny of stage-dependent expression of TJ components.[102,103] In the bovine model, we have also shown that environment factors (eg in vitro maturation or type of culture medium) influence the level of expression of TJ genes which may have important developmental and biotechnological implications. In the human embryo cultured in vitro, abnormality in gene expression for TJ constituents has also been identified and may impact on developmental potential. Two alternatively spliced isoforms of occludin, either with or without the fourth transmembrane domain have been identified in the human[104] but are absent in the mouse. Interestingly, ZO-1α+ and both occludin isoforms appear sensitive to loss of detectable expression in a number of human embryos.[103] In addition, failure in membrane assembly of occludin[103] and ZO-1 α+[104] protein at trophectoderm TJs has been identified in human embryos which may further compromise viability, and may indicate failure in normal intracellular signalling, for example, involving PKCs.[95,105,106]

Acknowledgements

We are grateful for funding from Medical Research Council, Biotechnology and Biological Sciences Research Council, and The Wellcome Trust for research in T.P. Fleming's laboratory.

References

1. Takeichi M. Cadherin cell adhesion receptors as a morphogenetic regulator. Science 1991; 251:1451-1455.
2. McCrea PD, Brieher WM, Gumbiner B. Induction of a secondary body axis in xenopus by antibodies to beta catenin. J Cell Biol 1993; 123:477-484.
3. Aberle H, Schwartz H, Kemler R. Cadherin-catenin complex: Protein interactions and their implications for cadherin function. J Cell Biochem 1996; 61(4):514-523.
4. Fleming TP, Papenbrock T, Fesenko I et al. Assembly of tight junctions during early vertebrate development. Semin Cell Dev Biol 2000; 11(4):291-299.
5. Johnson MH, Ziomek CA. The foundation of two distinct cell lineages within the mouse morula. Cell 1981; 24:71-80.
6. Fleming TP. A quantitative analysis of cell allocation to trophectoderm and inner cell mass in the mouse blastocyst. Dev Biol 1987; 119:520-531.
7. DiZio SM, Tasca RJ. Sodium-dependent aminoacid transport in preimplantation embryos. III. Na$^+$, K$^+$,-ATPase linked mechanism in blastocysts. Dev Biol 1977; 59:198-205.
8. Vorbrodt A, Konwinski M, Solter D et al. Ultrastructural cytochemistryof membrane-bound phosphatases in preimplantation mouse embryos. Dev Biol 1977; 55:117-134.
9. Watson AJ, Kidder GM. Immunofluorescence assessment of the timing of appearance and cellular distribution of Na/K-ATPase during mouse embryogenesis. Dev Biol 1988; 126:80-90.
10. MacPhee DJ, Barr KJ, DeSousa PA et al. Regulation of Na$^+$,K$^+$-ATPase alpha subunit gene expression during mouse preimplantation development. Dev Biol 1994; 162:259-266.
11. MacPhee DJ, Jones DH, Barr KJ et al. Differential involvement of Na(+), K(+)-ATPase isozymes in preimplantation development. Dev Biol 2000; 222:486-498.
12. Jones DH, Davies TC, Kidder GM. Embryonic expression of the putative gamma subunit of the sodium pump is required for aquisition of fluid transport capacity during mouse blastocyst development. J Cell Biol 1997; 139:1545-1552.
13. Fleming TP, Pickering SJ. Maturation and polarisation of the endocytotic system in outside blastomeres during mouse preimplantation embryos. J Embryol Exp Morphol 1985; 89:175-208.
14. Heyner S, Rao LV, Jarett L et al. Preimplantation mouse embryos internalize maternal insulin via receptor-mediated endocytosis: Pattern uptake and functional correlations. Dev Biol 1989; 134:48-58.
15. Manejwala FM, Cragoe EJ, Schultz RM. Blastocoel expansion in the preimplantation mouse embryo: A role for extracellular sodium and chloride and possible apical routes of their entry. Dev Biol 1986; 133:210-220.
16. Dardik A, Schultz RM. Protein secretion by the mouse blastocyst: Differences in the polypeptide composition secreted into the bastocoel and medium. Biol Reprod 1991; 45:328-333.
17. Wiley LM, Lever JE, Pape C et al. Antibodies to a renal Na$^+$/glucose cotransport systemlocalize to the apical membrane domain of polar mouse embryo blastomeres. Dev Biol 1991; 143:149-161.
18. Aghayan M, Rao LV, Smith RM et al. Developmental expression and cellular localization of glucose transporter molecules during mouse preimplantation development. Development 1992; 115:305-312.
19. Hewitson LC, Leese HJ. Energy metabolism of the trophectoderm and inner cell mass of the mouse blastocyst. J Exp Zool 1993; 267:337-343.
20. Brison DR, Hewitson LC, Leese HJ. Glucose, pyruvate and lactate concentrations in the blastocoel cavity of rat and mouse embryos. Mol Reprod Dev 1993; 35:272-232.
21. Smith RM, Garside WT, Aghayan M et al. Mouse preimplantation embryos exhibit receptor-mediated binding and transcytosis of maternal insulin-like growth factor I. Biol Reprod 1993; 49:1-12.
22. Barr KJ, Garrill A, Jones DH et al. Contributions of Na+/H+ exchanger isoforms to preimplantation development of the mouse. Reprod Dev 1998; 50:146-153.

23. Offenberg H, Barcroft LC, Caveney A et al. mRNAs encoding aquaporins are present during murine preimplantation development. Mol Reprod Dev 2000; 57(4):323-330.
24. Watson AJ, Barcroft LC. Regulation of blastocyst formation. Front Biosci 2001; 6:D708-D730.
25. Rossant J, Chazaud C, Yamanaka Y. Lineage allocation and asymmetries in the early mouse embryo. Philos Trans R Soc Lond B Biol Sci 2003; 358:1341-1349.
26. Ducibella T, Albertini DF, Anderson E et al. The preimplantation mammalian embryo: Characterization of intercellular junctions and their appearance during development. Dev Biol 1975; 45:231-250.
27. Magnuson T, Demsey A, Stackpole CW. Characterization of intercellular junctions in the preimplantation embryo by freeze-fracture and thin-section electron microscopy. Dev Biol 1977; 61:252-261.
28. Vestweber D, Gossler A, Boller K et al. Expression and distribution of cell adhesion molecule uvomorulin in mouse preimplantation embryos. Dev Biol 1987; 124:451-456.
29. Ohsugi M, Hwang S, Butz S et al. Expression and cell membrane localisationof catenins during mouse preimplantation development. Dev Dyn 1996; 206:391-402.
30. Handyside AH. Distribution of antibody- and lectin-binding sites on dissociated blastomeres of mouse morulae: Evidence for polarization at compaction. J Embryol Exp Morphol 1980; 60:99-116.
31. Ziomek CA, Johnson MH. Cell surface interactions induce polarisation of mouse 8-cell blastomeres at compaction. Cell 1980; 21:935-942.
32. Reeve WJ, Ziomek CA. Distribution of microvilli on dissociated blastomeres from mouse embryos: Evidence for surface polarisation at compaction. J Embryol Exp Morphol 1981; 62:339-350.
33. Johnson MH, Maro B. The distribution of cytoplasmic actin in mouse 8-cell blastomeres. J Embryol Exp Morphol 1984; 82:97-117.
34. Houliston E, Pickering SJ, Maro B. Redistribution of microtubules and pericentriole material during the development of polarity in mouse blastomeres. J Cell Biol 1987; 104:1299-1308.
35. Johnson MH, Maro B, Takeichi M. The role of cell adhesion in the synchronisation and orientation of polarisation in 8-cell mouse blastomeres. J Embryol Exp Morphol 1986; 93:239-255.
36. Matter K, Balda MS. Signalling to and from tight junctions. Nature Reviews 2003; 4:225-236.
37. D'Atri F, Citi S. Molecular complexity of vertebrate tight junctions. Mol Membr Biol 2002; 19:103-112.
38. Gonzalez-Mariscal L, Betanzos A, Nava P et al. Tight junction proteins. Prog Biophys Mol Biol 2003; 81:1-44.
39. Stevenson BR, Siliciano JD, Mooseker MS et al. Identification of ZO-1: A high molecular weight polypeptide associated with the tight junction (zonula occludens) in a variety of epithelia. J Cell Biol 1986; 103(3):755-766.
40. Anderson JM, Stevenson BR, Jesaitis LA et al. Characterization of ZO-1, a protein component of the tight junction from mouse liver and Madin-Darby canine kidney cells. J Cell Biol 1988; 106(4):1141-1149.
41. Itoh M, Nagafuchi A, Yonemura S et al. The 220-kD protein colocalizing with cadherins in nonepithelial cells is identical to ZO-1, a tight junction-associated protein in epithelial cells: cDNA cloning and immunoelectron microscopy. J Cell Biol 1993; 121(3):491-502.
42. Willott E, Balda MS, Fanning AS et al. The tight junction protein ZO-1 is homologous to the Drosophila discs-large tumor suppressor protein of septate junctions. Proc Natl Acad Sci USA 1993; 90(16):7834-7838.
43. Fleming TP, McConnell J, Johnson MH et al. Development of tight junctions de novo in the mouse early embryo: Control of assembly of the tight junction-specific protein, ZO-1. J Cell Biol 1989; 108(4):1407-1418.
44. Gonzalez-Mariscal L, Betanzos A, Avila-Flores A. MAGUK proteins: Structure and role in the tight junction. Semin Cell Dev Biol 2000; 11(4):315-324.
45. Willott E, Balda MS, Heintzelman M et al. Localization and differential expression of two isoforms of the tight junction protein ZO-1. Am J Physiol 1992; 262(5 Pt 1):C1119-C1124.
46. Balda MS, Anderson JM. Two classes of tight junctions are revealed by ZO-1 isoforms. Am J Physiol 1993; 264(4 Pt 1):C918-C924.
47. Sheth B, Fesenko I, Collins JE et al. Tight junction assembly during mouse blastocyst formation is regulated by late expression of ZO-1 alpha+ isoform. Development 1997; 124(10):2027-2037.
48. Novick P, Zettl KS. The diversity of rab proteins in vesicle transport. Curr Opin Cell Biol 1997; 9:496-504.
49. Chavrier P, Gour BJ. The role of ARF and rab GTPases in membrane transport. Curr Opin Cell Biol 1999; 11:466-475.
50. Zahraoui A, Louvard D, Galli T. Tight junction, a platform for trafficking and signaling protein complexes. J Cell Biol 2000; 151(5):F31-F36.
51. Zahraoui A, Joberty G, Arpin M et al. A small rab GTPase is distributed in cytoplasmic vesicles in non polarized cells but colocalizes with the tight junction marker ZO-1 in polarized epithelial cells. J Cell Biol 1994; 124(1-2):101-115.

52. Marzesco AM, Dunia I, Pandjaitan R et al. The small GTPase Rab13 regulates assembly of functional tight junctions in epithelial cells. Mol Cell Biol 2002; 13:1819-1831.
53. Sheth B, Fontaine JJ, Ponza E et al. Differentiation of the epithelial apical junctional complex during mouse preimplantation development: A role for rab13 in the early maturation of the tight junction. Mech Dev 2000; 97:93-104.
54. Joberty G, Tavitian A, Zahraoui A. Isoprenylation of rab proteins possessing a C-terminal CaaX motif. FEBS Lett 1993; 330:323-328.
55. Marzesco AM, Galli T, Louvard D et al. The rod cGMP phosphodiesterase δ subunit dissociates the small GTPase rab13 from membranes. J Biol Chem 1998; 273:22340-22345.
56 Bazzoni G. The JAM family of junctional adhesion molecules. Curr Opin Cell Biol 2003; 15:525-530.
57. Martin-Padura I, Lostaglio S, Schneemann M et al. Junctional adhesion molecule, a novel member of the immunoglobulin superfamily that distributes at intercellular junctions and modulates monocyte transmigration. J Cell Biol 1998; 142(1):117-127.
57a. Thomas FC, Sheth B, Eckert JJ et al. Contribution of JAM-1 to epithelial differentiation and tight junction biogenesis in mouse pre-implantation embryo. J Cell Sci 2004; 117:5599-5608.
58. Martinez-Estrada OM, Villa A, Breviario F et al. Association of junctional adhesion molecule with calcium/calmodulin-dependent serine protein kinase (CASK/Lin-2) in human epithelial Caco2 cells. J Biol Chem 2001; 276:9291-9296.
59. Bazzoni G, Martinez-Estrada OM, Orsenigo F et al. Interaction of junctional adhesion molecule with the tight junction components ZO-1, cingulin, and occludin. J Biol Chem 2000; 275(27):20520-20526.
60. Citi S, Sabanay H, Jakes R et al. Cingulin, a new peripheral component of tight junctions. Nature 1988; 333(6170):272-276.
61. Citi S, Sabanay H, Kendrick-Jones J et al. Cingulin: Characterization and localization. J Cell Sci 1989; 93(Pt 1):107-122.
62. Cordenonsi M, D'Atri F, Hammar E et al. Cingulin contains globular and coiled-coil domains and interacts with ZO-1, ZO-2 and myosin. J Cell Biol 1999; 147:1569-1582.
63. Fleming TP, Hay M, Javed Q et al. Localisation of tight junction protein cingulin is temporally and spatially regulated during early mouse development. Development 1993; 117(3):1135-1144.
64. Javed Q, Fleming TP, Hay M et al. Tight junction protein cingulin is expressed by maternal and embryonic genomes during early mouse development. Development 1993; 117(3):1145-1151.
65. Nowak RL, Sheth B, Fleming TP. Expression of the tight junction protein, ZO-2 in mouse preimplantation embryos. Mol Biol Cell 2000; 11S:512a.
66. Gumbiner B, Lowenkopf T, Apatira D. Identification of a 160-kDa polypeptide that binds to the tight junction protein ZO-1. Proc Natl Acad Sci USA 1991; 88(8):3460-3464.
67. Balda MS, Gonzalez-Mariscal L, Matter K et al. Assembly of the tight junction: The role of diacylglycerol. J Cell Biol 1993; 123(2):293-302.
68. Wittchen ES, Haskins J, Stevenson BR. Protein interactions at the tight junction. Actin has multiple binding partners, and ZO-1 forms independent complexes with ZO-2 and ZO-3. J Biol Chem 1999; 274(49):35179-35185.
69. Islas S, Vega J, Ponce L et al. Nuclear localization of the tight junction protein ZO-2 in epithelial cells. Exp Cell Res 2002; 274(1):138-148.
70. Traweger A, Fuchs R, Krizbai IA et al. The tight junction protein ZO-2 localizes to the nucleus and interacts with the heterogeneous nuclear ribonucleoprotein scaffold attachment factor-B. J Biol Chem 2003; 278(4):2692-2700.
71. Fleming TP, Sheth B, Fesenko I. Cell adhesion in the preimplantation mammalian embryo and its role in trophectoderm differentiation and blastocyst morphogenesis. Front Biosci 2001; 6:D1000-D1007.
72. Furuse M, Hirase T, Itoh M et al. Occludin: A novel integral membrane protein localising at tight junctions. Journal of Cell Biology 1993; 123(6(2)):1777-1788.
73. Morita K, Furuse M, Fujimoto K et al. Claudin multigene family encoding four-transmembrane domain protein components of tight junction strands. Proc Natl Acad Sci USA 1999; 96(2):511-516.
74. Tsukita S, Furuse M, Itoh M. Multifunctional strands in tight junctions. Nat Rev Mol Biol 2001; 2:285-293.
75. Sheth B, Moran B, Anderson JM et al. Post-translational control of occludin membrane assembly in mouse trophectoderm: A mechanism to regulate timing of tight junction biogenesis and blastocyst formation. Development 2000; 127(4):831-840.
76. Cordenonsi M, Mazzon E, De Rigo L et al. Occludin dephosphorylation in early development of xenopus laevis. J Cell Sci 1997; 110(Pt 24):3131-3139.
77. Sakakibara A, Furuse M, Saitou M et al. Possible involvement of phosphorylation of occludin in tight junction formation. J Cell Biol 1997; 137(6):1393-1401.
78. Wong V. Phosphorylation of occludin correlates with occludin localization and function at the tight junction. Am J Physiol 1997; 273(6 Pt 1):C1859-C1867.

79. Furuse M, Fujita K, Hiragi T et al. Claudin-1 and-2: Novel integral membrane proteins localising at tight junctions with no sequence similarity to occludin. Journal of Cell Biology 1998; 141:1539-1550.
80. Magnuson T, Jacobson JB, Stackpole CW. Relationship between intercellular permeability and junction organisation in the preimplantation mouse embryo. Developmental Biology 1978; 67:214-224.
81. Furuse M, Itoh M, Hirase T et al. Direct association of occludin with ZO-1 and its possible involvement in the localisation of occludin at tight junction. J Cell Biol 1994; 127:1617-1626.
82. Cordenonsi M, Turco F, D'Atri F et al. Xenopus laevis occludin. Identification of in vitro phosphorylation sites by protein kinase CK2 and its association with cingulin. Eur J Biochem 1999; 264:374-384.
83. Rajasekaran AK, Hojo M, Huima T et al. Catenins and zonula occludens-1 form a complex during early stages in the assembly of tight junctions. J Cell Biol 1996; 132(3):451-463.
84. Larue L, Ohsugi M, Hierchenhain J et al. E-cadherin null mutants fail to form a trophectoderm epithelium. Proc Natl Acad Sci USA 1994; 91:8263-8267.
85. Reithmacher DBV, Birchmeier C. A targeted mutation in the mouse E-cadherin gene results in defective preimplantation development. Proc Natl Acad Sci USA 1995; 92:855-859.
86. Torres M, Stoykova A, Huber O et al. An a-E-catenin gene trap mutation defines its function in preimplantation development. Proc Natl Acad Sci USA 1997; 94:901-906.
87. Ohsugi M, Larue L, Schwartz H et al. Cell-junctional and cytoskeletal organisation in mouse blastocysts lacking E-cadhern. Dev Biol 1997; 185:261-271.
88. Sefton M, Johnson MH, Clayton L. Synthesis and phosphorylation of uvomorulin during mouse early development. Development 1992; 115:313-318.
89. Winkel GK, Ferguson JE, Takeichi M et al. Activation of protein kinase C triggers premature compaction in the 4-cell stage mouse embryo. Dev Biol 1990; 138:1-15.
90. Pauken CM, Capco DG. Regulation of cell adhesion during embryonic compaction of mammalian embryos: Roles for PKC and β-catenin. Mol Reprod Dev 1999; 54:135-144.
91. Dodane V, Kachar B. Identification of isoforms of G proteins and PKC that colocalize with tight junctions. J Membr Biol 1996; 149(3):199-209.
92. Izumi Y, Hirose T, Tamai Y et al. An atypical PKC directly associates and colocalizes at the epithelial tight junction with ASIP, a mammalian homologue of Caenorhabditis elegans polarity protein PAR-3. J Cell Biol 1998; 143(1):95-106.
93. Balda MS, Anderson JM, Matter K. The SH3 domain of the tight junction protein ZO-1 binds to a serine protein kinase that phosphorylates a region C-terminal to this domain. FEBS Lett 1996; 399(3):326-332.
94. Pauken CM, Capco DG. The expression and stage-specific localisation of protein kinase C isotypes during mouse preimplantation development. Dev Biol 2000; 223:411-421.
95. Eckert JJ, McCallum A, Mears A et al. Specific PKC isoforms regulate blastocoel formation during mouse pre-implantation development. Dev Biol 2004; 274:384-401.
96. Etienne-Manneville S, Hall A. Rho-GTPases in cell biology. Nature 2002; 420:629-635.
97. Benais-Pont G, Punn A, Flores-Maldonado C et al. Identification of a tight junction-associated guanine nucleotide exchange factor that activates Rho and regulates paracellular permeability. J Cell Biol 2003; 160(5):729-740.
98. Reima I, Lehtonen E, Virtanen I et al. The cytoskeleton and associated proteins during cleavage, compaction and blastocyst differentiation in the pig. Differentiation 1993; 54:35-45.
99. Barcroft LC, Hay-Schmidt A, Caveney A et al. Trophectoderm differentiation in the bovine embryo: Characterization of a polarized epithelium. J Reprod Fertil 1998; 114(2):327-339.
100. Bloor DJ, Metcalfe AD, Rutherford A et al. Expression of cell adhesion molecules during human preimplantation embryo development. Mol Hum Reprod 2002; 8(3):237-245.
101. Nikas G, Ao A, Winston RM et al. Compaction and surface polarity in the human embryo. Biol Reprod 1996; 55:32-37.
102. Miller DJ, Eckert JJ, Lazzari G et al. Tight junction messenger RNA expression levels in bovine embryos are dependent upon the ability to compact and in vitro culture methods. Biol Reprod 2003; 68(4):1394-1402.
103. Ghassemifar MR, Eckert JJ, Houghton FD et al. Gene expression regulating epithelial intercellular junction biogenesis during human blastocyst development in vitro. Mol Hum Reprod 2003; 9(5):245-252.
104. Ghassemifar MR, Sheth B, Papenbrock T et al. Occludin TM4(-): An isoform of the tight junction protein present in primates lacking the fourth transmembrane domain. J Cell Sci 2002; 115(Pt 15):3171-3180.
105. Eckert JJ, Houghton FD, Fleming TP. Impaired assembly of junctional proteins in human embryos may be regulated by specific PKC isoforms. Theriogenology 2002; 57:635.
106. Eckert JJ, McCallum A, Rumsby MG et al. Protein kinase C isotypes are differentially involved in the biogenesis of the mouse blastocyst. Theriogenology 2001; 55:319.

CHAPTER 13

Tight Junctions and the Blood-Brain Barrier

Hartwig Wolburg, Andrea Lippoldt and Klaus Ebnet

Abstract

The blood-brain barrier protects the neural microenvironment from changes of the blood composition. It is located in the endothelium which is both seamless and interconnected by tight junctions. The restrictive paracellular diffusion barrier goes along with an extremely low rate of transcytosis and the expression of a high number of channels and transporters for such molecules which cannot enter or leave the brain paracellularly.

Many tight junction molecules have been identified and characterized including claudins, occludin, ZO-1, ZO-2, ZO-3, cingulin, 7H6, JAM and ESAM. Signaling pathways involved in tight junction regulation include G-proteins, serine-, threonine-, and tyrosine-kinases, extra- and intracellular calcium levels, cAMP levels, proteases, and cytokines Most of these pathways modulate the connection of the cytoskeletal elements to the tight junction transmembrane molecules. Additionally, crosstalk between components of the tight junction- and the cadherin-catenin-system of the adherens junction suggests a close functional interdependence of the two cell-cell-contact systems.

The blood-brain barrier endothelial cells are situated on top of a basal lamina which contains various molecules of the extracellular matrix. Pericytes and astrocytes directly contact this basal lamina; however, little is known about the signaling pathways between these cell types and the endothelium which possibly are mediated by components of the basal lamina. To understand the interplay between astrocytes, pericytes, the basal lamina and the endothelial cells is a big challenge for understanding the blood-brain barrier in the future.

Introduction

The blood-brain barrier (BBB) protects the microenvironment of the central nervous system from toxins and buffers fluctuations in blood composition.[1,2] The main structures responsible for the barrier properties are the tight junctions (Figs. 1, 2).[3-7] The most important cells responsible for the establishment of the barrier and interconnected by these tight junctions are the capillary endothelial cells in case of the BBB (Figs. 1A,B, 2B) and the epithelial (glial) cells in case of the blood-cerebrospinal fluid barrier (Fig. 2C).

The notion that endothelial cells form an efficient permeability barrier originally came from electron microscopical tracer experiments.[3] Then, Nagy et al[8] investigated the tight junctions of the BBB using the freeze-fracture method and found them the most complex ones in the whole vasculature of the body. In addition to the complexity of the tight junction network, the association of the tight junction particles with the inner (protoplasmatic: P-face) or outer (external: E-face) leaflet of the endothelial membrane has been described as a further parameter of the quality of the permeability barrier in the brain.[9] The BBB tight junctions are unique among all endothelial tight junctions by their specific rate of P-face association (Fig. 2B) which is slightly higher than the E-face association. Interestingly, the P-face/E-face-ratio of BBB tight junctions continuously increases during development[10] and is strongly reduced in

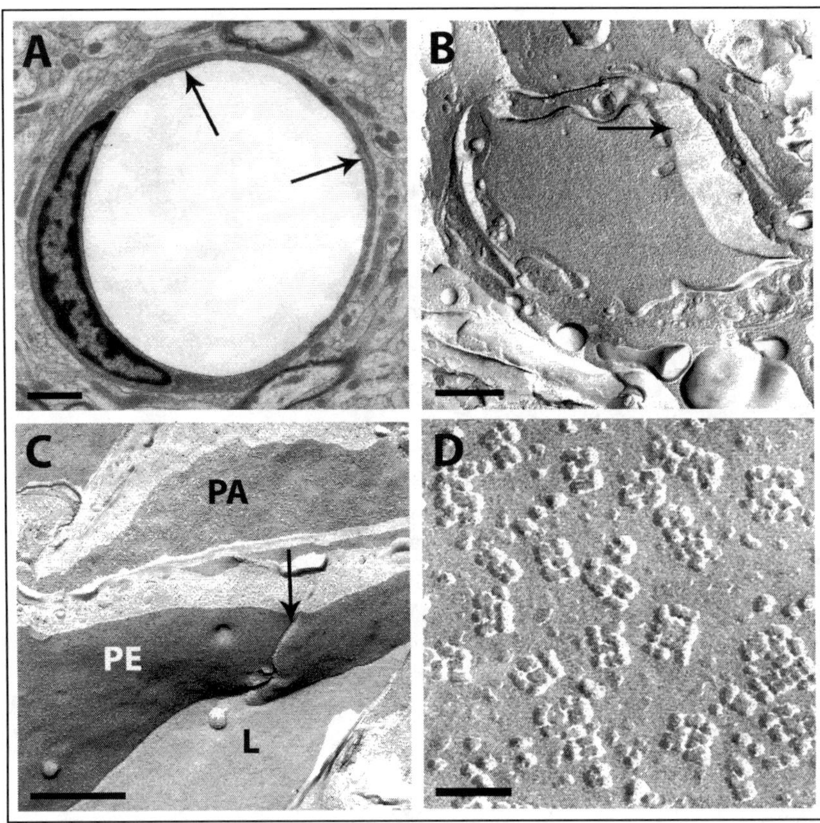

Figure 1. Electron microscopy of the blood-brain barrier. A) Ultrathin section of a mouse brain capillary. The arrows point to two tight junctions. The nucleus of the endothelial cell is at the left-hand side. Bar: 1 µm. B) Freeze-fracture replica of mouse brain capillary. The arrow points to the tight junction (for more details, see Fig. 2). Bar = 0.5 µm. C) Freeze-fracture replica of a rat brain capillary. L = lumen; PE = P-face of the endothelial cell; PA = P-face of the astrocytic endfoot membrane adjacent to the endothelial cells. The arrow points to a tight junction, which is hidden and not exposed as in (B). Bar = 0.5 µm. D) Freeze-fracture replica of a perivascular astrocytic endfoot membrane from rat brain. Many orthogonal arrays of intramembranous particles (OAPs) are a characteristic feature of astrocyte endfoot membranes. Bar = 50 nm.

cell culture (Fig. 2D).[9] The P-face/E-face-ratio of tight junctions of BBB endothelial cells in vitro is identical with that of nonBBB endothelial cells in vivo[11] indicating that the association of the strand particles with the membrane leaflets reflects the quality of the barrier and is under the control of the brain microenvironment (compare Figs. 2B, 2D).

In this chapter we will give a brief overview on structural and molecular features of the BBB tight junctions. Since most data on tight junction regulation were published from epithelial cells,[12] we will discuss some topics on whether these regulatory mechanisms may be valid for BBB endothelial cells as well.

Structure of Blood-Brain Barrier Tight Junctions

Tight junctions are domains of occluded (*"Zonula occludens"*) intercellular clefts[2,13] between endothelial and epithelial cells forming intramembrane networks of strands in freeze-fracture replicas (Fig. 2). If sectioned transversally, the tight junction appears as a system of fusion

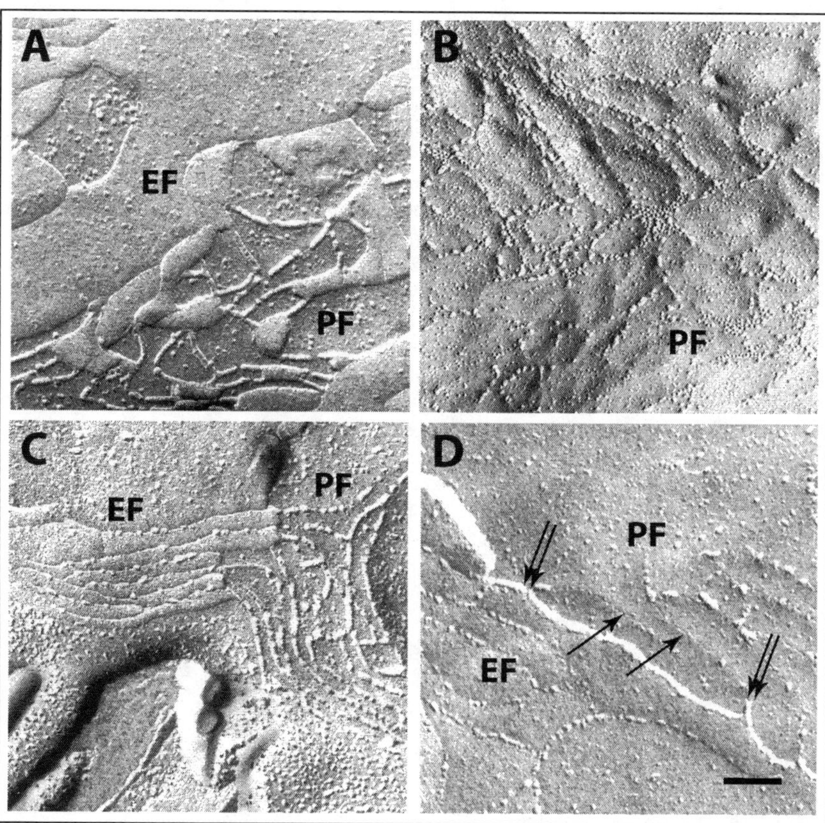

Figure 2. Freeze-fracture replicas of epithelial (A,C) and endothelial (B,D) tight junctions. EF = E-face; PF = P-face; Bar = 100 nm. A) Epithelial tight junctions of EPH4 cells. The strands are associated almost completely with the P-face. B) Endothelial blood-brain barrier tight junctions from capillary fragments freshly prepared from bovine brain. Most strand particles are associated with the P-face. C) Epithelial tight junction from the rat choroid plexus, the site of the blood-cerebrospinal fluid barrier. Most particles are associated with the P-face, forming a poorly anastomosing pattern of mostly parallel strands. This is characteristic of claudin-11 based tight junctions.[201] D) Rat brain capillary endothelial cells in culture have tight junctions almost completely associated with the E-face. This is also typical for nonBBB endothelial tight junctions. Arrows label particle-free ridges of tight junctions on the P-face, double-arrows point to strands which are interrupted where the fracture plane switches from the E- to the P-face underlining the observation that the anchorage of tight junction strands is stronger at the E- than at the P-face under culture conditions.

("kissing") points each of which represents a sectioned strand. Two parameters which can be visualized by freeze-fracture electron microscopy determine the functional quality of tight junctions: the complexity of strands and the association of the particles with the P- or E-face. Tight junction strands of most epithelial cells are associated with the P-face leaving grooves at the E-face which are occupied by only few particles (Fig. 2A,C; e.g., refs. 13-16). Epithelial cells in culture form tight junctions with identical properties as in vivo, namely high P-face association, high electrical resistance and low permeability.[16-18] The P-face association seems to be dependent on the energy metabolism of the cells: after ATP depletion, Madin-Darby canine kidney (MDCK) cells suffer from both deterioration of the paracellular barrier ("gate") function[19,20] and a decreased P-face association of the tight junctions. Also, tight junctions of low and high resistance MDCK cells differ by their association with the membrane leaflets. Low resistance tight junctions had a discontinuous and high resistance tight junctions a continuous particle association at their P-face.[21,22] However, the causal relationship between the degree of particle association to the P-face and the observed transepithelial resistance is imperfectly understood so far.

Tight junctions of endothelial cells have a much lower degree of P-face strand association than those of epithelial cells (compare Fig. 2A,C with Fig. 2B,D). The P-Face/E-face ratio of particle association depends on the vascular bed investigated. In peripheral endothelial cells, the particles of the tight junctions are predominantly associated with the E-face.[11,23] On the P-face, these tight junctions can be recognized only by particle-poor ridges (Fig. 2D). In contrast, the tight junctions of the BBB endothelial cells of mammalian species reveal the highest P-face association found in the vasculature of the body (Fig. 2B). This particle distribution is not maintained in vitro (Fig. 2D). Instead, the freeze-fracture morphology of cultured BBB endothelial cells is similar to nonBBB endothelial cells.[9] This shift in favor of the E-face association of tight junction particles seems to be the most sensitive parameter found so far to characterize the compromised barrier:[24] In asymptomatic adult stroke-prone spontaneously hypertensive rats (SHRSP) a strong reduction of the P-face/E-face ratio went along with a reduced polarity of BBB endothelial cells as assessed by the distribution of the glucose transporter isoform Glut-1. However, an increased permeability for lanthanum as well as an alteration of the immunoreactivities of tight junction proteins were not observed. In this case, the altered E-face/P-face ratio may reflect an alteration of the cytoskeletal anchorage of the tight junctional particles and/or the integrity of tight junctional strands without loss of tight junctional proteins which was described under pathological conditions such as glioblastoma and encephalomyelitis.[25] The importance of the cytoskeletal integrity for BBB maintenance in vivo was clearly demonstrated using cytochalasin B and colchicine treatments. Whereas cytochalasin infusions led to increased BBB permeability in acute hypertension, colchicine pretreatment of acutely hypertensive rats attenuated the permeability changes.[26]

In vitro models of the BBB have been established to investigate regulatory mechanisms.[9,27-37] Hamm et al[38] demonstrated an increased transendothelial permeability for horseradish peroxidase (HRP) after discontinued coculture with astrocytes, but this change in permeability was not paralleled by a change in tight junction protein expression. The authors concluded that loss of localization of tight junction associated proteins from the BBB tight junctions might be a relatively late event, which is not yet observed in the comparatively short term in vitro experiments. Consequently, early and more subtle changes at the tight junctions as discussed above in the SHRSP rat[24] being distinct from the complete loss of tight junction proteins may cause an increase in paracellular permeability towards small molecules.

Some studies have shown that astrocytes or related neuroepithelial cells participate in the induction of barrier properties in endothelial cells (for reviews see refs. 4,6). Humoral factors released from astrocytes were suggested to contribute to tight junction formation,[9,29,38-40] but are not sufficient to induce and maintain BBB characteristics.[9,28] Nevertheless, the glial cell line-derived neurotrophic factor (GDNF) seems to be necessary for BBB induction.[41-43] Recently, the src-suppressed C-kinase substrate (SSeCKS) in astrocytes has been reported to be responsible for the decreased expression of the angiogenic permeability factor vascular endothelial growth factor (VEGF) and the increased release of the anti-permeability factor angiopoietin-1 (Ang-1). SSeCKS overexpression was shown to increase the expression of tight junction molecules and to decrease the paracellular permeability in endothelial cells.[44,45]

Molecular Composition of Blood-Brain Barrier Tight Junctions

In the last ten years the knowledge of the molecular composition and regulation of the tight junctions has exploded (Table 1).[12,46-57] Occludin and the claudin family are the most important membranous components both of which are proteins with four transmembrane domains and two extracellular loops. It seems generally accepted that both loops in one membrane touch the appropriate loops in the adjacent membrane occluding the intercellular cleft.

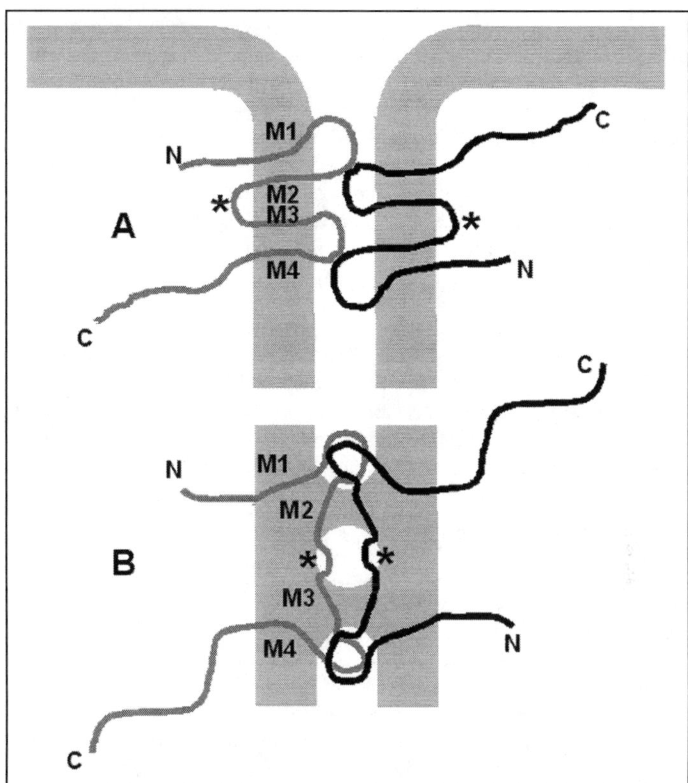

Figure 3. Hypothetical model of occludin/claudin folding in the tight junction. A) conventional protein model, showing both extracellular loops within the intercellular cleft. B) protein/lipid-model assuming an inverse lipidic micelle allowing the external leaflets of the membranes of two partner cells to be in continuity. The difference is that the protein domain between the transmembrane domains M2 and M3 (labelled by *) is in the cytosol (A) or in the micelle (B). Model B may explain the extremely high transcellular electrical resistance of some tight junctions, which are believed not to be realized by the pure protein model A. See text for details.

However, whether this simple molecular contact of peptide chains can serve as the basis of an extremely high transcellular electrical resistance, must seriously be questioned. Instead, there is unequivocal evidence for the continuation of outer membrane leaflets of two adjacent junction-connected partner cells[58,59] suggesting that lipid structures such as inverse micelles are involved in the formation of tight junctions. These micelles could be stabilized by proteins as was proposed recently by Wolburg et al,[60] the hydrophilic domain between transmembrane domain two and three (M2/3) could be located within the micelle instead of the cytoplasm (Fig. 3). In both models, the conventional protein model (Fig. 3A) and the lipid/protein-model (Fig. 3B), the extracellular loops have free access to the intercellular cleft. Indeed, experiments reported by Lacaz-Vieira et al[61] and Wong and Gumbiner[62] demonstrated that peptide molecules homologous to segments of the first external loop of occludin interfere with the resealing of tight junctions opened by Ca^{2+} removal. This increase of the intercellular permeability could be interpreted in favor of the assumption that the mode of protein folding can change and the equilibrium between two folding mechanisms is under the control of the cell. On the other hand, Colegio et al[63] were able to demonstrate recently that extracellular claudin domains determine the paracellular charge selectivity and resistance without changing the fibril architecture. This was done by interchanging the first extracellular loop between claudins-4 and −2. The way by which extracellular peptide loops can alter the paracellular permeability is not clear so far. In any case, combining the protein with the lipidic model of tight junctions seems necessary and overdue in order to explain the highly variable electrical resistances found in epithelial and endothelial cells. Indeed, Kan[64] demonstrated the presence of phospholipids in cylindrical tight junction strands using freeze-fracture labeling with gold-labeled phospholipase A_2. In addition, Yamagata et al[65] reported on the induction of tight junctions and increase of both the transendothelial electrical resistance and the expression level of occludin in brain capillary endothelial cell by polyunsaturated fatty acids.

Occludin

Occludin was the first tight junctional transmembrane molecule discovered.[47] The tight junctions in occludin-deficient mice[66] were not affected morphologically, and transepithelial resistance as measured in small and large intestine epithelial cells was not altered compared to wild-type mice. However, the mice developed chronic inflammation and hyperplasia of the gastric epithelium, calcifications in the brain and around brain vessels, thinning of bones, postnatal growth retardation, testicular atrophy and abnormalities in sexual behavior.[66] The authors concluded that occludin might have a function in tight junction modulation via the induction of intracellular signaling. Moreover, occludin is not required for the formation of tight junction strands. In a number of reports, posttranslational modifications of occludin such as phosphorylation of its cytoplasmic domains or binding to a ubiquitin-ligase[67] have been described as parts of tight junction regulation.[61,62,68-72] For example, DeMaio et al[73] reported on a clear reduction of occludin content in cultured aortic endothelial cells by shear stress (10 dyn/cm^2), but a time-dependent increase of occludin phosphorylation which could be attenuated by dibutyryl cAMP. As well, an increase of occludin phosphorylation has also been described after treatment with VEGF[74] suggesting that hormonal and mechanical changes are able to increase the paracellular permeability by an early increase of occludin phosphorylation and a subsequent decrease of the occludion content. Taken together, it seems that mature cells need occludin to regulate rather than establish their barrier properties.

The Claudin Family

The claudins are the tight junction molecules which seem to fulfill the task of establishing barrier properties (Table 1).[22,48,49,51,75] Claudins share with occludin the overall organization with four transmembrane domains, but have no sequence homology to occludin. The first claudins identified were isolated from chicken liver junctional fractions and were called claudin-1 and claudin-2.[48] Since then a number of related proteins has been identified and at present the claudin family contains more than 20 members.[76,77] It is now believed that claudins are responsible for the regulation of paracellular permeability through the formation of homotypic and heterotypic paired strands (for reviews, see refs. 51-53). In this model, ion selectivity is achieved through the selective expression and combination of distinct claudins in certain tissues.[52] Therefore, it is not surprising that the claudins are not randomly distributed throughout the organs, but at least in part show a tissue-specific expression pattern. For example, claudin-5 was originally described to be restricted to endothelial cells,[49] although it was recently also found in surface cells of the stomach and of the large and small intestine.[78] In addition, claudin-16 is selectively expressed in the thick ascending limb of Henle in the kidney where it regulates selectively the permeability for Mg^{2+} ions.[79]

Functional investigations support the view that the composition of the claudin species directly determines barrier function.[22] Tight junction-negative L-fibroblasts when transfected with claudin-1 or claudin-3 form tight junctions which appeared in freeze-fracture replicas associated with the P-face.[75] When transfected with claudin-2 or claudin-5, the cells form tight junctions associated with the E-face.[48,49] In contrast, occludin was found to be localized at both fracture faces.[80] Whereas occludin induced the formation of short strands, the claudin-induced strands were very long and branched resembling endogenous tight junctions.[51,75] Transfection of MDCK cells with claudin-1 increased the transepithelial resistance about 4-fold and reduced the paracellular flux.[81] Transfection with claudin-2 of high-resistant MDCK I cells that normally express claudin-1 and claudin-4, mimicked both the resistance behavior and the tight junction morphology of low-resistant MDCK II cells.[22] Claudin-4 was formerly known as the *Clostridium perfringens* enterotoxin receptor (CPE-R). By treatment of MDCK I cells, *Clostridium perfringens* enterotoxin (CPE) selectively removed claudin-4 from the tight junctions. Tight junctions were disintegrated to a simple network with only few anastomosing strands, and the transepithelial electrical resistance (TER) was decreased. After CPE removal, barrier properties were reestablished.[82] These results suggested that the combination and stoichiometry of the claudins might be responsible for the outcome of a given resistance or permeability.

The claudins detected in endothelial cells were initially claudin-1 and claudin-5.[25,83-85] However, soon it turned out that the anti-claudin-1 antibody employed recognized claudin-3 as well (for methodological aspects of antibodies against claudins, see ref. 25). The specificity of the anti-claudin antibodies was verified by immunofluorescence staining and western blot analysis of transfected mouse L cells expressing either murine claudin-1 or murine claudin-3 (Engelhardt, Wolburg, Furuse, unpublished). Thus, claudin-3 has been recognized as a novel molecule of BBB endothelial tight junctions.[25] A novel anti-claudin-1 antibody also stained BBB endothelial cells, but the staining was not restricted to the tight junctions[25] suggesting that the molecule is present in the cell but not targeted to the junction. Additionally, claudin-12 has been described in the BBB endothelial cell,[86] so that at least claudin-3, -5 and -12 are the BBB-specific claudins.

Concerning claudin-5, the deficiency of this molecule as reported by the Tsukita group[86] could be expected to be compromising the quality of the BBB and therefore the viability of the animal. However, astonishingly the newborn knockout animals did not differ from the wildtype in terms of macroscopic morphology, and even the electron microscopy of the tight junctions was normal. However, since a low molecular weight tracer (e.g., Hoechst dye) and not the higher molecular weight tracer microperoxidase extravasated in the claudin-5 knock out mouse out of the brain vessels, it was concluded that the claudin-5 based tight junctions restrict the permeability for small molecules < 800 Da.[86] Claudin-12 which was detected as a constituent of the BBB tight junctions, was postulated to be responsible to be selectively restrictive for molecules > 800 Da. Thus, the concept of BBB tight junctional components as selectively regulated molecular sieves is arising which may be important for therapeutic procedures in the future.

As mentioned above, claudin-3 if transfected into cultured fibroblasts is associated with the P-face, whereas claudin-5 is associated with the E-face.[48,82] BBB endothelial cells in vivo reveal a P-face/E-face ratio of about 55/45%;[10] as well, claudin-3 and claudin-5 are well expressed. In nonBBB endothelial cells, tight junctions are almost completely associated with the E-face, and claudin-3 is rarely or not expressed. BBB endothelial cells cultured in vitro develop tight junctions which are associated with the E-face[9] and express less claudin-1;[85] however, an antibody was used now known to recognize claudin-3 as well. Under pathological conditions such as malignant glioma or experimental allergic encephalomyelitis, claudin-1/3 was found to be lost and/or the tight junctions were E-face associated.[25,84]

Immunoglobulin-Like Proteins at Tight Junctions

Almost simultaneously with the identification of claudins, junctional adhesion molecule (JAM) has been reported as the first member of the immunogloblin (Ig)-superfamily to be present at tight junctions (Table 1).[87] JAM, which is now called JAM-A,[88] localizes at homotypic cell-cell contacts of endothelial and epithelial cells and is highly enriched at tight junctions.[87,89] Two Ig-like proteins closely related to JAM-A, JAM-B and JAM-C, have been identified recently.[90-94] With respect to multicellular tissues, JAM-B and JAM-C are restricted to endothelial cells and are largely absent from epithelial cells. Although ultrastructural analyses for JAM-B and JAM-C in endothelial cell are still missing, the colocalization of JAM-C with occludin and ZO-1 upon ectopic expression in MDCK epithelial cells suggests its localization at tight junctions.[93]

Table 1. Molecular composition of tight junctions

Integral Membrane Proteins	Adaptor Proteins 1st Order (Direct)	Adaptor Proteins 2nd Order (Indirect)	Signaling Proteins	Regulatory Proteins
Tight junction strand proteins				
Occludin	**ZO-1, -2, -3**	AF-6, *cingulin*	PI3-kinase, CK2, PKC	*Itch*
Claudins (**Cl-3, -5, -12**)	**ZO-1**	AF-6, *cingulin*	c-Yes, $G\alpha_{12}$	ZONAB
	ZO-2	*cingulin*		c-Jun, c-Fos, C-EBP, SAF-B
	ZO-3 MUPP1	PATJ, *cingulin*		
Ig-superfamily members				
JAM-A	AF-6, **ZO-1** PAR-3, MUPP1	PAR-6	aPKC, PP2A Cdc42	
JAM-C	**ZO-1**, PAR-3	PAR-6	aPKC, PP2A Cdc42	
CAR	?	**ZO-1**		
ESAM	?		Rap-GEF	
JAM4	MAGI-1			
CLMP	?			
Crumbs homologues	Pals-1	PATJ, PAR-6	aPKC, PP2ACdc42	
	Pals-1, PAR-6	PATJ, PAR-6	aPKC, PP2A Cdc42	
CRB1 CRB3				
Unknown				
?	MAGI-2		PTEN	
?	MAGI-3		PTEN, RPTPβ	
—	?		CGEF-H1	
—	?		Rab3B, Rab13	
—	—			huASH1 symplekin
—	—			
—	—		Gas	
—	—		Gai2	
—	—		Ga0	
—	—		RGS5	

The molecular components identified at tight junctions are grouped into different classes based on their structures and functions. The first column displays integral membrane proteins. The second column displays adaptor proteins and distinguishes between 1st order and 2nd order adaptors based on their direct or indirect association with the integral membrane proteins. The third column contains signaling proteins which include tyrosine and serine/threonine kinases as well as heterotrimeric G-proteins, small GTP-binding proteins and guanine-nucleotide exchange factors. Molecules in the fourth column include regulatory proteins such as transcription factors, transcription regulatory proteins or proteins regulating posttranslational modifications. The inclusion of the molecules listed in the third and fourth column in the table is based on either their association with adaptor proteins or their localization at tight junctions. Note that the mere association with a tight junction associated protein does not necessarily mean that the two proteins are associated at tight junctions. Actin-binding proteins such as cortactin, synaptopodin or α-actinin which associate with tight junction proteins and which link the tight junctions to actin cytoskeleton are not included in the table. Molecules which are written in **bold and italics** were described to occur in the blood-brain barrier.

continued on next page

Table 1. Continued

Abbreviations are: AF-6 = ALL-1 fusion partner from chromosome 6; aPKC = atypical PKC; ASH1 = absent, small, or homeotic 1; CAR = coxsackie- and adenovirus receptor; C/EBP = CCAAT/enhancer binding protein; CK2 = casein kinase 2; CLMP = coxsackie- and adenovirus receptor-like membrane protein; CRB = Crumbs; ESAM = endothelial cell-selective adhesion molecule; GEF = guanine nucleotide exchange factor; JAM = junctional adhesion molecule; MAGI = membrane-associated guanylate kinase inverted; MUPP1 = multi-PDZ-domain protein; Pals = protein associated with Lin-7; PAR = partitioning defective; PATJ = Pals-1-associated tight junction protein/protein associated with tight junctions; PI = phosphatidyl-inositol; PKC = protein kinase C; PP2A = protein phosphatase 2A; PTEN = phosphatase and tensin homologue deleted on chromosome 10; RGS = regulator of G-protein signaling; RPTP = receptor protein tyrosine phosphatase; SAF-B = scaffold attachment factor-B; ZO = zonula occludens; ZONAB = ZO-1-associated nucleic acid-binding protein

One role of JAMs in endothelial cells might be related to their predicted function in regulating leukocyte-endothelial cell interaction during inflammation through homophilic and heterophilic interactions. This role is based (i) on the findings that anti-JAM-A and anti-JAM-C antibodies reduce transendothelial migration of leukocytes, and (ii) on the identification of integrin ligands expressed on leukocytes for all three JAMs (see refs. 95,96 for recent reviews). However, additional evidence suggests a role for JAMs in the formation of tight junctions. This is based on the observation that anti-JAM-A antibodies negatively affect the formation of functional tight junctions after Ca^{2+}-switch-induced cell-cell contact formation[89,97] and on the identification of cytosolic proteins which associate with JAMs and which are implicated in the formation/function of tight junctions.

All three JAMs associate with ZO-1 in a PDZ domain-dependent manner.[98-100] ZO-1 most likely acts as a scaffolding protein that helps to organize large protein complexes consisting of transmembrane and cytosolic proteins at tight junctions (see Chapter 6 by A. Fanning), and the direct assocation with JAMs might serve to localize these at the subapical complex. JAM-A, JAM-B and JAM-C also associate with ASIP/PAR-3,[100-102] and this interaction might be important for tight junction biogenesis. PAR-3 is part of a ternary protein complex which is conserved through evolution and which localizes to tight junctions of vertebrate epithelial cells.[103] This complex consists of PAR-3, atypical PKC (aPKC) and PAR-6. Overexpression of dominant-negative mutants of all three components during cell-cell contact formation results in the mislocalization of tight junction proteins as well as in functional defects of tight junctions such as a decrease in TER, increase in paracellular permeability and a loss of membrane polarity.[104-107] The findings that the dominant-negative mutants in all cases are effective when overexpressed in the process of cell contact formation but not in fully polarized epithelial cells implicates a role of the PAR-3-aPKC-PAR-6 complex for the formation of tight junctions rather than for their maintenance.[103] The recruitment of the PAR-3-aPKC-PAR-6 complex to early cell-cell contact sites and the subsequent activation of aPKC by Cdc42/Rac-1 might promote the formation of tight junctions from primordial adherens junctions and thus the polarization of the cell. In accordance with this, primordial, spot-like adherens junctions can be formed in the presence of dominant-negative aPKC, but their maturation into adherens junctions and tight junctions is blocked.[108] JAM-A which is present early at cell-cell contact sites at primordial, spot-like adherens junctions could serve to recruit the PAR-3-aPKC-PAR-6 complex to early sites of cell adhesion.[101,108]

The roles of JAM-B and JAM-C in tight junction formation is less well understood. As mentioned above, both are predominantly expressed by endothelial cells. Like JAM-A, both associate with PAR-3 through PDZ domain 1 of PAR-3 and JAM-C recruits endogenous PAR-3 to cell-cell contact sites in CHO cells[100] suggesting that JAM-B and JAM-C contribute to the recruitment of the PAR-3-aPKC-PAR-6 complex in endothelial cells. It is striking that among all integral membrane proteins analyzed so far, including JAM-A, JAM-B, JAM-C, occludin, claudin-1, claudin-4, claudin-5, CAR and ESAM, PAR-3 associates exclusively with JAM-A, JAM-B and JAM-C[100,101] suggesting a specific role for JAMs in tight junction formation.

More recently, four additional Ig-superfamily members have been identified at tight junctions (Table 1). These include the coxsackie- and adenovirus receptor (CAR,[109]), endothelial cell-selective adhesion molecule (ESAM,[110]), junctional adhesion molecule (JAM) 4[111] (see also Chapter 4) and coxsackie- and adenovirus receptor-like membrane protein (CLMP).[112] They share with JAM-A, JAM-B and JAM-C a similar organization with two Ig-like domains. However, they are more closely related among each other than to the three JAMs and thus form a subfamily within tight junction-associated Ig-superfamily members.[96] Interestingly, CAR, ESAM and JAM4 end in a type I PDZ domain-binding motif whereas JAM-A, JAM-B and JAM-C end in a type II motif which suggests functional differences between the two subfamilies. The function of these four Ig-superfamily members at tight junctions is not clear. CAR, JAM4 and CLMP are predominantly expressed by epithelial cells, whereas ESAM is expressed exclusively in endothelial cells including those in brain capillaries.[110,113] Endothelial cells derived from ESAM-deficient mice display defects in endothelial tube formation suggesting a role for ESAM in endothelial cell contact formation.[114] How this function relates to its specific localization at tight junctions is not yet clear.

Peripheral Membrane Components at Tight Junctions

The transmembrane proteins associate in the cytoplasm with peripheral membrane components which form large protein complexes, the cytoplasmic "plaque" (Table 1). One type of plaque proteins consists of adaptors, proteins with multiple protein-protein interaction domains such as SH-3 domains, guanylate kinase (GK) domains and PDZ domains.[115,116] The adaptor proteins include members of the MAGUK (membrane-associated guanylate kinase,[117] and MAGI (membrane-associated guanylate kinase with an inverted orientation of protein-protein interaction domains[118] families, such as ZO-1, -2, -3, Pals-1, MAGI-1, -2, -3, as well as proteins with one or several PDZ domains such as PAR-3, PAR-6, PATJ and MUPP1.[119] The latter two proteins contain ten and 13 PDZ domains,[120,121] respectively, and thus seem to be particularly well suited to assemble large protein complexes. The adaptor proteins serve as scaffolds to organize the close proximity of the second type of plaque proteins, the regulatory and signalling proteins. These include small GTPases like Ras, Rab13 or Cdc42,[122] and their regulators, e.g., guanine nucleotide exchange factors,[123] protein kinases and phosphatases such as aPKC, PP2A and PTEN[124-127] as well as transcriptional regulators like ZONAB, huASH1, Jun, Fos and the CCAAT/enhancer binding protein (C/EBP).[128-130] In many cases, the role of the regulatory and signaling proteins in tight junction biology is still poorly understood, and it is to be expected that they are involved in completely different aspects of tight junction biology. Some proteins might be involved in the regulation of tight junction formation from preexisting primordial adherens junctions, for example the proteins of the PAR-3-aPKC-PAR-6.[103] Some proteins might be necessary for signaling to the cell interior, for example the transcription factors ZONAB and huAsh1. Information on the maturation state of cell-cell contacts is required for many cellular events which are regulated by cell density (e.g., proliferation), and transcription factors associated with the cytoplasmic plaque at tight junction provide a direct link between tight junctions and the nucleus.[12] In summary, the large number of scaffolding and signaling proteins at tight junctions points to a highly complex protein machinery that is required to regulate the formation and maintenance of tight junctions; it also points to tight junctions as signalling centers that influence various cellular processes.

Regulatory Mechanisms in the Blood-Brain Barrier

Tight junctions have long been considered as static barriers responsible for both the compartmentalization of the intercellular cleft (gate function) and the polarity of cell (fence function). But in the last years it came out that the tight junctions are under the control of multiple regulatory systems in which many molecular systems such as adhesion molecules, extracellular matrix components and signal transduction pathways are involved.[12,131] Most data were gathered in epithelial cells. In endothelial cells, in particular in those of the BBB, less data are available, probably due to the fact that the system is considerably more complex as it includes pericytes and astrocytes and a special composition of the extracellular matrix, and not well established in vitro.

G-Protein Signaling

G-proteins play an essential role in maintaining barrier integrity of epithelial cells.[12] The G-proteins involved are the classical heterotrimeric G-proteins and the small G-proteins/ small GTPases (Ras superfamily). Several $G\alpha$ subunits have been localized within the tight junctions of cultured epithelial cells, such as $G\alpha_{i2}$,[132] $G\alpha s$,[133] $G\alpha_{12}$[134] and $G\alpha_O$, when transfected into MDCK cells.[129] Moreover, $G\alpha_O$ could be coprecipitated with ZO-1.[132] The activation of heterotrimeric G-proteins leads to activation of second messengers like cAMP/cGMP or Ca^{++} and to an increased transepithelial resistance of the cell monolayer. Additionally, $G\alpha_O$ accelerated tight junction biogenesis, whereas $G\alpha_{i2}$ was important for development and maintenance of the tight junctions.[133,135] The inhibition of the protein kinase A (PKA) has been shown to result in the preservation of tight junctions and low permeability in MDCK cells during removal of calcium,[136] suggesting that in epithelial cells the PKA could be involved in the destabilization of tight junctions.[137] As activated $G\alpha_i$ evokes a decrease in cAMP and thus probably a decrease in the amount of activated PKA, G-protein signaling could influence epithelial permeability via inhibiting the cAMP/PKA-pathway. In contrast, in endothelial cells elevation of cAMP by forskolin resulted in a stabilization of tight junctions and decrease of permeability.[9,28,138] Accordingly, elevation of cAMP by forskolin or cholera toxin was able to reverse the permeability increasing effect of pertussis toxin (PTX) on cerebral endothelial cells in vitro.[139] PTX is known to inhibit G-protein signaling by ADP-ribosylation of $G\alpha_i$ proteins. The pathway by which PTX permeabilized the barrier was shown to include the protein kinase C (PKC),[139] probably by operating via extracellular signal-regulated kinase (ERK)-activation.[140]

Recently, a family of G-protein regulating proteins has been identified. These so called RGS proteins (regulators of G-protein signaling) interact with the $G\alpha$ subunit of heterotrimeric G-proteins. They inactivate the $G\alpha$ subunit by accelerating GTP hydrolysis.[141] Interestingly, a genomic suppression subtractive hybridization (SSH) approach has been used to identify specific genes expressed at the BBB.[142] Among other genes, RGS5 was found to be expressed at the BBB but also in other tissues. By using the same approach in hypertensive rats, the RGS5 mRNA was identified in isolated brain capillaries and localized in the BBB endothelial cells.[143] Immunocytochemical detection of RGS5 at the light and electron microscope level revealed its occurrence in tight junctions of BBB endothelial cells (Lippoldt et al, in preparation). However, the function of RGS5 in endothelial cells is not yet clear, nor the signaling pathway RGS5 is acting on.

Small GTPases

The RhoA and Rac1 small GTPases were shown to play a promoting role in the regulation of tight and adhesion junction structure and function.[12,144,145] In MDCK-cells expressing RhoA and Rac1 mutants, the organization of tight junctions is disturbed and the permeability for inulin, anionic or neutral dextran increased and the TER decreased.[146,147] In T84 cells, the inhibition of the Rho pathway by *Clostridium botulinum* toxin C3 transferase resulted in a disorganization of the perijunctional actin ring and ZO-1 distribution, whereas the transient expression of RhoC resulted in actin concentration at intercellular contacts.[148] On the other hand, *E. coli* cytotoxic necrotizing factor-1 (CNF-1) activated the Rho pathway, but reduced the gate function of tight junctions in T84 cells and impaired tight junction assembly in the calcium switch assay.[149] In addition, a guanine nucleotide exchange factor (GEF) has been identified which activated Rho, associated with tight junctions, and increased the paracellular permeability in MDCK cells.[123]

Regarding cerebral endothelial cells, the activation of the Rho pathway in vitro by lysophosphatidic acid (LPA)[150] disrupted the paracellular barrier.[151] Accordingly, the inhibition of the Rho pathway prevented the LPA-induced increase in permeability.[71] Inflammation, as caused by bradykinin, thrombin, histamin cytokines, matrix metalloproteinases,[152-157] or by bacterial toxins[144,158-160] increases the transendothelial permeability by affecting tight and adherens junctions via reorganization of the actin cytoskeleton and formation of intercellular gaps. As well, the transendothelial migration of lymphocytes was shown to be dependent on Rho signaling: when activation of Rho as a consequence of leukocyte binding to the adhesion receptor ICAM-1 was blocked by C3-transferase the leukocyte transmigration was compromised.[160] Whereas the extracellular domain of endothelial ICAM-1 suffices to mediate T cell adhesion, the cytoplasmic domain is required to mediate transmigration of T cells probably by inducing Rho-signaling within the endothelial cells.[161] In human umbilical vein endothelial cells (HUVECs), affection of the permeability requires the activation of the Cdc42-, rac-, Rho-cascade following stimulation by TNF-α.[152] Furthermore, in vitro studies demonstrated that a Rho/Rho kinase-dependent pathway is a central target for an increase of vascular permeability by inactivating the myosin light chain (MLC) phosphatase, thus enhancing MLC phosphorylation. This leaded to endothelial cell contraction and increased permeability.[159]

The Putative Role of the Extracellular Matrix on BBB Regulation

The BBB is under the strict control of the brain microenvironment which is composed of neuronal and glial cells, pericytes, and the extracellular matrix (ECM). The basal lamina of cerebral endothelial cells is very complex and consists of various collagens, laminin, fibronectin, entactin, thrombospondin, as well as heparan and chondroitin sulfate proteoglycans.[162-166] Many informations on the functional impact of the ECM on the vasculature in the brain came from cell culture studies or those investigating conditions during aging, tumor growth or inflammation (e.g., see refs. 165,167-171). For example, the matrix metalloproteinase (MMP)-9 deficient mice were protected from transient focal ischemia by attenuation of serum extravasation via the BBB and reduction of the lesion volume demonstrating the beneficial role of ECM components on the integrity of the BBB under these conditions.[172]

The heparan sulfate proteoglycan agrin originally characterized as the essential molecule for clustering acetylcholine receptors at the motor endplate,[173,174] has also been described as important within the CNS and for the integrity of the BBB.[162,175,176] The agrin isoform Y0Z0 was reported to be specifically present in the CNS capillary endothelial cell basal lamina.[177] Agrin binds to α-dystroglycan,[178] but also to some integrins and heparin.[174] α-dystroglycan is a member of the dystrophin-dystroglycan complex which localizes at glial endfeet membranes.[179] Under conditions of BBB disruption, agrin is lost.[175,180] The possible implication of this agrin loss for the integrity of the BBB will be discussed in the following paragraph of this chapter.

The Astroglial Cell as an Organizer of the BBB

It is generally accepted now that the astrocytes play a decisive role in the maintenance if not induction of the BBB.[5,7,38,45,46,138,181-183] Unfortunately, there is little known about the molecules that are involved in this induction process. An interesting correlation exists between astroglial differentiation and BBB maturation. The astroglial cells form processes to many compartments of the CNS including synapses, Ranvier nodes, and neural-mesenchymal borders at the surface of the CNS and around vessels (Fig. 1C). These extensions at the glial limiting border are called "endfeet". Their membranes are characterized by a very special molecular architecture. They carry plenty of transporters and ion channels, but also the dystrophin-dystroglycan complex,[179] aquaporin-4, the so-called orthogonal arrays of particles (OAPs) (Fig. 1D),[184] and a recently identified member of the immunoglobulin superfamily, limitrin.[185] Within the neuropil, the parenchymal astroglial membranes do not express these molecules and membrane structures or only at a very low level. This polarization of astrocytes arises concomitantly with the maturation of the BBB[183,185-187] and is not maintained by cultured astrocytes.[184] Regarding the OAPs, it is well-known that they consist of the water channel protein aquaporin-4 (AQP4). Aquaporins mediate water movements between the intracellular, interstitial, vascular and ventricular compartments which are under the strict control of osmotic and hydrostatic pressure gradients.[188,189] The involvement of AQP4 in the OAP formation was demonstrated by the absence of OAPs in astrocytes of the AQP4-deficient mouse,[190] by formation of OAPs in chinese hamster ovary cells stably transfected with AQP4 cDNA,[191] and by the immunogold fracture-labeling technique showing that AQP4 is a component of the arrays.[192] Moreover, Nielsen et al[193] were able to demonstrate by immunogold immunocytochemistry that the distribution of the AQP4-related immunoreactivity was identical to that of the OAPs. It should be stressed that AQP4 is the only member of the aquaporin family which is associated with a membrane structure demonstrable by electron microscopy (Fig. 1D).

The OAP/AQP4-related polarity of glial cells has been demonstrated to decrease under brain tumor conditions.[194] The density of OAPs in membranes of glioma cells was found to be extremely low,[194] whereas the AQP4 content as detected by immunocytochemistry was increased.[195] These findings, if taken together, suggest that under glioma conditions AQP4 exists separated from the OAP in the membrane and is no more restricted to the glial membranes contacting the basal membrane. There was a clear positive correlation between AQP4 restriction at the endfoot membrane and presence of agrin in the vessel basal lamina.[196] Importantly, agrin does not bind to AQP4, but to α-dystroglycan.[178] AQP4 is connected to the dystrophin-dystroglycan complex via a PDZ-binding domain in the C-terminus of α-syntrophin.[197] Therefore, we believe that loss of agrin reduces the OAP/AQP4-related polarity of astrocytes leading to a redistribution of "free" AQP4 outside the OAP structure, and that this may be one reason for the loss of capability of the glial cell to maintain the BBB properties within the endothelial cell. Indeed, loss of agrin has been demonstrated to correlate with loss of different tight junction molecules in glioma capillary endothelial cells.[180]

Conclusions

It has been 120 years since the discovery of the BBB by Paul Ehrlich.[198] In the following period blood-brain barrier research has focused on the morphological description of the barrier using mainly conventional histological and electron microscopical methods e.g[1-3] as well as methods to demonstrate the tightness of the barrier against a variety of low and high molecular weight substances[4] (for review see ref. 199). Tight junctions have been described as a network of protein particles using freeze-fracture electron microscopy. But, it needed almost 30 years and a great methodological advance in molecular biology to discover protein components of the tight junctional complex and to identify the protein particles within the replicas. Moreover, due to the advances in molecular biological methods like knock out and GFP-technologies it is now possible to recognize the tight junctional complex as a very dynamic structure. Using these approaches[200] it turned out that paired claudin strands are the molecular equivalents of the so called "kissing points" seen in transmission electron microscopy. In vitro time laps studies of annealing tight junction strands possessing different types of claudins for the first time demonstrated the high dynamic organization potential of the tight junctional complex.[200] Now, it might be possible to go more into detail to characterize the interplay of these molecules and their regulation by and association with cellular signaling cascades and cytoskeletal components. These data then will allow us to use this knowledge for therapeutic interventions in pathologies influencing blood-brain barrier integrity.

The enormous complexity of the endothelial barrier in the brain and its regulation seems to be clearly different from that in epithelial barriers. The simple observation that epithelial, but not endothelial cells, are able to form a high resistant and low permeability barrier in vitro, sheds light to the importance of the brain microenvironment in the formation and maintenance of the barrier in vivo. This microenvironment consists of a cocktail of neuronal, astroglial, microglial, pericytic factors and those of the extracellular matrix which itself forms a microcosmos of its own. All components may or may not operate together or at certain periods during development or during pathological derangements in certain combinations which are not recognizable to us so far. It will be a great challenge in the future to characterize the mechansims involved in BBB differentiation and derangement and to understand the interplay of the BBB components in the adult CNS.

References

1. Brightman MW, Reese TS. Junctions between intimately apposed cell membranes in the vertebrate brain. J Cell Biol 1969; 40:648-677.
2. Begley DJ, Brightman MW. Structural and functional aspects of the blood-brain barrier. Progr Drug Res 2003; 61:39-78.
3. Reese TS, Karnovsky MJ. Fine structural localization of a blood-brain barrier to exogenous peroxidase. J Cell Biol 1967; 34:207-217.
4. Mollgard K, Saunders NR. The development of the human blood-brain and blood-CSF barriers. Neuropathol appl Neurobiol 1986; 12:337-358.
5. Wolburg H, Lippoldt A. Tight junctions of the blood-brain barier: Development, composition and regulation. Vascular Pharmacol 2002; 38:323-337.
6. Vorbrodt AW, Dobrogowska DH. Molecular anatomy of intercellular junctions in brain endothelial and epithelial barriers: Electron microscopist's view. Brain Res Rev 2003; 42:221-242.
7. Engelhardt B. Development of the blood-brain barrier. Cell Tissue Res 2003; 314:119-129.
8. Nagy Z, Peters H, Hüttner I. Fracture faces of cell junctions in cerebral endothelium during normal and hyperosmotic conditions. Lab Invest 1984; 50:313-322.
9. Wolburg H, Neuhaus J, Kniesel U et al. Modulation of tight junction structure in blood-brain barrier ECs. Effects of tissue culture, second messengers and cocultured astrocytes. J Cell Sci 1994; 107:1347-1357.
10. Kniesel U, Risau W, Wolburg H. Development of blood-brain barrier tight junctions in the rat cortex. Dev Brain Res 1996; 96:229-240.
11. Mühleisen H, Wolburg H, Betz E. Freeze-fracture analysis of endothelial cell membranes in rabbit carotid arteries subjected to short-term atherogenic stimuli. Virch Arch B Cell Pathol 1989; 56:413-417.
12. Matter K, Balda MS. Signalling to and from tight junctions. Nature Rev Mol Biol 2003; 4:225-236.
13. Farquhar MG, Palade GE. Junctional complexes in various epithelial. J Cell Biol 1963; 17:375-412.
14. Bentzel CJ, Hainau B, Ho S et al. Cytoplasmic regulation of tight junction permeability: Effect of plant cytokinins. Am J Physiol 1980; 239:C75-C89.
15. Martinez-Palomo A, Meza I, Beaty G et al. Experimental modulation of occluding junctions in a cultured transporting epithelium. J Cell Biol 1980; 87:736-745.
16. Madara JL, Dharmsathaphorn K. Occluding junction structurefunctiuon relationships in a cultured epithelial monolayer. J Cell Biol 1985; 101:2124-2133.
17. González-Mariscal L, Chavez de Ramirez B, Cereijido M. Tight junction formation in cultured epithelial cells (MDCK). J Membrane Biol 1985; 86:113-121.
18. Gumbiner B, Simons K. A functional assay for proteins involved in establishing an epithelial occluding barrier: Identification of an uvomorulin-like polypeptide. J Cell Biol 1986; 102:457-468.
19. Mandel LJ, Bacallao R, Zampighi G. Uncoupling of the molecular "fence" and paracellular "gate" functions in epithelial tight junctions. Nature 1993; 361:552-555.
20. Bacallao R, Garfinkel A, Monke S et al. ATP-depletion: A novel method to study junctional properties in epithelial tissues. I. Rearrangement of the actin cytoskeleton. J Cell Sci 1994; 107:3301-3313.
21. Stevenson BR, Anderson JM, Goodenough DA et al. Tight junction structure and ZO-1 content are identical in two strains of Madin-Darby canine kidney cells which differ in transepithelial resistance. J Cell Biol 1988; 107:2401-2408.
22. Furuse M, Furuse K, Sasaki H et al. Conversion of Zonulae occludentes from tight to leaky strand type by introducing claudin-2 into Madin-Darby canine kidney I cells. J Cell Biol 2001; 153:263-272.
23. Simionescu M, Simionescu N, Palade GE. Segmental differentiations of cell junctions in the vascular endothelium. Arteries and veins. J Cell Biol 1976; 68:705-723.
24. Lippoldt A, Kniesel U, Liebner S et al. Structural alterations of tight junctions are associated with loss of polarity in stroke-prone spontaneously hypertensive rat blood-brain barrier endothelial cells. Brain Res 2000; 885:251-261.
25. Wolburg H, Wolburg-Buchholz K, Kraus J et al. Localization of claudin-3 in tight junctions of the blood-brain barrier is selectively lost during experimental autoimmune encephalomyelitis and human glioblastoma multiforme. Acta Neuropathol 2003; 105:586-592.

26. Nag S. Role of the endothelial cytoskeleton in blood-brain barrier permeability to protein. Acta Neuropathol 1995; 90:454-460.
27. Méresse S, Dehonk M-P, Delorme P et al. Bovine brain ECs express tight junctions and monoamine oxidase activity in long-term culture. J Neurochem 1989; 53:1363-1371.
28. Rubin LL, Hall DE, Porter S et al. A cell culture model of the blood-brain barrier. J Cell Biol 1991; 115:1725-1736.
29. Tontsch U, Bauer HC. Glial cells and neurons induce blood brain barrier related enzymes in cultured cerebral ECs. Brain Res 1991; 539:247-253.
30. Abbott NJ, Hughes CCW, Revest PA et al. Development and characterization of a rat capillary endothelial culture. Towards an in vitro BBB. J Cell Sci 1992; 103:23-38.
31. Pekny M, Stanness KA, Eliasson C et al. Impaired induction of blood-brain barrier properties in aortic endothelial cells by astrocytes from GFAP-deficient mice. Glia 1998; 22:390-400.
32. Stanness KA, Neumaier JF, Sexton TJ et al. A new model of the blood-brain barrier: Coculture of neuronal, endothelial and glial cells under dynamic conditions. NeuroReport 1999; 10:3725-3731.
33. Franke H, Galla H-J, Beuckmann CT. Primary cultures of brain microvessel endothelial cells: A valid and flexible model to study drug transport through the blood-brain barrier in vitro. Brain Res Protocols 2000; 5:248-256.
34. Gaillard PJ, Voorwinden LH, Nielsen JL et al. Establishment and functional characterization of an in vitro model of the blood-brain barrier, comprising a coculture of brain capillary endothelial cells and astrocytes. Europ J Pharmac Sci 2001; 12:215-222.
35. Cucullu L, McAllister MS, Kight K et al. A new dynamic in vitro model for the multidimensional study of astrocyte-endothelial cell interactions at the blood-brain barrier. Brain Res 2002; 951:243-254.
36. Nitz T, Eisenblätter T, Psathaki K et al. Serum-derived weaken the barrier properties of cultured porcine brain capillary endothelial cells in vitro. Brain Res 2003; 981:30-40.
37. Parkinson FE, Friesen J, Krizanac-Bengez L et al. Use of a three-dimensional in vitro model of the rat blood-brain barrier to assay nucleoside efflux from brain. Brain Res 2003; 980:233-241.
38. Hamm S, Dehouck B, Kraus J et al. Astrocyte mediated modulation of blood-brain barrier permeability does not correlate with a loss of tight junction proteins from the cellular contacts. Cell Tissue Res 2004; 315:157-166.
39. Arthur FE, Shivers RR, Bowman PD. Astrocyte-mediated induction of tight junctions in brain capillary endothelium: An efficient in vitro model. Dev Brain Res 1987; 36:155-159.
40. Dehouck B, Dehouck M-P, Fruchart J-C et al. Upregulation of the low density lipoprotein receptor at the blood-brain barrier: Intercommunications between brain capillary ECs and astrocytes. J Cell Biol 1994; 126:465-474.
41. Igarashi Y, Utsumi H, Chiba H et al. Glial cell line-derived neurotrophic factor (GDNF) enhances barrier function of endothelial cells forming the blood-brain barrier. Biochem Biophys Res Comm 1999; 261:108-112.
42. Utsumi H, Chiba H, Kamimura Y et al. Expression of GFRa-1, receptor for GDNF, in rat brain capillary during postnatal development of the BBB. Am J Physiol 2000; 279:C361-C368.
43. Yagi T, Jikihara I, Fukumura M et al. Rescue of ischemic brain injury by adenoviral gene transfer of glial cell line-derived factor after transient global ischemia in gerbils. Brain Res 2000; 885:273-282.
44. Rieckmann P, Engelhardt B. Building up the blood-brain barrier. Nature Med 2003; 9:828-829.
45. Lee S-W, Kim W-J, Choi YK et al. SSeCKS regulates angiogenesis and tight junction formation in blood-brain barrier. Nature Med 2003; 9:900-906.
46. Ando-Akatsuka Y, Saitou M, Hirase T et al. Interspecies diversity of the occludin sequence: cDNA cloning of human, mouse, dog, and rat-kangaroo homologues. J Cell Biol 1996; 133:43-48.
47. Furuse M, Hirase T, Itoh M et al. Occludin: A novel integral membrane protein localizing at tight junctions. J Cell Biol 1993; 123:1777-1788.
48. Furuse M, Fujita K, Hiiragi T et al. Claudin-1 and -2: Novel integral membrane proteins localizing at tight junctions. J Cell Biol 1998; 141:1539-1550.
49. Morita K, Furuse M, Fujimoto K et al. Claudin multigene family encoding four-transmembrane domain protein components of tight junction strands. Proc Natl Acad Sci USA 1999; 96:511-516.
50. Balda MS, Matter K. Transmembrane proteins of tight junctions. Sem Cell Develop Biol 2000; 11:281-289.
51. Tsukita S, Furuse M. Occludin and claudins in tight-junction strands: Leading or supporting players? Trends Cell Biol 1999; 9:268-273.
52. Tsukita S, Furuse M. Pores in the wall: Claudins constitute tight junction strands containing aqueous pores. J Cell Biol 2000; 149:13-16.
53. Tsukita S, Furuse M, Itoh M. Multifunctional strands in tight junctions. Nature Reviews Mol Cell Biol 2001; 2:285-293.

54. Heiskala M, Peterson PA, Yang Y. The roles of claudin superfamily proteins in paracellular transport. Traffic 2001; 2:92-98.
55. Huber JD, Egleton RD, Davis TP. Molecular physiology and pathophysiology of tight junctions in the blood-brain barrier. Trends Neurosci 2001; 24:719-725.
56. D'Atri F, Citi S. Molecular complexity of vertebrate tight junctions. Mol Membr Biol 2002; 19:103-112.
57. González-Mariscal L, BetanzosA, Nava P et al. Tight junction proteins. Progr Biophys Mol Biol 2003; 81:1-44.
58. Hein M, Madefessel C, Haag B et al. Reversible modulation of transepithelial resistance in high and low resistance MDCK-cells by basic amino acids, Ca^{2+}, protamine and protons. Chem Phys Lipids 1992; 63:223-233.
59. Grebenkämper K, Galla H-J. Translational diffusion measurements of a fluorescent phospholipid between MDCK-1 cells support the lipid model of the tight junctions. Chem Phys Lipids 1994; 71:133-143.
60. Wolburg H, Liebner S, Lippoldt A. Freeze-fracture studies of cerebral endothelial tight junctions. Meth Mol Med 2003; 89:51-66.
61. Lacaz-Vieira F, Jaeger MMM, Farshori P et al. Small synthetic peptides homologous to segments of the first loop of occludin impair tight junction resealing. J Membrane Biol 1999; 168:289-297.
62. Wong V, Gumbiner BM. A synthetic peptide corresponding to the extracellular domain of occludin perturbs the tight junctioin permeability barrier. J Cell Biol 1997; 136:399-409.
63. Colegio OR, Van Itallie C, Rahner C et al. Claudin extracellular domains determine paracellular charge selectivity and resistance but not tight junction fibril architecture. Am J Physiol Cell Physiol 2003; 284:C1346-C1354.
64. Kan FWK. Cytochemical evidence for the presence of phospholipids in epithelial tight junction strands. J Histochem Cytochem 1993; 41:649-656.
65. Yamagata K, Tagami M, Takenaga F et al. Polyunsaturated fatty acids induce tight junctions to form in brain capillary endothelial cells. Neuroscience 2003; 116:649-656.
66. Saitou M, Furuse M, Sasaki H et al. Complex phenotype of mice lacking occludin, a component of tight junction strands. Mol Biol Cell 2000; 11:4131-4142.
67. Traweger A, Fang D, Liu Y-C et al. The tight junction specific protein occludin is a functional target of the E3 ubiquitin-protein ligase Itch. J Biol Chem 2002; 277:10201-10208.
68. Sakakibara A, Furuse M, Saitou M et al. Possible involvement of phosphorylation of occludin in tight junction formation. J Cell Biol 1997; 137:1393-1401.
69. Chen Y, Merzdorf C, Paul DL et al. COOH terminus of occludin is required for tight junction barrier function in early Xenopus embryos. J Cell Biol 1997; 138:891-899.
70. Hirase T, Kawashima S, Wong EYM et al. Regulation of tight junction permeability and occludin phosphorylation by RhoA-p160ROCK-dependent and –independent mechanisms. J Biol Chem 2001; 276:10423-10431.
71. Balda MS, Flores-Maldonado C, Cereijido M et al. Multiple domains of occludin are involved in the regulation of paracellular permeability. J Cell Biochem 2000; 78:85-96.
72. Huber D, Balda MS, Matter K. Occludin modulates transepithelial migration of neutrophils. J Biol Chem 2000; 275:5773-5778.
73. DeMaio L, Chang YS, Gardner TW et al. Shear stress regulates occludin content and phosphorylation. Am J Physiol 2001; 281:H105-113.
74. Antonetti DA, Barber AJ, Hollinger LA et al. Vascular endothelial growth factor induces rapid phosphorylation of tight junction proteins occludin and zonula occludens 1. A potential mechnism for vascular permeability in diabetic retinopathy and tumors. J Biol Chem 1999; 174:23463-23467.
75. Furuse M, Sasaki H, Tsukita S. Manner of interaction of heterogenous claudin species within and between tight junction strands. J Cell Biol 1999; 147:891-903.
76. Morita K, Furuse M, Fujimoto K et al. Claudin multigene family encoding four-transmembrane domain protein components of tight junction strands. Proc Natl Acad Sci USA 1999; 96:511-516.
77. Mitic LC, Van Itallie CM, Anderson JM. Molecular physiology and pathophysiology of tight junctions. I. Tight junction structure and function: Lessons from mutant animals and proteins. Am J Physiol 2000; 279:G250-G254.
78. Rahner C, Mitic LL, Anderson JM. Heterogeneity in expression and subcellular localization of claudins 2, 3, 4, and 5 in rat liver, pancreas, and gut. Gastroenterology 2001; 120:411-422.
79. Simon DB, Lu Y, Choate KA et al. Paracellin-1, a renal tight junction protein required for paracellular Mg^{2+} resorption. Science 1999; 285:103-106.
80. Hirase T, Staddon JM, Saitou M et al. Occludin as a possible determinant of tight junction permeability in ECs. J Cell Sci 1997; 110:1603-1613.

81. Inai T, Kobayashi J, Shibata Y. Claudin-1 contributes to the epithelial barrier function in MDCK cells. Europ J Cell Biol 1999; 78:849-855.
82. Sonoda N, Furuse M, Sasaki H et al. Clostridium perfringens enterotoxin fragment removes specific claudins from tight junction strands: Evidence for direct involvement of claudins in tight junction barrier. J Cell Biol 1999; 147:195-204.
83. Morita K, Sasaki H, Furuse M et al. Endothelial claudin: Claudin-5/TMVCF constitutes tight junction strands in endothelial cells. J Cell Biol 1999; 147:185-194.
84. Liebner S, Fischmann A, Rascher G et al. Claudin-1 expression and tight junction morphology are altered in blood vessels of human glioblastoma multiforme. Acta Neuropathol 2000; 100:323-331.
85. Liebner S, Kniesel U, Kalbacher H et al. Correlation of tight junction morphology with the expression of tight junction proteins in blood-brain barrier endothelial cells. Europ J Cell Biol 2000; 79:707-717.
86. Nitta T, Hata M, Gotoh S et al. Size-selective loosening of the blood-brain barrier in claudin-5-deficient mice. J Cell Biol 2003; 161:653-660.
87. Martin-Padura I, Lostaglio S, Schneemann M et al. Junctional adhesion molecule, a novel member of the immunoglobulin superfamily that distributes at intercellular junctions and modulates monocyte transmigration. J Cell Biol 1998; 142:117-127.
88. Muller WA. Leukocyte-endothelial-cell interactions in leukocyte transmigration and the inflammatory response. Trends Immunol 2003; 24:327-334.
89. Liu Y, Nusrat A, Schnell FJ et al. Human junction adhesion molecule regulates tight junction resealing in epithelia. J Cell Sci 2000; 113:2363-2374.
90. Palmeri D, van Zante A, Huang CC et al. Vascular endothelial junction-associated molecule, a novel member of the immunoglobulin superfamily, is localized to intercellular boundaries of endothelial cells. J Biol Chem 2000; 275:19139-19145.
91. Cunningham SA, Arrate MP, Rodriguez JM et al. A novel protein with homology to the Junctional Adhesion Molecule: Characterization of leukocyte interactions. J Biol Chem 2000; 275:34750-34756.
92. Aurrand-Lions MA, Duncan L, Du Pasquier L et al. Cloning of JAM-2 and JAM-3: An emerging junctional adhesion molecular family? Curr Top Microbiol Immunol 2000; 251:91-98.
93. Aurrand-Lions MA, Duncan L, Ballestrem C et al. JAM-2, a novel Immunoglobulin Superfamily Molecule, expressed by endothelial and lymphatic cells. J Biol Chem 2001; 276:2733-2741.
94. Arrate MP, Rodriguez JM, Tran TM et al. Cloning of human junctional adhesion molecule 3 (JAM3) and its identification as the JAM2 counter-receptor. J Biol Chem 2001; 276:45826-45832.
95. Bazzoni G. The JAM family of junctional adhesion molecules. Curr Opin Cell Biol 2003; 15:525-530.
96. Ebnet K, Suzuki A, Ohno S et al. Junctional adhesion molecules (JAMs): More molecules with dual functions? J Cell Sci 2004; 117:19-29.
97. Liang TW, DeMarco RA, Mrsny RJ et al. Characterization of huJAM: Evidence for involvement in cell-cell contact and tight junction regulation. Am J Physiol Cell Physiol 2000; 279:C1733-C1743.
98. Bazzoni G, Martinez-Estrada OM, Orsenigo F et al. Interaction of junctional adhesion molecule with the tight junction components ZO-1, cingulin, and occludin. J Biol Chem 2000; 275:20520-20526.
99. Ebnet K, Schulz CU, Meyer Zu et al. Junctional adhesion molecule interacts with the PDZ domain-containing proteins AF-6 and ZO-1. J Biol Chem 2000; 275:27979-27988.
100. Ebnet K, Aurrand-Lions M, Kuhn A et al. The junctional adhesion molecule (JAM) family members JAM-2 and JAM-3 associate with the cell polarity protein PAR-3: A possible role for JAMs in endothelial cell polarity. J Cell Sci 2003; 116:3879-3891.
101. Ebnet K, Suzuki A, Horikoshi Y et al. The cell polarity protein ASIP/PAR-3 directly associates with junctional adhesion molecule (JAM). EMBO J 2001; 20:3738-3748.
102. Itoh M, Sasaki H, Furuse M et al. Junctional adhesion molecule (JAM) binds to PAR-3: A possible mechanism for the recruitment of PAR-3 to tight junctions. J Cell Biol 2001; 154:491-498.
103. Ohno S. Intercellular junctions and cellular polarity: The PAR - aPKC complex, a conserved core cassette playing fundamental roles in cell polarity. Curr Opin Cell Biol 2001; 13:641-648.
104. Suzuki A, Yamanaka T, Hirose T et al. Atypical protein kinase C is involved in the evolutionary conserved PAR protein complex and plays a critical role in establishing epithelia-specific junctional structures. J Cell Biol 2001; 152:1183-1196.
105. Yamanaka T, Horikoshi Y, Suzuki A et al. Par-6 regulates aPKC activity in a novel way and mediates cell-cell contact-induced formation of epithelial junctional complex. Genes Cells 2001; 6:721-731.
106. Gao L, Joberty G, Macara IG. Assembly of epithelial tight junctions is negatively regulated by Par6. Curr Biol 2002; 12:221-225.

107. Nagai-Tamai Y, Mizuno K, Hirose T et al. Regulated protein-protein interaction between aPKC and PAR-3 plays an essential role in the polarization of epithelial cells. Genes Cells 2002; 7:1161-1171.
108. Suzuki A, Ishiyama C, Hashiba K et al. aPKC kinase activity is required for the asymmetric differentiation of the premature junctional complex during epithelial cell polarization. J Cell Sci 2002; 115:3565-3573.
109. Cohen CJ, Shieh JT, Pickles RJ et al. The coxsackievirus and adenovirus receptor is a transmembrane component of the tight junction. Proc Natl Acad Sci USA 2001; 98:15191-15196.
110. Nasdala I, Wolburg-Buchholz K, Wolburg H et al. A transmembrane tight junction protein selectively expressed on endothelial cells and platelets. J Biol Chem 2002; 277:16294-16303.
111. Hirabayashi S, Tajima M, Yao I et al. JAM4, a junctional cell adhesion molecule interacting with a tight junction protein, MAGI-1. Mol Cell Biol 2003; 23:4267-4282.
112. Raschperger E, Engstrom U, Pettersson RF et al. CLMP, a novel member of the CTX family and a new component of epithelial tight junctions. J Biol Chem 2004; 279:796-804.
113. Hirata K, Ishida T, Penta K et al. Cloning of an immunoglobulin family adhesion molecule selectively expressed by endothelial cells. J Biol Chem 2001; 276:16223-16231.
114. Ishida T, Kundu RK, Yang E et al. Targeted disruption of endothelial cell-selective adhesion molecule inhibits angiogenic processes in vitro and in vivo. J Biol Chem 2003; 278:34598-34604.
115. Harris BZ, Lim WA. Mechanism and role of PDZ domains in signaling complex assembly. J Cell Sci 2001; 114:3219-3231.
116. Pawson T, Nash P. Assembly of cell regulatory systems through protein interaction domains. Science 2003; 300:445-452.
117. Anderson JM. Cell signalling: MAGUK magic. Curr Biol 1996; 6:382-384.
118. Dobrosotskaya I, Guy RK, James GL. MAGI-1, a membrane-associated guanylate kinase with a unique arrangement of protein-protein interaction domains. J Biol Chem 1997; 272:31589-31597.
119. Roh MH, Margolis B. Composition and function of PDZ protein complexes during cell polarization. Am J Physiol Renal Physiol 2003; 285:F377-387.
120. Roh MH, Makarova O, Liu CJ et al. The Maguk protein, Pals1, functions as an adapter, linking mammalian homologues of Crumbs and Discs Lost. J Cell Biol 2002; 157:161-172.
121. Ullmer C, Schmuck K, Figge A et al. Cloning and characterization of MUPP1, a novel PDZ domain protein. FEBS Lett 1998; 424:63-68.
122. Hopkins AM, Li D, Mrsny RJ et al. Modulation of tight junction function by G protein-coupled events. Adv Drug Deliv Rev 2000; 41:329-340.
123. Benais-Pont G, Punn A, Flores-Maldonado C et al. Identification of a tight junction-associated guanine nucleotide exchange factor that activates Rho and regulates paracellular permeability. J Cell Biol 2003; 160:729-740.
124. Avila-Flores A, Rendon-Huerta E, Moreno E et al. Tight-junction protein zonula occludens 2 is a target of phophorylation by protein kinase C. Biochem J 2001; 360:295-304.
125. Izumi Y, Hirose T, Tamai Y et al. An atypical PKC directly associates and colocalizes at the epithelial tight junction with ASIP, a mammalian homologue of Caenorhabditis elegans polarity protein PAR-3. J Cell Biol 1998; 143:95-106.
126. Nunbhakdi-Craig V, Machleidt T, Ogris E et al. Protein phosphatase 2A associates with and regulates atypical PKC and the epithelial tight junction complex. J Cell Biol 2002; 158:967-978.
127. Wu Y, Dowbenko D, Spencer S et al. Interaction of the tumor suppressor PTEN/MMAC with a PDZ domain of MAGI3, a novel membrane-associated guanylate kinase. J Biol Chem 2000; 275:21477-21485.
128. Balda MS, Garrett MD, Matter K. The ZO-1-associated Y-box factor ZONAB regulates epithelial cell proliferation and cell density. J Cell Biol 2003; 160:423-432.
129. Nakamura T, Blechman J, Tada S et al. huASH1 protein, a putative transcription factor encoded by a human homologue of the Drosophila ash1 gene, localizes to both nuclei and cell-cell tight junctions. Proc Natl Acad Sci USA 2000; 97:7284-7289.
130. Betanzos A, Huerta M, Lopez-Bayghen E et al. The tight junction protein ZO-2 associates with Jun, Fos and C/EBP transcription factors in epithelial cells. Exp Cell Res 2004; 292:51-66.
131. Sheth P, Basuroy, Li C et al. Role of phosphatidylinositol 3-kinase in oxidative stress-induced disruption of tight junctions. J Biol Chem 2003; 278:49239-49245.
132. Denker BM, Saha C, Khawaja S et al. Involvement of a heterotrimeric G protein alpha subunit in tight junction biogenesis. J Biol Chem 1996; 271:25750-25753.
133. Saha C, Nigam SK, Denker BM. Expanding role of G proteins in tight junction regulation: Gαs stimulates tight junction assembly. Biochem Biophys Res Commun 2001; 285:250-256.
134. Dodane V, Kachar B. Identification of isoforms of G proteins and PKC that colocalize with tight junctions. J Membr Biol 1996; 149:199-209.

135. Denker BM, Nigam SK. Molecular structure and assembly of the tight junction. Am J Physiol 1998; 274:F1-F9.
136. Klingler C, Kniesel U, Bamforth S et al. Disruption of epithelial tight junctions is prevented by cyclic nucleotide-dependent protein kinase inhibitors. Histochem Cell Biol 2000; 113:349-361.
137. Balda MS, Gonzalez-Mariscal L, Contreras RG et al. Assembly and sealing of tight junctions: Possible participation of G-proteins, phospholipase C, protein kinase C and calmodulin. J Membr Biol 1991; 122:193-202.
138. Raub TJ. Signal transduction and glial cell modulation of cultured brain microvessel endothelial cell tight junctions. Am J Physiol 1996; 271:C495-C503.
139. Brückener KE, El Baya A, Galla H-J et al. Permeabilization in a cerebral endothelial barrier model by pertussis toxin involves the PKC effector pathay and is abolished by elevated levels of cAMP. J Cell Sci 2003; 116:1837-1846.
140. Garcia JGN, Wang P, Liu F et al. Pertussis toxin directly activates endothelial cell p42/p44 MAP kinases via a novel signaling pathway. Am J Physiol 2001; 280:C1233-C1241.
141. De Vries L, Farquahr MG. RGS proteins: More than just GAPs for heterotrimeric G proteins. Trends Cell Biol 1999; 9:138-144.
142. Li JY, Boado RJ, Pardridge WM. Blood-brain barrier genomics. J Cerebr Blood Flow Metab 2001; 21:61-68.
143. Kirsch T, Wellner M, Haller H et al. Altered gene expression in cerebral capillaries of stroke-prone spontaneously hypertensive rats. Brain Res 2001; 910:106-115.
144. Adamson RH, Curry FE, Adamson G et al. Rho and rho kinase modulation of barrier properties: Cultured endothelial cells and intact microvessels of rats and mice. J Physiol 2002; 539:295-308.
145. Van Hinsbergh VWM, Van Nieuw Amerongen GP. Intracellular signalling involved in modulating human endothelial barrier function. J Anat 2002; 200:549-560.
146. Jou TS, Nelson WJ. Effects of regulated expression of mutant rhoA and rac1 small GTPases on the development of epithelial (MDCK) cell polarity. J Cell Biol 1998; 142:85-100.
147. Jou TS, Schneeberger EE, Nelson WJ. Structural and functional regulation of tight junctions by rhoA and rac1 small GTPases. J Cell Biol 1998; 142:101-115.
148. Nusrat A, Giry M, Turner JR et al. Rho protein regulates tight junctions and perijunctional actin organization in polarized epithelia. Proc Natl Acad Sci USA 1995; 92:10629-10633.
149. Hopkins AM, Walsh SV, Verkade P et al. Constitutive activation of Rho proteins by CNF-1 influences tight junction structure and epithelial barrier function. J Cell Sci 2002; 116:725-742.
150. van Leeuwen FN, Giepmans BN, van Meeteren LA et al. Lysophosphatidic acid: Mitogen and motility factor. Biochem Soc Trans 2003; 31:1209-1212.
151. Schulze C, Smales C, Rubin LL et al. Lysophophatidic acid increases tight junctional permeability in cultured brain ECs. J Neurochem 1997; 68:991-1000.
152. Lum H, Malik AB. Mechanisms of increased endothelial permeability. Can J Physiol Pharmacol 1996; 74:787-800.
153. Wójciak-Stothard B, Entwistle A, Garg R et al. Regulation of TNF-α-induced reorganization of the actin cytoskeleton and cell-cell junctions by rho, rac, and cdc42 in human endothelial cells. J Cell Physiol 1998; 176:150-165.
154. Johnson-Léger C, Aurrand-Lions M, Imhof BA. The parting of the endothelium: Miracle, or simply a junctional affair? J Cell Sci 2000; 113:921-933.
155. Mayhan WG. Leukocyte adherence contriburtes to disruption of the blood-brain barrier during activation of mast cells. Brain Res 2000; 869:112-120.
156. Liu F, Verin AD, Borbiev T et al. Role of cAMP-dependent protein kinase A activity in endothelial cell cytoskeleton rearrangement. Am J Physiol 2001; 280:L1309-L1317.
157. Petty MA, Lo EH. Junctional complexes of the blood-brain barrier: Permeability changes in neuroinflammation. Progr Neurobiol 2002; 68:311-323.
158. Essler M, Amano M, Kruse HJ et al. Thrombin inactivates myosin light chain phosphatase via rho and its target rho kinase in human endothelial cells. J Biol Chem 1998; 273:21867-21874.
159. Essler M, Staddon JM, Weber PC et al. Cyclic AMP blocks bacterial lipopolysaccharide-induced myosin light chain phosphorylation in endothelial cells through inhibition of rho/rho kinase signaling. J Immunol 2000; 164:6543-6549.
160. Adamson P, Etienne S, Couraud PO et al. Lymphocyte migration through brain endothelial cell monolayers involves signaling through endothelial ICAM-1 via a rho-dependent pathway. J Immunol 1999; 162:2964-2973.
161. Etienne S, Adamson P, Greenwood J et al. ICAM-1 signaling pathways associated with rho activation in microvascular brain endothelial cells. J Immunol 1998; 161:5755-5761.
162. Barber AJ, Lieth E. Agrin accumulates in the brain microvascular basal lamina during development of the blood-brain barrier. Dev Dyn 1997; 208:62-74.

163. Gladson CL. The extracellular matrix of gliomas: Modulation of cell function. J Neuropathol exp Neurol 1999; 58:1029-1040.
164. Savettieri G, Di Liegro I, Catania C et al. Neurons and ECM regulate occludin localization in brain endothelial cells. NeuroReport 2000; 11:1081-1084.
165. Sixt M, Engelhardt B, Pausch F et al. Endothelial cell laminin isoforms, laminins 8 and 10, play decisive roles in T cell recruitment across the blood-brain barrier in experimental autoimmune encephalomyelitis. J Cell Biol 2001; 153:933-945.
166. Alexander JS, Elrod JW. Extracellular matrix, junctional integrity and matrix metalloproteinase interactions in endothelial permeability regulation. J Anat 2002; 200:561-574.
167. Robert AM, Robert L. Extracellular matrix and blood-brain barrier function. Pathol Biol 1998; 46:535-542.
168. Sobel RA, Hinojoza JR, Maeda A et al. Endothelial cell integrin laminin receptor expression in multiple sclerosis lesions. Am J Pathol 1998; 153:405-415.
169. Rascher G, Wolburg H. The blood-brain barrier in the aging brain. In: J de Vellis, ed. Neuroglia in the Aging Brain. Totowa: Humana Press, 2001:305-320.
170. Tilling T, Engelbertz C, Decker S et al. Expression and adhesive properties of basement membrane proteins in cerebral capillary endothelial cell cultures. Cell Tissue Res 2002; 310:19-29.
171. Jiang X, Couchman JR. Perlecan and tumor angiogenesis. J Histochem Cytochem 2003; 51:1393-1410.
172. Asahi M, Wang X, Mori T et al. Effects of matrix metalloproteinase-9 gene knockout on the proteolysis of blood-brain barrier and whize mater components after cerebral ischemia. J Neurosci 2001; 21:7724-7732.
173. McMahon UJ. The agrin hypothesis. Cold Spring Harb Symp Quant Biol 1990; 55:407-418.
174. Bezakova G, Ruegg MA. New insights into the roles of agrin. Nature Rev Mol Cell Biol 2003; 4:295-308.
175. Berzin TM, Zipser BD, Rafii MS et al. Agrin and microvascular damage in Alzheimer's disease. Neurobiol Aging 2000; 21:349-355.
176. Smith MA, Hilgenberg LGW. Agrin in the CNS: A protein in search of a function? NeuroReport 2002; 13:1485-1495.
177. Stone DM, Nikolics K. Tissue- an age-specific expression patterns of alterantively spliced agrin mRNA transcripts in embryonic rat suggest novel developmental roles. J Neurosci 1995; 15:6767-6778.
178. Gee SH, Montanaro F, Lindenbaum MH et al. Dystroglycan-α: A dystrophin-associated glyoprotein, is a functional agrin receptor. Cell 1994; 77:675-686.
179. Blake DJ, Kröger S. The neurobiology of Duchenne muscular dystrophy: Learning lessons from muscle? Trends Neurosci 2000; 23:92-99.
180. Rascher G, Fischmann A, Kröger S et al. Extracellular matrix and the blood-brain barrier in glioblastoma multiforme: Spatial segregation of tenascin and agrin. Acta Neuropathologica 2002; 104:85-91.
181. Janzer RC, Raff MC. Astrocytes induce blood-brain barrier properties in endothelial cells. Nature 1997; 325:253-257.
182. Abbott NJ. Astrocyte-endothelial interactions and blood-brain barrier permeability. J Anat 2002; 200:629-638.
183. Brillault J, Berezowski V, Cecchelli R et al. Intercommunications between brain capillary endothelial cells and glial cells increase the transcellular permeability of the blood-brain barrier during ischemia. J Neurochem 2002; 83:807-817.
184. Wolburg H. Orthogonal arrays of intramembranous particles. A review with special reference to astrocytes. J Brain Res 1995; 36:239-258.
185. Yonezawa T, Ohtsuka A, Yoshitaka T et al. Limitrin, a novel immunoglobulin superfamily protein localized to glia limitans formed by astrocyte endfeet. Glia 2003; 44:190-204.
186. Wolburg H. Glia-neuronal and glia-vascular interrelations in blood-brain barrier formation and axon regeneration in vertebrates. In: Vernadakis A, Roots B, eds. Neuro-Glial Interactions during Phylogeny. Totowa: Humana Press Inc., 1995:479-510.
187. Nico B, Frigeri A, Nicchia GP et al. Role of aquaporin-4 water channel in the development and integrity of the blood-brain barrier. J Cell Sci 2001; 114:1297-1307.
188. Papadopoulos MC, Krishna S, Verkman AS. Aquaporin water channels and brain edema. Mount Sinai J Med 2002; 69:242-248.
189. Badaut J, Lasbennes F, Magistretti PJ et al. Aquaporins in brain: Distribution, physiology and pathophysiology. J Cer Blood Flow Metabol 2002; 22:367-378.
190. Verbavatz J-M, Ma T, Gobin R et al. Absence of orthogonal arrays in kidney, brain and muscle from transgenic knockout mice lacking water channel aquaporin-4. J Cell Sci 1997; 110:2855-2860.

191. Yang B, Brown D, Verkman AS. The mercurial insensitive water channel (AQP-4) forms orthogonal arrays in stably transfected chinese hamster ovary cells. J Biol Chem 1996; 271:4577-4580.
192. Rash JE, Yasumura T, Hudson CS et al. Direct immunogold labeling of aquaporin-4 in square arrays of astrocyte and ependymocyte plasma membranes in rat brain and spinal cord. Proc Natl Acad Sci USA 1998; 95:11981-11986.
193. Nielsen S, Nagelhus EA, Amiry-Moghaddam M et al. Specialized membrane domains for water transport in glial cells: High-resolution immunogold cytochemistry of aquaporin-4 in rat brain. J Neurosci 1997; 17:171-180.
194. Neuhaus J. Orthogonal arrays of particles in astroglial cells: Quantitative analysis of their density, size, and correlation with intramembranous particles. Glia 1990; 3:241-251.
195. Saadoun S, Papadopoulos MC, Davies DC et al. Aquaporin-4 expression is increased in oedematous human brain tumours. J Neurol Neurosurg Psychiatry 2002; 72:262-265.
196. Warth A, Kröger S, Wolburg H. Redistribution of aquaporin-4 in human glioblastoma correlates with loss of agrin immunoreactivity from brain capillary basal laminae. Acta Neuropathol 2004; 107:311-318.
197. Neely JD, Amiry-Moghaddam M, Ottersen OP et al. Syntrophin-dependent expression and localization of aquaporin-4 water channel protein. Proc Natl Acad Sci USA 2001; 98:14108-14113.
198. Ehrlich P. Das Sauerstoff-Bedürfnis des Organismus. eine farbenanalytische studie. Herschwald, Berlin: PhD thesis, 1885:69-72.
199. Nag S. Blood-brain barrier permeability using tracers and immunohistochemistry. Methods Mol Med Humana Press, 2003:89.
200. Sasaki H, Matsui C, Furuse K et al. Dynamic behavior of paired claudin strands within apposing plasma membranes. Proc Natl Acad Sci USA 2003; 100:3971-3976.
201. Wolburg H, Wolburg-Buchholz K, Liebner S et al. Claudin-1, claudin-2 and claudin-11 are present in tight junctions of choroid plexus epithelium of the mouse. Neurosci Letters 2001; 307:77-80.

CHAPTER 14

Tight Junctions in CNS Myelin

Jeff M. Bronstein and Seema Tiwari-Woodruff

Abstract

Myelin in the central nervous system (CNS) is composed of a complex multilamellar sheath wrapped around axons facilitating axonal conduction. Tight junctions (TJs) between plasma membrane layers of myelin were recognized over 40 years ago but only recently has the molecular composition been determined. CNS myelin TJs stain with anti-OSP/claudin-11 antibodies and are absent when the OSP/claudin-11 gene is disrupted. These TJs, along with PLP/DM20, serve an adhesive function maintaining myelin compaction. In addition to its structural role in mature myelin, OSP/claudin-11 associates with a member of the tetraspanin superfamily and $\beta 1$ integrin, and regulates proliferation and migration of oligodendrocyte progenitor cells.

Introduction

Myelin is composed of a complex multilamellar sheath essential for normal neuronal function. The formation of myelin occurs early in development and requires oligodendrocyte and Schwann cell proliferation, and migration followed by the wrapping and compaction of the plasma membrane around axons. This myelin sheath completely covers the axon except at specialized gaps called the nodes of Ranvier. Myelin is essential for normal axonal function by increasing the resistance facilitating saltatory conduction at the nodes and its multilaminar structural organization is similar in the peripheral and central nervous system (PNS and CNS) with some important differences. In the CNS, oligodendrocytes extend processes that myelinate several axons whereas Schwann cells in the PNS only myelinate a single axon. Oligodendrocytes do not have a basal lamina but do have a distinctive structure not seen in the PNS called the radial component (see below). The protein composition of CNS and PNS myelin also differs. In the CNS, 80-90% of total myelin protein corresponds to myelin basic proteins (MBPs) and proteolipid proteins (PLP and DM20) and 7% corresponds to OSP/claudin-11 (Fig. 1).[1,2] There are also several other minor proteins such as 2'3'-cyclic nucleotide-phosphodiesterase (CNPase), myelin/oligodendrocyte glycoprotein (MOG), and myelin-associated glycoprotein (MAG) but little is known about their structure and function. In the peripheral nervous system (PNS), protein zero (P_0) and peripheral myelin protein 22 (PMP-22) are highly expressed specifically in Schwann cells; along with the MBPs, they form the bulk of peripheral myelin protein.

The Radial Component of CNS Myelin Are Tight Junctions

The 1st report of a radial structure in myelin was made by Peters in 1961.[3] Using electron microscopy (EM), he described a series of plaque-like thickenings of the intraperiod line and later confirmed that these thickenings run parallel to the length of the axon in a spiral manner.[4] Dermietzel and Reale et al used freeze-fracture to demonstrate that these thickenings observed in CNS myelin were in fact zona occludens or TJs.[5-7] Evidence for a barrier function of the

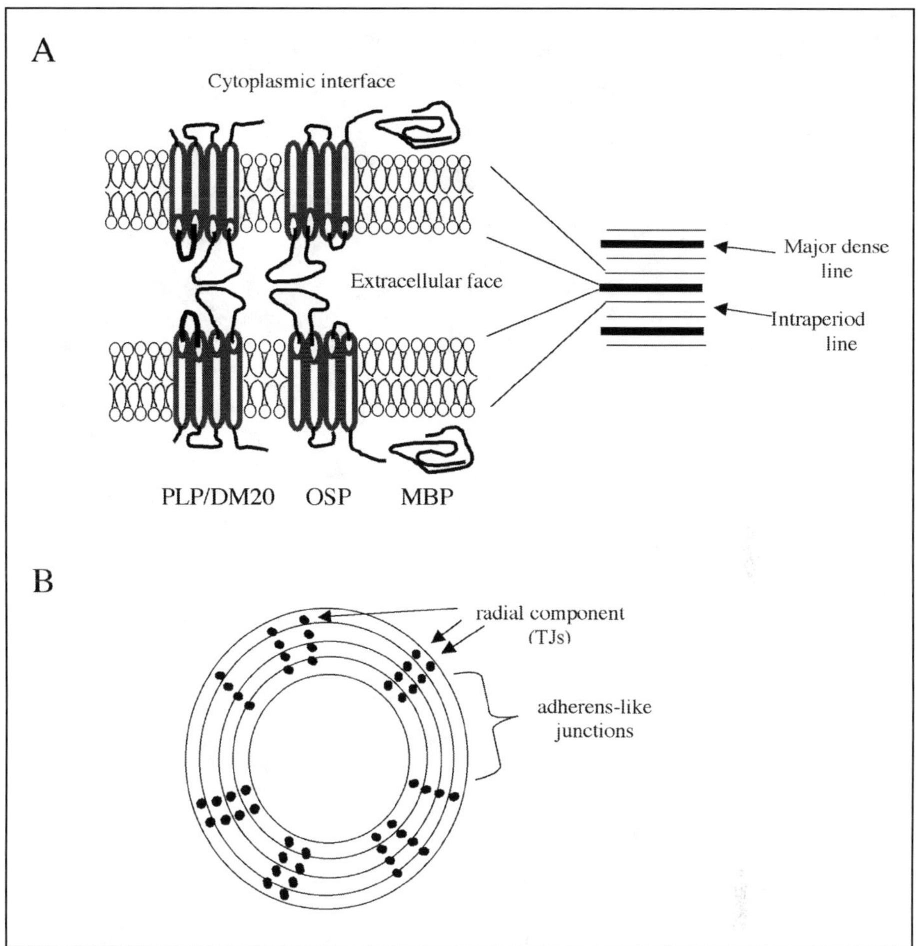

Figure 1. Schematic representation of the predicted structure and orientation of CNS myelin proteins. The proposed composition of the major dense and intraperiod lines observed by electron microscopy is shown on the right. B) Diagram of the cross section of a myelinated axon in the CNS showing the orientation of the radial component and the adherens-like junctions between laminar sheaths.

radial component was later described by Tabira et al who demonstrated that the junctions provided a diffusion barrier between the lamellar components of the myelin sheaths.[8] Furthermore, they demonstrated that when hexachlorophene was administered to the animals, intramyelinic vacuolation occurred but the myelin splitting was limited by the TJs. These early reports suggested that the radial component of CNS myelin were composed of TJs that provided structural support as well as a diffusion barrier.

OSP/Claudin-11 Is Essential for the Formation of CNS Myelin TJs

Oligodendroycte-Specific Protein (OSP) was first identified from a subtractive hybridization library from mouse spinal cord enriched with cDNAs coding for proteins expressed during myelination.[9] OSP was further characterized because sequence and structural homology was found to PMP-22 suggesting that it may be the CNS homologue to this important peripheral

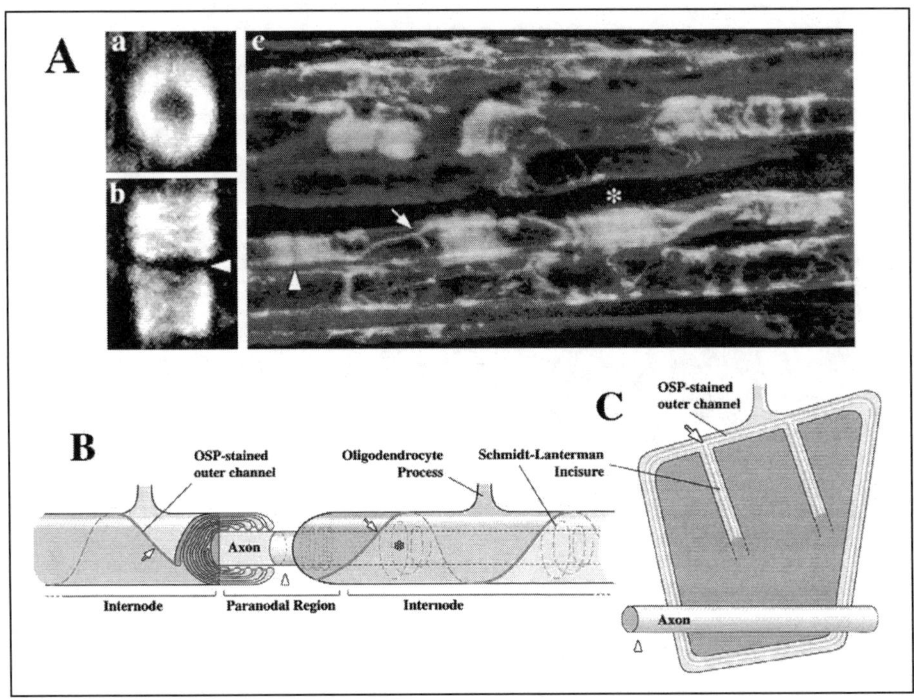

Figure 2. OSP/claudin-11 is localized to tight junctions in CNS Myelin Sheaths. A) Confocal images of monkey CNS. a) extended-focus image of a paranodal OSP-stained channel viewed along the axon. b) The data set in (a) rotated 90 degrees to show the node of Ranvier and paranodal OSP staining. c) extended focu image of monkey myelinated fibers from brainstem stained for PLP (red) and OSP (green). An OSP-stained channel extends away from the paranode and bifurcates near apparent Schmidt-Lanterman incisures (asterisk). Magnification (a) and (b) 2,600 x; (c) 1,500 x. B,C) Schematic representations of salient features of OSP-stained channels derived from the data in (A). Arrow, bifurcating OSP-stained channel, arrowhead, node of Ranvier, asterisk; Schmidt-Lanterman incisure. From Gow et al, Cell 1999; 99(6):649-659,[15] ©1999 with permission from Elsevier. A color version of this figure is available online at www.Eurekah.com.

myelin protein. Mutations in the PMP-22 gene cause Charcot Marie-Tooth Disease 1A, a common form of peripheral neuropathy.[10] OSP was found to be expressed in high concentrations in CNS myelin (7% of total myelin protein) and testes in the adult but was more widely distributed during development.[1,11] Developmental expression of OSP includes premyelinating oligodendrocyte progenitor cells (OPC). During the time we were making the OSP knockout (KO) mouse to help determine the function of OSP, Furuse et al reported the discovery of two 22-kD proteins (claudin-1 and 2) that were capable of forming TJs in cultured epithelial cells.[12] Soon later, database searches revealed other cDNAs with similar sequences suggesting that they belonged to a new gene family called the claudin family.[13] OSP was quickly identified by Morita et al and us as a distant member of this new family of TJ-associated proteins and was renamed OSP/claudin-11.[14,15]

Several lines of evidence support the concept that OSP/claudin-11 plays an instrumental functional and structural roles in CNS myelin. Anti-OSP/claudin-11 antibodies stained a linear structure in CNS myelin reminiscent of the radial component previously described by Peters[14,15] (Fig. 2). Furthermore, freeze fracture replicas of CNS myelin TJs stained specifically with anti-OSP/claudin-11 antibodies (Fig. 3).[14] The most compelling evidence that OSP/claudin-11 is an essential component of CNS TJs came from the creation of the OSP/claudin-11

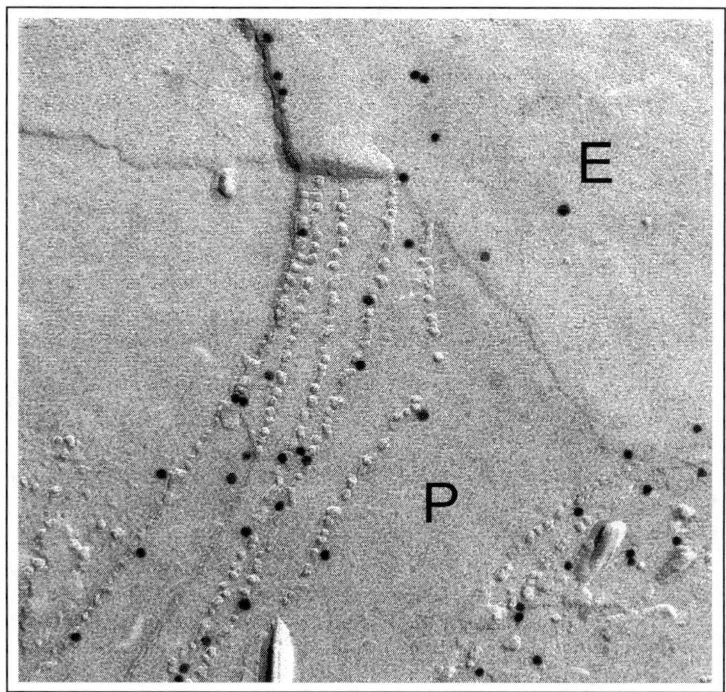

Figure 3. Freeze-fracture immunolabeling of CNS myelin with an antibody against OSP/claudin-11. Slices of mouse brainstem were rapidly frozen, digested with SDS buffer and the anti-OSP/claudin-11 antibodies were visualized with 10 nM diameter gold-conjugated Protein A. The gold particles labeled specifically the rows of intramembane particles representing the tight junctions. The protoplasmic surface (P) contained larger number of gold particles than the external face (E). Image produced by Dr. Guido Zampighi.

KO mouse. These mice did not form TJs in CNS myelin or in the testes (Fig. 4).[15] In fact, this was the first demonstration that claudins were necessary to form TJs.

The OSP/claudin-11 KO mouse offered an important opportunity to study its function. The phenotype of these mice was relatively subtle and consisted of mild hindlimb weakness, sterility of males, and deafness (unpublished data). The weakness is likely due to slow conduction velocities although the myelin appeared normal except for the absence of radial component and TJs noted by EM.[15] On the other hand, the structure of the testes was abnormal in the OSP/claudin-11 KO mice. Seminiferous tubules were narrow, lumina were ill defined, and spermatozoa were never found. Interestingly, some of the components of the testes TJ plaques were still targeted to the proper place. Specifically, ZO-1 and occludin staining was present in a parallel array in Sertoli cells suggesting that OSP/claudin-11 is not necessary for targeting of these TJ-associated proteins. ZO-1 and occludin are not expressed by oligodendrocytes and therefore are not OSP/claudin-11-associated proteins or components of CNS myelin TJs. The neurological phenotype of the OSP/claudin-11 KO mouse supports the hypothesis that TJs serve as a diffusion barrier but does not support a mechanical structural role.

Myelin Compaction Is Dependent on Both OSP/Claudin-11 and PLP/DM20

Like OSP/claudin-11, PLP/DM20 is an integral four transmembrane domain protein that is expressed in high levels by maturing oligodendrocytes. It contributes up to 60% of total

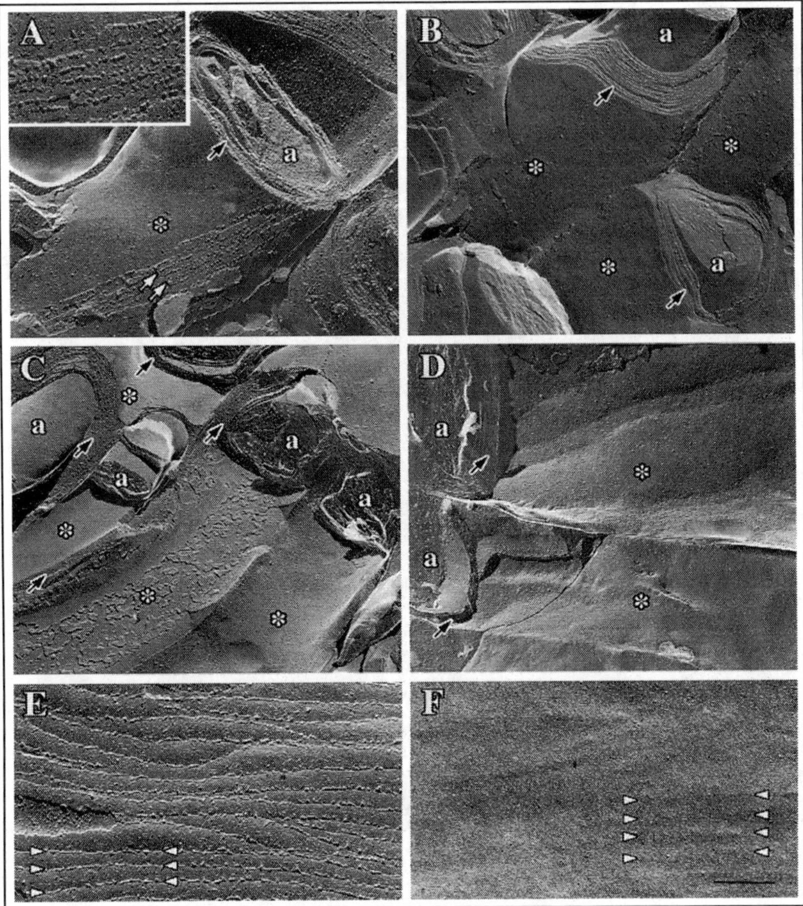

Figure 4. Freeze fracture of tight junction strands in wild-type and knockout mice. A) Optic nerve from a wild-type mouse. Black arrow, internodal myelin; white arrows, tight junction strands; asterisk, outer wrapping of the sheath; a, axon. Inset: the strands are comprised of linear arrays of intramembranous particles. B) Optic nerve from a knockout mouse. Tight junction strands are absent. C,D) Outer wraps of myelin sheaths in spinal cords of knockout mice. Tight junction strands are absent. E) Testis from an adult wild-type mouse. Arrowheads: parallel tight junction strands similar to those in CNS myelin. F) Testis from an adult knockout mouse. Tight junction strands are absent. Arrowheads: ripples in the replica surface exhibit an organization similar to normal Sertoli cell junctions. Bar = 0.2 μm. From Gow et al, Cell 1999; 99(6):649-659,[15] ©1999 with permission from Elsevier.

myelin protein in the CNS but is not present in significant quantities in the PNS. The PLP gene gives rise to two alternatively spliced transcripts that encode for PLP and a smaller isoform DM20.[16] It has been proposed that PLP/DM20 plays an important structural role in holding the myelin sheaths together in a manner similar to adherens junctions even though there is no cadherin present.[17] Insight into the function of PLP/DM20 has come from the targeted disruption of the gene and spontaneous deletions or mutations that cause a stop codon resulting in complete lack of protein expression. Surprisingly, these mutations result in relatively mild disease both in mice and man.[18] Compact myelin formed normally in the PLP/DM20-null mice but ultrastructural examination revealed condensed intraperiod lines and reduced stability and susceptibility to fixation artifacts.[19,20] There were no motor abnormalities in one study[20] while mice demonstrated impairment in motor coordination in another.[19] The differences between these two models may be due to the fact that the later study produced a RNA splice defect similar to that of the *jimpy* mouse, which causes an abnormal protein that is toxic to oligodendrocytes.

Given that the lack of expression of the two best candidates for the adhesive function maintaining myelin compaction (OSP/claudin-11 and PLP/DM20) failed to result in significant abnormalities, we hypothesized that these proteins could be compensating for each other. This idea is supported by their similar structure, localization, and high abundance in CNS myelin. To test this hypothesis, we created an OSP/PLP double KO (dKO) transgenic mouse. Interestingly, when either of these proteins was absent, the expression of the other was up-regulated while MBP expression remained unchanged. The dKO mice developed early hindlimb weakness followed by forelimb weakness after a few months. Conduction velocities, measured by visual evoked potentials (VEPs), were markedly delayed compared to the normal values measured in the wild-type and PLP KO animals, and the mildly delayed VEPs measured in the OSP/claudin-11 KO mice. Most strikingly, CNS myelin compaction was severely disrupted and no normal appearing myelin was observed in the OSP/PLP dKO brains.[21] We conclude from these experiments both OSP/claudin-11 and PLP/DM20 play a redundant mechanical role in maintaining myelin compaction.

Regulation of Migration and Proliferation; OSP/Claudin-11 and Associated Proteins

We first hypothesized that OSP/claudin-11 may have more than simply a structure role based on its homology with PMP-22.[9] PMP-22 is a four transmembrane domain protein that appears to regulate proliferation of Schwann cells.[22-24] As with PMP-22, transfection of cell lines with OSP/claudin-11 constructs resulted in marked changes in rates of proliferation.[9,25] This potential role of OSP/claudin-11 in growth regulation is further supported by the fact that OSP/claudin-11 expression is up-regulated in mesenchymal, meningeal, and testicular cells during times of active growth, as well as being expressed during proliferation of OPC.[11,15,26] Interestingly, OSP/claudin-11 expression in meningeal cells in the adult is low but is markedly increased in meningeal tumors (Fig. 5).[27] Furthermore, expression of claudin-3 and 4 were also found to be highly up-regulated in some cancers[28] and claudin-7 is down regulated in some carcinomas.[29] Further support for a role of TJ proteins and growth regulation comes from transgenic flies and mice.[30,31] Mutations in *dlg* or TamA, components of Drosophila septate junctions (functional equivalent to vertebrate TJs) result in abnormal proliferation.[32] Mice lacking occludin, another TJ-associated integral membrane protein, have a complex phenotype that includes gastric hyperplasia.[33] It is unclear whether the association of OSP/claudin-11 expression with proliferation is dependent on TJ formation or if it exerts a proliferative effect in the absence of TJ formation. In vivo, OSP/claudin-11 is expressed in OPC before myelin TJ form and within mature myelin sheaths, much of the claudin pool localizes to TJ strands but some does not (Fig. 3).[14] It is clear that in cell cultures grown at low density, OSP/claudin-11 can have a powerful effect on cell growth rates in the absence of TJ formation.[25]

Regulation of growth, migration, and the cytoskeleton by TJs presumably would occur via interacting proteins on the cytoplasmic face. Several signaling molecules have been implicated including ZO-1, 2, and 3, MUPP1, protein kinase C, phospholipase C, Cdc42, Par3, Par6, Symplekin, Rab8, Rab13, and Rab3b. Direct interactions have been found between the C-terminus of claudins 1-8 (ending in YV) and the PDZ domain containing proteins ZO-1, ZO-2, and ZO-3.[34] MUPP1 is another PDZ domain containing protein that has been found to interact with claudins 1, 5, and 8. OSP/claudin-11 ends in AHV and thus contains the second class of PDZ binding domains on the C-terminus but has not been shown to directly bind to ZO 1,2, and 3. The lack of interaction with PDZ domain binding proteins and OSP/claudin-11 is supported by the fact that ZO 1 is targeted to TJ region in Sertoli cells even in the absence of OSP/claudin-11.

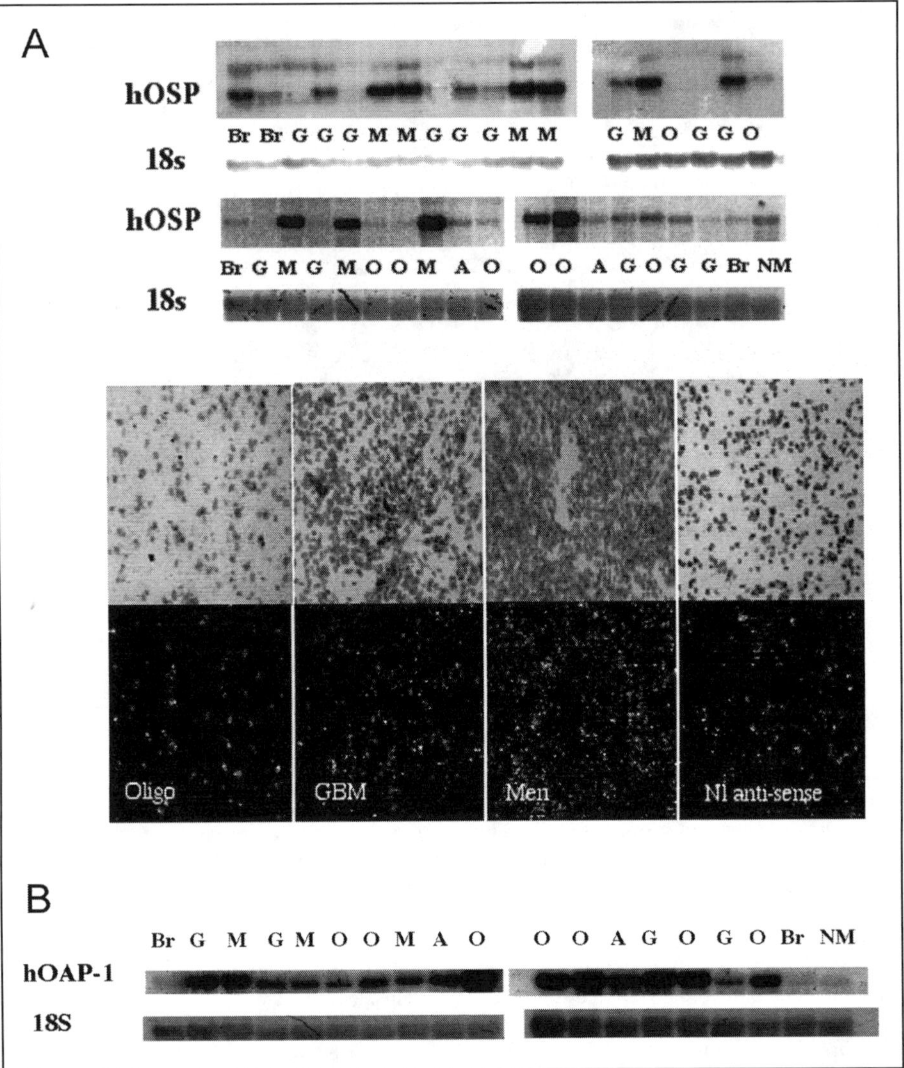

Figure 5. OSP/claudin-11 and OAP-1/Tspan-3 expression in brain tumors. A) Human OSP/claudin-11 (hOSP) cDNA was used to probe Northern blots and tumor sections. Tumors were quickly frozen following resection and prepared for RNA isolation and in situ hybridization. Top) Northern blot analysis of tumors and nontumor brain. Each lane represents tissue from different patients. Br = nontumor brain and NM = nontumor meninges obtained from an autopsy less than 6 hrs postmortem; O and Oligo = oligodendroglioma; G and GBM = glioblastoma multiforme; M and Men = meningioma. Note the high level of expression of OSP/claudin-11 in meningiomas compared to normal brain (Br) and meninges (NM). Bottom) Representative micrographs of tumors following in situ hybridization with ^{35}S-hOSP cRNA. Bright field images are shown in the top row and the corresponding dark field images are shown below. There was excellent correlation between expression levels measured using Northern blot analysis and in situ hybridization. B) Northern blot analysis of human OAP-1/Tspan-3 (hOAP-1) expression in brain tumors. A = astrocytoma. Note the high level of expression in all tumors compared to normal brain and meninges.

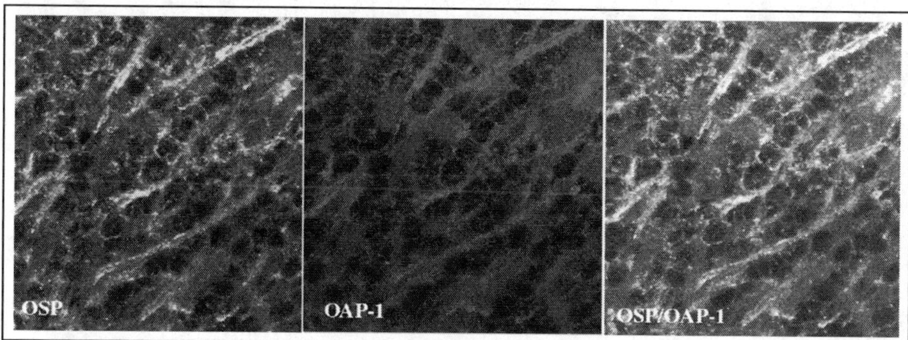

Figure 6. Confocal images of mouse brainstem section stained with anti-OSP/claudin-11 (green) and anti-OAP-1/Tspan-3 (red) antibodies. The images were fused and areas of colocalization are shown in yellow. Note the spiral pattern of staining for both OSP/claudin-11 and OAP-1/Tspan-3 (arrows) in the myelin sheaths suggesting that OAP-1/Tspan-3 localizes to tight junctions. A color version of this figure is available online at www.Eurekah.com.

In order to determine the pathway that OSP/claudin-11 may exert its effect on proliferation, migration and the cytoskeleton, we used the yeast-two hybrid system to identify associated proteins. Two OSP-interacting proteins (OAP) were identified using a cDNA that coded for the 2nd extracellular domain and intracellular C-terminus of OSP/claudin-11 as bait but none were identified using the 1st extracellular domain.[25] OAP-1 was the 1st clone to be characterized and was found to code for a novel member of a recently identified family of four transmembrane domain proteins called the tetraspanin super family (TM4SF). Given the high homology of OAP-1 with a human EST in GenBank termed Tspan-3, we refer to this protein as OAP-1/Tspan-3.[35] Members of TM4SF are widely expressed membrane proteins that are implicated in a number of basic biological phenomena including proliferation, migration, signal transduction, cell activation, and tumor invasion.[36-39] Most, if not all, TM4SF proteins associate with integrins but they may also associate with other TM4SF members (e.g., CD4, CD8, Peta-3, NAG-2, and CR2/CD19).[40]

The association of OSP/claudin-11 with OAP-1/Tspan-3 was confirmed using coimmunoprecipitation (IP) under relatively stringent conditions and by double-labeled immunohistochemistry (IH).[25] This interaction surprisingly dose not occur at the intracellular C-termini of OSP/claudin-11 but at the 2nd extracellular domain and is dependent on a 7 amino acid peptide. Substitution of two of these results in loss of OSP-OAP-1 interactions as determined by the yeast-two hybrid system (unpublished data). TM4SF members interact with integrins at their large 2nd extracellular loop in a similar manner as with OSP/claudin-11.[41] When we performed IP with either anti-OSP/claudin-11 or OAP-1/Tsapn-3 antibodies, we were able to bring down β1-integrin as well. These 3 proteins also colocalized using IH suggesting that they form a complex and biotinylation experiments confirmed the surface localization of this complex.[25] OSP/claudin-11 and OAP-1/Tspan-3 not only colocalize in cell culture, but also in CNS myelin (Fig. 6). We still have to confirm that OAP-1/Tspan-3 localizes to TJs using freeze fracture IH but these preliminary results suggest that it may represent another TJ-associated protein. Based on these results and others in cell culture, we hypothesize that OSP/claudin-11 targets OAP-1/Tspan-3 and other associated proteins to TJs.

The identification of a TM4SF member as an OSP-interacting protein added additional support for the concept that OSP/claudin-11 regulates OPC migration and proliferation. Functional experiments confirm this. OPC from OSP/claudin-11 KO mice lose integrin-dependent migration. Antibodies that bind to OSP/claudin-11, OAP-1/Tspan-3 or β1-integrin dramatically inhibit migration OPC. Finally, over-expression of either OSP/

claudin-11 or OAP-1/Tspan-3 promotes cell growth in a cell line. The precise signaling pathways by which this complex exerts its effects is not known but some good candidates have been identified for other members of the TM4SF. Interestingly, PKC and tyrosine kinases have been reported to be associated with TM4SF members and TJs.[40,42] Phosphatidylinositide 4-kinase (PI 4-kinase) has also been reported to interact directly with TM4SF members.[43] Several signal transduction molecules have also been reported to be associated with integrin signaling, the 3rd member of the OSP/OAP-1/integrin complex and are also good candidates.[44]

The yeast-two hybrid screen resulted in the identification of two OSP/claudin-11-interacting proteins. The 2nd cDNA identified coded for the voltage dependent potassium channel Kv3.1. Recent studies have demonstrated that this channel is expressed in OPC but not in significant amounts in mature oligodendrocytes and therefore does not appear to be part of the TJ. Kv3.1 appears to regulate OPC proliferation and migration and the association with OSP/claudin-11 alters its current.[45]

Summary

OSP/claudin-11 localizes to TJs and is essential for TJ formation and function in the CNS and testes. OSP/claudin-11, along with PLP/DM20, provide the adhesive component between myelin sheaths and are necessary for normal myelin compaction. OSP-associated proteins have been identified and appear to regulated oligodendrocyte progenitor cell proliferation and migration. Identification of other OSP/claudin-11-associated protein and signal transduction pathways will lead to a more comprehensive understanding of the function of CNS myelin TJs and their components.

References

1. Bronstein JM, Micevych PE, Chen K. Oligodendrocyte-specific protein (OSP) is a major component of CNS myelin. J Neurosci Res 1997; 50(5):713-720.
2. Campagnoni AT. Molecular biology of myelin proteins from the central nervous System. J Neurochem 1988; 51:1-14.
3. Peters A. A radial component of central myelin sheaths. J Biophys Biochem Cytol 1961; 11:733-735.
4. Peters A. Further observations on the structure of myelin sheaths in the central nervous system. J Cell Biol 1964; 20:281-296.
5. Dermietzel R. Junctions in the central nervous system of the cat. II. A contribution to the tertiary structure of the axonal-glial junctions in the paranodal region of the node of Ranvier. Cell Tissue Res 1974; 148(4):577-586.
6. Dermietzel R. Junctions in the central nervous system of the cat. I. Membrane fusion in central myelin. Cell Tissue Res 1974; 148(4):565-576.
7. Reale E, Luciano L, Spitznas M. Zonulae occludentes of the myelin lamellae in the nerve fibre layer of the retina and in the optic nerve of the rabbit: A demonstration by the freeze-fracture method. J Neurocytol 1975; 4(2):131-140.
8. Tabira T, Cullen MJ, Reier PJ et al. An experimental analysis of interlamellar tight junctions in amphibian and mammalian C.N.S. myelin. J Neurocytol 1978; 7(4):489-503.
9. Bronstein JM, Popper P, Micevych PE et al. Isolation and characterization of a novel oligodendrocyte-specific protein. Neurology 1996; 47(3):772-778.
10. Roa BB, Dyck PJ, Marks HG et al. Dejerine-Sottas syndrome associated with point mutation in the peripheral myelin protein 22 (PMP22) gene. Nat Genet 1993; 5(3):269-273.
11. Bronstein JM, Chen K, Tiwari-Woodruff S et al. Developmental expression of OSP/claudin-11. J Neurosci Res 2000; 60(3):284-290.
12. Furuse M, Fujita K, Hiiragi T et al. Claudin-1 and -2: novel integral membrane proteins localizing at tight junctions with no sequence similarity to occludin. J Cell Biol 1998; 141(7):1539-1550.
13. Morita K, Furuse M, Fujimoto K et al. Claudin multigene family encoding four-transmembrane domain protein components of tight junction strands. Proc Natl Acad Sci USA 1999; 96(2):511-516.
14. Morita K, Sasaki H, Fujimoto K et al. Claudin-11/OSP-based tight junctions of myelin sheaths in brain and sertoli cells in testis. J Cell Biol 1999; 145(3):579-588.
15. Gow A, Southwood CM, Li JS et al. CNS myelin and sertoli cell tight junction strands are absent in Osp/claudin-11 null mice. Cell 1999; 99(6):649-659.
16. Nave KA, Lai C, Bloom FE et al. Splice site selection in the proteolipid protein (PLP) gene transcript and primary structure of the DM-20 protein of central nervous system myelin. Proc Natl Acad Sci USA 1987; 84(16):5665-5669.

17. Sinoway MP, Kitagawa K, Timsit S et al. Proteolipid protein interactions in transfectants: Implications for myelin assembly. J Neurosci Res 1994; 37(5):551-562.
18. Garbern J, Cambi F, Shy M et al. The molecular pathogenesis of Pelizaeus-Merzbacher disease. Arch Neurol 1999; 56(10):1210-1214.
19. Boison D, Stoffel W. Disruption of the compacted myelin sheath of axons of the central nervous system in proteolipid protein-deficient mice. Proc Natl Acad Sci USA 1994; 91(24):11709-11713.
20. Klugmann M, Schwab MH, Puhlhofer A et al. Assembly of CNS myelin in the absence of proteolipid protein. Neuron 1997; 18(1):59-70.
21. Chow E, Mottahedeh J, Prins M. Disrupted compaction of CNS myelin in an OSP/Claudin-11 and PLP/DM20 double knockout mouse. Mol Cell Neurosci 2005; 29(3):405-413.
22. Jetten AM, Suter U. The peripheral myelin protein 22 and epithelial membrane protein family. Prog Nucleic Acid Res Mol Biol 2000; 64:97-129.
23. Zoidl G, Blass-Kampmann S, D'Urso D et al. Retroviral-mediated gene transfer of the peripheral myelin protein PMP22 in Schwann cells: Modulation of cell growth. EMBO J 1995; 14(6):1122-1128.
24. Welcher AA, Suter U, De Leon M et al. A myelin protein is encoded by the homologue of a growth arrest-specific gene. Proc Natl Acad Sci USA 1991; 88:7195-7199.
25. Tiwari-Woodruff SK, Buznikov AG, Vu TQ et al. OSP/claudin-11 forms a complex with a novel member of the tetraspanin super family and beta1 integrin and regulates proliferation and migration of oligodendrocytes. J Cell Biol 2001; 153(2):295-305.
26. Hellani A, Ji J, Mauduit C et al. Developmental and hormonal regulation of the expression of oligodendrocyte-specific protein/claudin 11 in mouse testis. Endocrinology 2000; 141(8):3012-3019.
27. Bronstein JM, Buznikov A, Liau LM et al. Is oligodendrocyte-specific protein (OSP) a proto-oncogene? Soc Neuroscience 1997; 43:42.
28. Hough CD, Sherman-Baust CA, Pizer ES et al. Large-scale serial analysis of gene expression reveals genes differentially expressed in ovarian cancer. Cancer Res 2000; 60(22):6281-6287.
29. Kominsky SL, Argani P, Korz D et al. Loss of the tight junction protein claudin-7 correlates with histological grade in both ductal carcinoma in situ and invasive ductal carcinoma of the breast. Oncogene 2003; 22(13):2021-2033.
30. Hoskins R, Hajnal AF, Harp SA et al. The C. elegans vulval induction gene lin-2 encodes a member of the MAGUK family of cell junction proteins. Development 1996; 122(1):97-111.
31. Willott E, Balda MS, Fanning AS et al. The tight junction protein ZO-1 is homologous to the Drosophila discs- large tumor suppressor protein of septate junctions. Proc Natl Acad Sci USA 1993; 90(16):7834-7838.
32. Woods DF, Bryant PJ. ZO-1, DlgA and PSD-95/SAP90: Homologous proteins in tight, septate and synaptic cell junctions. Mech Dev 1993; 44(2-3):85-89.
33. Saitou M, Furuse M, Sasaki H et al. Complex phenotype of mice lacking occludin, a component of tight junction strands. Mol Biol Cell 2000; 11(12):4131-4142.
34. Itoh M, Furuse M, Morita K et al. Direct binding of three tight junction-associated MAGUKs, ZO-1, ZO-2, and ZO-3, with the COOH termini of claudins. J Cell Biol 1999; 147(6):1351-1363.
35. Todd SC, Doctor VS, Levy S. Sequences and expression of six new members of the tetraspanin/ TM4SF family. Biochim Biophys Acta 1998; 1399(1):101-104.
36. Hemler ME, Mannion BA, Berditchevski F. Association of TM4SF proteins with integrins: Relevance to cancer. Biochim Biophys Acta 1996; 1287(2-3):67-71.
37. Hemler ME. Specific tetraspanin functions. J Cell Biol 2001; 155(7):1103-1107.
38. Berditchevski F. Complexes of tetraspanins with integrins: More than meets the eye. J Cell Sci 2001; 114(Pt 23):4143-4151.
39. Sugiura T, Berditchevski F. Function of alpha3beta1-tetraspanin protein complexes in tumor cell invasion. Evidence for the role of the complexes in production of matrix metalloproteinase 2 (MMP-2). J Cell Biol 1999; 146(6):1375-1389.
40. Berditchevski F, Odintsova E. Characterization of integrin-tetraspanin adhesion complexes. Role Of tetraspanins in integrin signaling. J Cell Biol 1999; 146(2):477-492.
41. Yauch RL, Kazarov AR, Desai B et al. Direct extracellular contact between integrin alpha(3)beta(1) and TM4SF protein CD151. J Biol Chem 2000; 275(13):9230-9238.
42. Zhang XA, Bontrager AL, Hemler ME. TM4SF proteins associate with activated PKC and Link PKC to specific beta1 integrins. J Biol Chem 2001; 26:26.
43. Berditchevski F, Tolias KF, Wong K et al. A novel link between integrins, transmembrane-4 superfamily proteins (CD63 and CD81), and phosphatidylinositol 4-kinase. J Biol Chem 1997; 272(5):2595-2598.
44. Aplin AE, Howe AK, Juliano RL. Cell adhesion molecules, signal transduction and cell growth. Curr Opin Cell Biol 1999; 11(6):737-744.
45. Tiwari-Woodruff SK, Bronstein JM. OSP/claudin-11 associates with a voltage-sensitive K+ channel Kv3.1 in oligodendrocytes. Soc Neurosci 2001.

CHAPTER 15

Tight Junction Modulation and Its Relationship to Drug Delivery

Noha N. Salama, Natalie D. Eddington and Alessio Fasano

Abstract

In order for therapeutic agents to exert their pharmacological effects, they have to cross the biological membranes into the systemic circulation and reach the site of action. Drugs cross the membranes by one of two pathways; paracellular or transcellular. Most drugs are transported transcellularly depending on their physiocochemical properties, however the paracellular route is usually the main route of absorption for hydrophilic drugs (proteins, peptides, etc.). The paracellular pathway is governed by the tight junctions (TJs). The modulation of the TJs by absorption enhancers for paracellular drug transport enhancement and hence drug delivery improvement has been hampered for so many years by lack of comprehensive understanding of the structure and function of the TJs. The TJs are a multiple unit structure composed of multiprotein complex that affiliates with the underlying apical actomyosin ring. TJ proteins identified include transmembrane proteins occludin and claudin and cytoplasmic plaque proteins ZO-1, ZO-2, ZO-3, cingulin, and 7H6. Among the new absorption enhancers that evolved in the past few years is Zonula Occludens toxin, Zot. In vivo and in vitro studies have shown that Zot and its biologically active fragment ΔG could be effectively used to increase the transport/absorption of paracellular markers and low bioavailable drugs across the intestinal epithelium and the blood brain barrier endothelium. Zot was also shown to be an excellent mucosal adjuvant. Above all, the transient opening of the TJs by Zot suggests that it could be used as a novel approach for the safe drug delivery of therapeutic agents.

Introduction

The identification of therapeutic agents has been sometimes compromised by their biological behavior following administration. For drugs to be therapeutically effective, they have to possess favorable characteristics to cross the biological membranes into the systemic circulation and reach the site of action. Drugs cross the membranes via the transcellular or the paracellular routes (Fig. 1). The transcellular pathway involves the passage of the drug across the cells, while the paracellular pathway refers to the passage of drugs in between the adjacent cells. The major pathway for absorption or transport of a drug depends on its physicochemical characteristics as well as the membrane features. In general, lipophilic drugs cross the biological membrane transcellularly while hydrophilic drugs cross the membrane paracellularly. In order to ameliorate drug absorption via the transcellular pathway, the physicochemical features of the drug have to be manipulated (lipophilicity, pKa, conformation, H-bond characteristics, etc.) or the membrane characteristics have to be altered.

Agents used to increase the penetration or absorption of drugs are called absorption/penetration enhancers. In most cases, those that act via alteration of the characteristics of the

Tight Junctions, edited by Lorenza Gonzalez-Mariscal. ©2006 Landes Bioscience and Springer Science+Business Media.

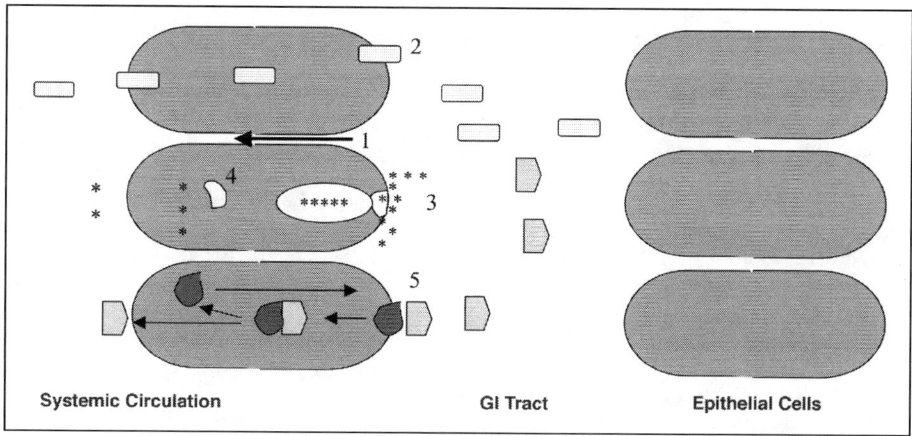

Figure 1. A schematic representation of the paracellular transport (1) and the transcellular transport of drugs (▢,▢) or solutes (* *) across the epithelial cells of the GI tract into the systemic circulation. 2) transcellular passive diffusion, 3) transcellular endocytosis followed by exocytosis (4), 5) carrier-mediated transport processes between a specific carrier (▮) and a drug(▢).

membrane to be more permeable, tend to compromise cell viability. On the other hand, manipulation of the paracellular pathway could be used to increase the transport of hydrophilic drugs and modify the absorption route of the fraction absorbed paracellularly for other drugs. The manipulation of the paracellular route has only been explored recently because the structural features of the TJs governing the permeation via this route have been partially unraveled in the past few years. In this chapter, we will focus on the use of the paracellular pathway to increase the bioavailability of therapeutic agents. Novel absorption enhancers will be discussed with emphasis on Zonula occludens toxin.

The Paracellular Route

Paracellular transport is the transport of drugs through the intercellular spaces. The paracellular pathway is governed by TJs. TJ or Zonula occludens constitute the major rate-limiting barrier towards the paracellular transport for permeation by ions and larger solutes.[1] The dimensions of the paracellular space lie between 10 and 30-50 °A, suggesting that solutes with a molecular radius exceeding 15 °A (~3.5 kDa) will be excluded from this uptake route.[2] TJs are dynamic structures, which normally regulate the trafficking of nutrients, medium sized compounds (≤ 15 °A) and relatively large amounts of fluids between the intestinal lumen and the submucosa.[3] TJs serve two primary functions in epithelia and endothelia. They form a regulated barrier in the spaces between cells (the paracellular space), restricting the movement of molecules as small as ions across cell sheets and they act as a boundary within the plasma membrane itself, separating the compositionally unique apical and basolateral cell surface domains.[4] TJs play a role in the transduction of signals in one or more directions across cell membranes and in regulating links to the cytoskeleton of the cells.[5]

Structural Characteristics of Tight Junctions

The TJs are a multiple unit structure composed of multiprotein complex that affiliates with the underlying apical actomyosin ring. Among the TJ proteins identified so far are transmembrane proteins occludin and claudin and cytoplasmic plaque proteins ZO-1, ZO-2, ZO-3, cingulin, and 7H6.[6] In vitro, the electrical resistance is a measure of charge flow across the membrane and thus reflects the permeability of the paracellular shunt pathway[7] and the tightness of the tight junctions.

TJ assembly is the result of a complex cascade of biochemical events that ultimately lead to the formation of an organized network of TJ elements, the composition of which has been only partially characterized. Two candidates for the transmembrane protein strands, occludin and claudins, have recently been identified.[8]Several proteins have been identified in a cytoplasmic submembraneous plaque underlying membrane contacts, but their function remains to be established. Zonula occludens-1 (ZO-1) and Zonula occludens-2 (ZO-2) are the best characterized proteins within the cytoplasmic plaque of the TJ. Most immunoelectron microscopic studies have localized ZO-1 to precisely beneath membrane contacts.[9] Both ZO-1 and ZO-2 belong to the membrane-associated guanylate kinase (MAGUK) family of proteins.[10] They are bound to each other[11] and to 130-kDa protein (ZO-3). ZO-1 and ZO-2 each exists as a heterodimer[11] in a detergent-stable complex with ZO-3. ZO-1 is a phosphoprotein having a species-dependent relative mass between 210 and 225 K.[12,13] ZO-1 binds directly to the COOH-terminal cytoplasmic tail of occludin.[14] ZO-2 is a 160K phosphoprotein peripherally associated with the cytoplasmic surface of the TJ.[11,15] ZO-2 shares strong sequence homology with ZO-1.[15] The expression of ZO-2 is restricted exclusively to TJs, whereas ZO-1 is also observed at some types of cadherin-based junctions, such as the intercalated discs of cardiac myocytes.[16] ZO-1 and ZO-2 are involved in creating the proper organization of proteins within the tight junctional plaque so that signaling events can be propagated.[17] ZO-3 (130K, p130), shares strong sequence homology with both ZO-1 and ZO-2, and directly interacts with ZO-1 but not ZO-2.[18]

Several other peripheral membrane proteins have been localized to the TJs, including cingulin,[19] 7H6,[20] rab13,[21] $G\alpha_{I-2}$[22] and PKC.[23] Actin , although not exclusively localized to the TJs, was actually the first protein shown by morphological techniques to be associated with the junctional membrane.[24] The architecture of the actin cytoskeleton appears to be critical for TJ function. Most of the actin is positioned under the apical junctional complex where myosin II and several actin binding proteins, including α-catenin, vinculin and radixin have been identified.[25] These actin binding proteins, concentrated directly under the adherens junction, could serve as links to the plasma membrane.[17] Myosin movement along actin filaments is regulated by ATP and phosphorylation of the regulatory light chain by Ca^{2+}-calmodulin-activated myosin light chain kinase.[26] Increases in intracellular Ca^{2+} can affect phosphorylation of myosin regulatory light chain contraction of perijunctional actin and cause increased paracellular permeability.[27]

Occludin is a transmembrane phosphoprotein of TJs of ~65kDa.[17] It is expressed in the TJs of both epithelial and endothelial cells. The proposed folding topology places both NH_2 and COOH terminals within the cytoplasm, allowing the polypeptide to pass out and back inside twice within the NH_2 terminal half.[17] The two extracellular loops, 44 and 45 residues in length, lack consensus sites for glycosylation and although lacking obvious sequence homology with each other, share an extremely high hydrophobic residue and glycine content (25% tyrosine and 36% glycine in the first loop). The cytoplasmic domains are heavily charged, whereas the two extracellular loops contain a total of only one positive and three negative side chains.[17] Occludin has recently been shown to function as a cell-cell adhesion molecule and appears to participate in maintaining the intramembrane diffusion barrier.[28] Occludin is likely involved in establishing the seal at the sites of junctional strands.[29]

Claudins interact with ZO-1, ZO-2 and ZO-3 and are able to polymerize and form TJs strands in the absence of their ZO binding region. Two additional cytoplasmic plaque proteins, cingulin (140kDa phosphoprotein) and 7H6 antigen (155kDa-175 K polypeptide) have been localized precisely to TJs using immuno-electron microscopic techniques. Two others are Symplekin (126.5kDa) and ZA-1TJ. The novel protein, Symplekin[30] has been described to associate with TJ and to the nucleus. Similar to ZO-1, Symplekin is also expressed by cells that do not form TJs, where it appears to be only in the nucleus. ZO-1 also can be localized to the nucleus, but unlike Symplekin, only in growing but not in differentiated epithelial cells.[31] This dual localization for these TJ components suggests that TJs might also be involved in the regulation of gene expression, cell growth, and differentiation.[32] Another set of proteins has been localized only at the resolution of the light microscope to the apical junction complex and includes the small GTP-binding proteins Rab13 and Rab3B (25kD), the tyrosine kinase protooncogenes c-Src and c-Yes, and the Src substrate p120, which can bind cadherin/β-catenin complex.[17] These proteins are presumably involved in signal transduction and cell adhesion, although implications for regulation of the tight junction remain unclear. Beside rab 13, the other small GTP-binding proteins are known to regulate the cortical cytoskeleton; Rho regulates actin polymerization and focal adhesion formation.[33] In polarized epithelial cells, Rho also regulates TJ organization and permeability.[34] Other proteins, such as Rac and focal adhesion kinase (FAK), play a role in plasma membrane ruffling and focal adhesion formation.[35]

Enhancing Drug Delivery via Modulation of the TJs

Absorption enhancers are compounds capable of increasing the absorption of therapeutic agents and hence improve therapeutic effectiveness. Many studies focused on evaluating absorption enhancers based on the extent of bioavailability enhancement achieved, the influence of formulation and physiological variables, toxicity associated with permeation enhancers and the mechanism of permeation enhancement.[36] Many studies focused on the investigation of absorption enhancers with compounds acting via one or more mechanisms, including influence on the thermodynamic activity of the drug in solution, alteration of the molecular structure of the cell membrane ranging from temporary membrane pore formation to complete membrane destruction, loosening of the TJs between epithelial cells, inhibition of the protease activity present in the mucosa, and alteration of the characteristics of the mucus in ways which reduce its diffusion barrier properties.[37] Ideally, the action of an absorption enhancer should be immediate and should coincide with the presence of the drug at the absorption site.[36,38] For a thorough evaluation of absorption enhancing compounds, it is important to study the extent of improved drug absorption and the absorption-enhancement–time profile as well as the toxicity of the enhancer.[39] In most cases, drug absorption enhancement is accompanied by mucosal damage induced by the enhancer.[36,40] Numerous classes of compounds with diverse chemical properties, including detergents, surfactants, bile salts, Ca^{2+} chelating agents, fatty acids, medium chain glycerides, acyl carnitine, alkanoyl cholines, N-acetylated α-amino acids, N-acetylated non α-amino acids, chitosans, mucoadhesive polymers, and phospholipids have been reported to enhance the intestinal absorption of small drug molecules and large polypeptide drugs.[36,41-50] Many of these absorption enhancers act as detergents /surfactants to increase the transcelluar transport of drugs by disrupting the structure of the lipid bilayer and rendering the cell membrane more permeable and/or by increasing the solubility of insoluble drugs.[51] Others act as Ca^{2+} chelators and improve the paracellular transport of hydrophilic drugs by disrupting the TJs after the removal of extracellular Ca^{2+} from the medium or upregulation of intracellular Ca^{2+}. Ca^{2+} depletion induces global changes in the cells, including disruption of actin filaments, disruption of adherent junctions, and diminished cell adhesion.[52] In the case of surfactants, the potential lytic nature of these agents may cause exfoliation of the intestinal epithelium, irreversibly compromising its barrier functions.[53] Reports about some enhancers including fatty acid sodium caprate and long chain acylcarnitines, have been shown to improve absorption without obvious harmful effects to the intestinal mucosa.[53] Sakai and coworkers tested the cytotoxicity of three absorption enhancers; sodium caprate, dipotassium glycyrrhizinate and sodium deoxycholate by the trypan blue exclusion test, the protein-release test, the neutral-red assay, the DNA-propidium iodide staining assay and the test for recovery of transepithelial electrical resistance (TEER).[54] Among the three enhancers, dipotassium glycyrrhizinate was found not to be cytotoxic. They reported that sodium deoxycholate (0.1%w/ v) and sodium caprate (0.5%w/v) were cytotoxic to the plasma membrane and nuclear membrane as indicated by the trypan blue exclusion, the protein release tests and the DNA-propidium iodide staining test. Sodium glycolate (1%) has been shown to increase the absorption of insulin[55] and glucagons[56] from the nasal cavity. Sodium glycolate induces slight cellular disruption in both in vitro and in vivo tests.[37] With respect to the mucus and mucous membranes, enhancers may act by alteration of the properties of the mucus layer, by opening the TJs between epithelial cells or by increasing membrane fluidity, either by creating disorders in the phospholipid domain in the membrane or by facilitating the leaching of proteins and lipids from the membrane.[57] Sodium lauryl sulfate is known to be capable of lysing biological membranes by a mechanism which appears to be a stepwise process involving both lipid solubilization and subsequent protein denaturation and solubilization.[58,59] Sodium lauryl sulfate was chosen as a positive control that promotes serious skin damage.[60]

Based on these studies, it would appear that a transient opening of TJs would seem less damaging than a disruption of cell membrane structure.[36] Absorption enhancers include palmitoyl carnitine. It has been one of the most extensively studied and used absorption enhancers, which increases the intracellular concentration of Ca^{2+} ions however the exact mechanism of action is yet to be understood.[48,53,61] Some studies reported no separation between toxic and effective concentrations.[51] Duizer et al conducted absorption enhancement studies for palmitoyl carnitine with paracellular markers, mannitol and PEG4000, along with cytotoxicity studies (LDH, TEER reversibility). Their studies showed that palmitoyl carnitine expressed clear signs of cell damage at all effective (mannitol or PEG4000 transport enhancing) concentrations with LDH leakage and inability of the cells to recover full viability after exposure to the enhancer.[62]

Several studies introduced sodium dodecyl sulfate, sodium caprate, and long-chain acylcarnitines to increase permeability through the paracellular pathways.[53] Sodium caprate (10-13mM) increased the permeability of mannitol, PEGs, argovasopressin, and FITC-dextrans of 4000 and 20,000 molecular weight (MW) and compounds with MW< 1200 across Caco2 cells.[63] Tomita et al[64] and Lindmark et al[49] proposed that the mechanism of paracellular transport enhancement by sodium caprate was via phospholipase C activation and upregulation of intracellular Ca^{2+} leading to contraction of calmodulin dependent actin-myosin filaments and opening of TJs. Dodecylphosphocholine increases the permeability of hydrophilic compounds across Caco2 cells by modulation of the TJs directly or through perturbation in the apical membrane and not by disruption of the cell membrane.[51] The enhancing effects by Quillaja saponin, dipotassium glycyrrhizinate, 18 beta-glycyrrhetinic acid, sodium caprate and taurine were determined by changes in TEER and the amount of heparin disaccharide transported across Caco2 cells. The results showed that these absorption enhancers can widen TJs and improve the transport of macromolecules and hydrophilic drugs.[65]

An approach built on understanding the mechanisms by which TJ solute permeation is naturally regulated may provide insights into simple everyday problems, including improved bioavailability of orally administered drugs.[1] The discovery of an absorption enhancer which functionally and immunologically mimics an endogenous modulator of the TJs appears to be the ideal approach to use for the manipulation of TJs as a means to improve drug delivery. Among the recent absorption enhancers displaying this principle and exhibiting the most safe and effective promising results in enhancing drug delivery is Zonula Occludens Toxin or Zot.

Discovery of Zonula Occludens Toxin

Pathogenic *Vibrio cholerae* (*V. cholerae*) strains produce several endotoxins, e.g., cholera toxin (CT), zonula occludens toxin (Zot), and accessory cholera enterotoxin (Ace), as well as a haemolysin that may have enterotoxic effects.[66] Zot is a single polypeptide chain of 44.8 kDa, 399 amino acids (AA) in length, with a predicted pI of 8.5, of bacteriophage origin, present in toxigenic stains of *V. cholerae*[67] with the ability to reversibly alter intestinal epithelial TJs, allowing the passage of macromolecules through mucosal barriers.[68] It was first identified by Fasano et al in the outer membrane of *V. cholerae* while searching for other factors responsible for the residual diarrhea observed in CT- *V. cholerae* vaccine candidates.[69] Zot possess multiple domains that allow a dual function as a morphogenetic phage protein and as an enterotoxin.[70] After cleavage at AA residue 287, a carboxyl terminal fragment of 12 kDa is excreted, that is probably responsible for the biological effect of the toxin.[70] The ~33kDa N- terminal left after cleavage, is probably involved in CTXΦ assembly[70] and remains associated to the bacterial membrane. Studies in Ussing chambers have shown that the activity of Zot is reversible, heat-labile, sensitive to protease digestion, and found in culture supernatant fractions containing molecules between 10 and 30kDa in size.[68] To identify the Zot domain(s) directly involved in the protein permeating effect, several Zot gene deletion mutants were constructed and tested for their biological activity in Ussing chamber assay and their ability to bind to the target receptor on intestinal epithelial cell cultures.[71] With these studies, the Zot biologically active domain was localized toward the carboxyl terminus of the protein and coincided with the predicted cleavage product generated by *V. cholerae*. It was hypothesized that Zot may mimic the effect of a functionally and immunologically related endogenous modulator of epithelial TJs.[72] The combination of affinity-purified anti-Zot antibodies and the Ussing chamber assay revealed an intestinal Zot analogue that was named Zonulin.[73]

Insights into the Zonulin System

When Zonulin was studied in a nonhuman primate model, it reversibly opened intestinal TJs after engagement to the same receptor activated by Zot and therefore acts with the same effector mechanism described for the toxin.[73] Comparison of the amino termini of the secreted Zot fragment (AA288-399) and its eukaryotic analogue, zonulin, governing the permeability of intercellular TJ,[71] revealed an 8-amino acid shared motif corresponding to 291-298 AA region, identified by the binding experiments as the putative binding domain. Zonulin has a molecular weight of 47 kDa, an N-terminal receptor binding motif that is structurally and functionally similar to the Zot binding motif[73] and a C-terminal domain probably involved in the rearrangement of the cytoskeletal elements functionally connected to the intercellular TJs.[74] Zonulin was detected in the human intestine, heart and brain.[69] The physiological role of the zonulin system remains to be established but it is likely that this system is involved in several functions, including TJ regulation during developmental, physiological, and pathological processes, including tissue morphogenesis, protection against microorganisms' colonization, the movement of fluid, macromolecules, and leucocytes between the bloodstream and the intestinal lumen and vice versa.[74] Dysregulation of this conceptual zonulin model may contribute to disease states that involve disordered intercellular communication including developmental and intestinal disorders leading to autoimmune disease (celiac disease and type 1 diabetes), tissue inflammation, malignant transformation, and metastasis.[75] It was hypothesized that zonulin has multiple functions with the adult form mainly in charge of the regulation of the paracellular permeability and its fetal counterpart possibly more involved in the regulation of molecules trafficking between body compartments during embryogenesis.[73] The fact that the interaction of bacteria with the intestinal mucosa induces zonulin release, irrespective of their pathogenic traits or viability, can be interpreted as a bacteria independent mechanism of defense of the host that reacts to the abnormal presence of microorganisms on the surface of the small intestine. Following zonulin-induced opening of TJs, water is secreted into the intestinal lumen following hydrostatic pressure gradients[72] and bacteria are flushed out from the small intestine.[74] The increase in intraluminal zonulin was found to be age related, to be detectable only in the small intestine (jejunum and ileum) but not in the colon, to correlate with an increase in intestinal permeability of the intestine, to precede the onset of diabetes by at least 3-4 weeks, to remain high in these diabetic prone rats, and to correlate with progression towards full-blown diabetes.[76]

Identification of the Zot Receptors

Zot binding to the surface of rabbit intestinal epithelium has been shown to vary along the different regions of the intestine.[72] Zot increases tissue permeability in the ileum but not in the colon where the presence of the colonic microflora and /or their bio-products could be harmful if the mucosal barrier is compromised.[68] This binding distribution coincides with the regional effect of Zot on the intestinal permeability and with the preferential F-actin redistribution induced by Zot in the mature cells of the villi.[72,77]

The identification of Zot receptor was initiated by binding studies using [^{125}I] maltose binding protein (Mbp)-Zot in different epithelial cell lines, including IEC6 (rat, intestine), Caco2 (human, intestine), T84 (human intestine), MDCK (canine, kidney), and BPA endothelial cells (bovine, pulmonary artery).[78] The Zot receptor was purified by ligand-affinity chromatography.[78] Only IEC6 cells (derived from rat crypt cells) and Caco2 cells (that resemble mature absorptive enteric cell of the villi), but not T84 (crypt-like cells) or MDCK, express receptors on their surface. Zot action is tissue specific, it is active on epithelial cells of the small intestine (jejunum and distal ileum) and on endothelial cells but not on colonocytes or renal epithelial cells.[77] Moreover, Zot is active only on the mucosal side.[77] Zot showed differential sensitivity among the enterocytes with the mature cells of the tip of the villi being more sensitive than the less mature crypt cells.[77] The number of Zot receptors may decrease along the villous crypt axis, and this explains why the effect of Zot on IEC6 cells requires longer exposure time compared to whole tissue.[77] Fluorescence staining of Zot binding to the intestinal epithelium is maximal on the surface of the mature absorptive enterocytes at the tip of the villi and completely disappears along the surface of the immature crypt cells, suggesting that the expression of Zot/Zonulin receptor is up-regulated during enterocyte differentiation.[72] It was proposed that the 66-kDa Zot/Zonulin putative receptor identified in intestinal cell lines is an intracellular modulator of the actin cytoskeleton and of the tight junctional complex, whose activation is triggered by the ligand engagement and subsequent internalization of the ligand/receptor complex.[78] Earlier studies discussed the partial characterization of the Zonulin receptor in the brain and revealed it is a 45 kDa glycoprotein containing multiple sialic acid residues with structural similarities to myeloid-related protein, a member of the calcium-binding protein family possibly linked to cytoskeletal rearrangements.[79] The functional significance of glycosylation of the Zonulin/Zot receptor remains unknown, however, sialic acid residues may contribute to the stability (protection from proteolysis) or assist in translocation to the cell surface during synthesis.[80] Affinity column purification revealed another ~55 kDa Zot binding protein in the human brain plasma membrane preparation, which was identified as tubulin. Pretreatment with a specific protein kinase C inhibitor, CGP41251, completely abolished the Zot effects on both tissue permeability and actin polymerization.[77] Zot decreases the soluble G-actin pool with a reciprocal increase in the filamentous F-actin pool, thus exerting an effect on actin filaments (polymerization).[77] Thus, Zot triggers a cascade of intracellular events that lead to a protein kinase C (PKC) α dependent polymerization of actin microfilaments strategically localized to regulate the paracellular pathway,[77] and consequently leading to opening of the TJs at a toxin concentration as low as 1.1×10^{-13} M.[72] So far, the receptor was identified in the small intestine,[72] brain,[79] nasal epithelium[81] and possibly the heart[79] (since Zonulin was identified there).[82]

Zot as a TJ Modulator for Improved Drug Delivery

Extensive studies of the biological activity of Zot were triggered by its ability to allow for the oral administration of insulin, an agent exclusively administered subcutaneously[83] to diabetic rats. None of the animals treated with insulin and Zot experienced diarrhea, fever, or other systemic symptoms, and no structural changes were demonstrated in the small intestine on histological examination.[83] The observation that in rabbit small intestine, Zot does not affect Na$^+$-glucose coupled active transport, not cytotoxic, fails to completely abolish the transepithelial resistance and induces a reversible increase of tissue permeability,[68] made it a potential tool for studying intestinal TJ regulation.[77] This information is of valuable importance in cases of clinical conditions affecting the gastrointestinal system involving alteration in intestinal TJ function, including food allergies,[84] malabsorption syndromes[85] and inflammatory bowel diseases.[86] In vitro experiments in the rabbit ileum demonstrated that Zot reversibly increased intestinal absorption of insulin (MW 5733 Da) by 72% and immunoglobulin G (140-160kDa) by 52% in a time dependent manner.[83] In vivo[83,87] and in vitro[68,87] studies showed that the effect of Zot on tissue permeability occurs within 20 mins of the addition of the protein to the intestinal mucosa and reaches a peak effect in 80 mins. Zot was also tested in an in vivo primate model of diabetes mellitus. Insulin was intragastrically administered to diabetic monkeys either alone or in combination with increasing amounts of Zot. Measurements of blood insulin levels revealed that insulin bioavailability increased from 5.4% in controls to 10.7% and 18% when Zot 2 μg/kg and 4 μg/kg were administered, respectively.[88]

Karyekar et al has recently reported that Zot increases the permeability of molecular weight markers (sucrose, inulin) and chemotherapeutic agents (paclitaxel and doxorubicin) across the bovine brain microvessel endothelial cells in a reversible and concentration dependant manner and without affecting the transcellular pathway as indicated by the unaltered transport of propranolol in the presence of Zot.[89] A good correlation was established between the transport enhancement and the molecular size with r^2 of 0.875. Since many therapeutic agents including anticancer drugs have molecular weights ranging from 300 to 1000 Da, these studies introduced Zot as a potential absorption enhancer for the effective delivery of therapeutic agents to the central nervous system.

Moreover, studies have shown that Zot enhances the transport of drug candidates of varying molecular weights (mannitol, PEG4000, inulin) or low bioavailability (Doxorubicin, paclitaxel, acyclovir, cyclosporin A, anticonvulsant enaminones) up to 30 fold as seen with paclitaxel across Caco2 cell monolayers without modulating the transcellular transport.[90,91] In addition, the transport enhancing effect of Zot is reversible and nontoxic.[68,91] Recent studies have identified a smaller 12 kDa fragment of Zot, referred to as ΔG.[71] These studies focused on identifying the Zot domain(s) directly involved in the protein permeating effect. ΔG results in a cascade of intracellular events leading to actin cytoskeletal rearrangement, disengagement of junctional complex proteins and finally opening of TJs.

ΔG displayed significant potential as a TJ modulator, which could be used to improve the bioavailability of drugs. In vitro studies showed that it is capable of significantly increasing the apparent permeability coefficients for a wide variety of therapeutic agents and markers across the Caco2 cell model[92-94] (Fig. 2). In addition, ΔG improved the bioavailability

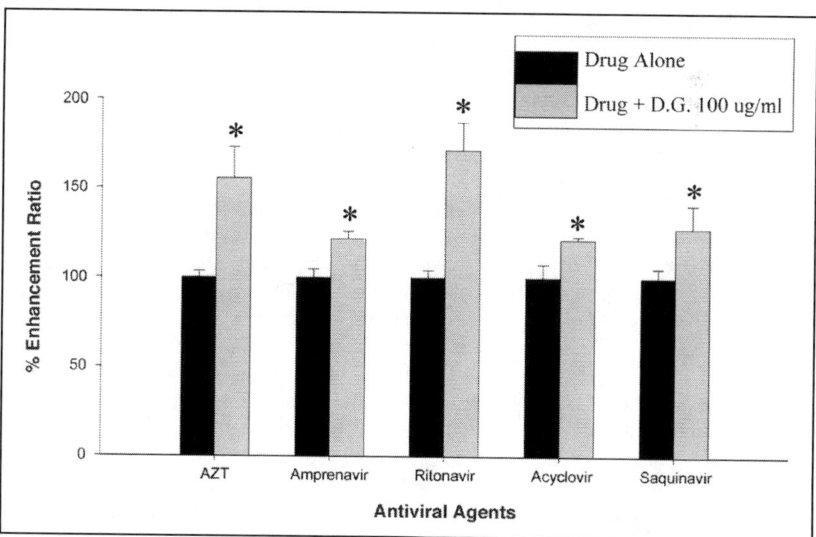

Figure 2. The percent fold enhancement in the apparent permeability coefficient (Papp) of [³H]Amprenavir, [³H]Ritonavir, [³H]Saquinavir, [¹⁴C]Zidovudine (AZT) and [³H]Acyclovir in the presence of ΔG at 0, and 100 μg/ml across Caco2 cell monolayers (n = 3). * Significant at $p<0.05$ compared to control. Data presented as mean (±SD).

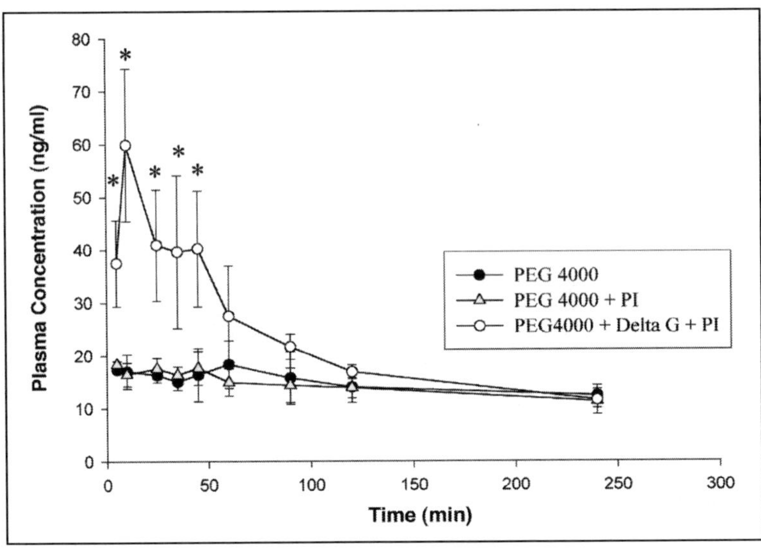

Figure 3. Average Plasma concentration versus time profile for [^{14}C] PEG4000 alone and with ΔG (720 µg/kg) and/or PI administered ID to jugular vein cannulated Sprague Dawley rats. * Significant at $p < 0.05$ compared to PEG4000. Data presented as Mean $\pm$ SD (N = 3-5/group). Republished with permission from: © copyright J Pharm Sci 2004; 93(5):1310-1319.

of paracellular markers, mannitol, inulin and PEG4000 (Fig. 3). after intraduodenal (ID) administration to rats in the presence of peptidase inhibitors (PI).[92,93] The transport/absorption of different therapeutic agents exhibiting different physicochemical properties (lipophilicity/hydrophilicity, molecular weights, efflux properties, structural differences) and belonging to a wide range of drug classes was tested with ΔG. For example, when the anti-HIV protease inhibitors were administered orally to Sprague Dawley rats with ΔG, the rate and/or extent of drug absorption were significantly ameliorated, with the exception of drugs such as saquinavir, whose low oral bioavailability is mainly attributed to extensive first pass metabolism.[95] The in vivo studies with ΔG displayed up to 57 and 50 fold increases in drug bioavailability parameters; Cmax and AUC, respectively, as seen with cyclosporin A (Fig. 4)[95] after metabolic protection was provided.

Moreover, Zot is a good mucosal adjuvant, considering its ability to interfere with the suppression of specific cell mediated immunity, probably as a result of the increased dose and/ or altered processing of antigen at the mucosal level.[96] To further characterize the role of Zot as an adjuvant, its ability to abrogate nasal tolerance to an unrelated protein as gliadin was examined.[96] When mice were given intranasally Zot with gliadin, the cytokine pattern showed reduced down-regulation of IL-2 and IFN-γ secretions, together with a significantly less suppression in T-cell proliferation. This suggested that the mechanism of adjuvanticity mediated by Zot preferentially leads to a functional activation of Th1-cell differentiation

Recent studies have shown that Zot could be exploited to deliver soluble antigens (Ag) through the nasal mucosa for the induction of Ag-specific systemic and mucosal immune responses.[81] The coadministration of Zot and ovalbumin (Ova) was found to induce anti-Ova serum immunoglobulin G (IgG) titers that were approximately 40-fold higher than those induced by immunization with Ag alone. Zot also stimulated Ag-specific IgA titres in vaginal and intestinal mucosa.

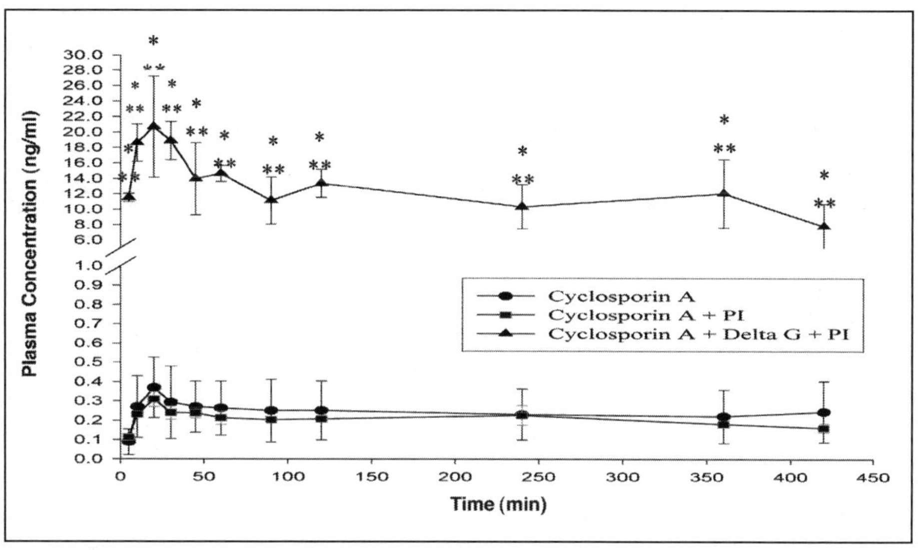

Figure 4. Average Plasma concentration versus time profile for [^{3}H] cyclosporin A alone and with ΔG (720 μg/kg) and/or PI administered ID to jugular vein cannulated Sprague Dawley rats. * Significant at $p < 0.05$ compared to cyclosporine A. ** Significant at $p < 0.05$ compared to cyclosporine A/PI. Data presented as Mean ± SD (N = 3-6/group). Republished with permission from: J Pharmacol Exp Ther 2005; 312(1):199-205.

Conclusion

The low bioavailability of drugs remains to be an active area of research. Novel approaches to improve drug delivery have been introduced lately. Among which, the modulation of TJs to improve paracellular drug transport appears to be a very appealing and attractive solution. Transient opening of TJs would be beneficial to the therapeutic effect because it avoids entry of metabolic waste as well as leakage of important proteins and nutrients. Zot was identified as a potential modulator of the TJ, which initiates a cascade of intracellular events upon binding to its receptor, leading to opening of the TJs. Extensive in vivo and in vitro studies have identified Zot receptors in the small intestine, the nasal epithelium, the heart and the brain endothelium. The discovery of Zot triggered studies to identify possible eukaryotic analogues. Comparison of the amino termini of the secreted Zot fragment (AA288-399) and its eukaryotic analogue, Zonulin, governing the permeability of intercellular TJ,[71] revealed an octapeptide amino acid shared motif corresponding to 291-298 AA region. The physiological role of the Zonulin system remains to be established but it is likely that this system is involved in several functions, including TJ regulation during developmental, physiological, and pathological processes, including tissue morphogenesis, protection against microorganisms' colonization, as well as the movement of fluid, macromolecules, and leucocytes between the blood stream and the intestinal lumen and vice versa.[74] In vitro and in vivo studies across the blood brain barrier have shown that Zot increases the transport and tissue accumulation of therapeutic agents when administered concurrently. In addition, Zot and its biologically active fragment ΔG displayed significant potential in enhancing the oral bioavailability of anticancer, immunosuppressant, and antiretroviral drugs in Caco2 cells and Sprague Dawley rats. Moreover, toxicity studies have shown that Zot and ΔG do not compromise cell viability or cause membrane toxicity contrary to other absorption enhancers. Collectively, this chapter presents modulation of the TJs as a promising route for enhanced drug delivery approaches as evidenced by the novel modulator, Zot.

References

1. Madara JL. Regulation of the movement of solutes across tight junctions. Annu Rev Physiol 1998; 60:143-159.
2. Rubas W, Cromwell ME, Shahrokh Z et al. Flux measurements across Caco2 monolayers may predict transport in human large intestinal tissue. J Pharm Sci 1996; 85(2):165-169.
3. Hollander D. The intestinal permeability barrier: A hypothesis as to its regulation and involvement in Crohn's Disease. Scand J Gastroenterol 1992; 27:721-726.
4. Stevenson BR. The tight junction: Morphology to molecules. Annu Rev Cell Dev Biol 1998; 14:89-109.
5. Dunina-Barkovskaya A. Tight junctions: Facts and models. Membr Cell Biol 1998; 11(5):555-589.
6. Nusrat A, Turner JR, Madara JL. Molecular physiology and pathophysiology of tight junctions. IV. Regulation of tight junctions by extracellular stimuli: Nutrients, cytokines, and immune cells. Am J Physiol - Gastrointest Liver Physiol 2000; 279(5):G851-857.
7. Fromter E, Diamond J. Route of passive ion permeation in epithelia. Nature New Biol 1972; 53:9-13.
8. Tsukita S, Furuse M. Occludin and claudins in tight junction strands: Leading or supporting players. Trends Cell Biol 1999; 9:268-273.
9. Stevenson BR, Anderson JM, Bullivant S. The epithelial tight junction: Structure, function, and preliminary biochemical characterization. Mol Cell Biochem 1988; 83:129-145.
10. Anderson JM. Cell signalling: MAGUK magic. Curr Biol 1996; 6:382-384.
11. Gumbiner B, Lowenkopf T, Apatira D. Identification of a 160-kDa polypeptide that binds to the tight junction protein ZO-1. Proc Natl Acad Sci USA 1991; 88(8):3460-3464.
12. Stevenson BR, Siliciano JD, Mooseker MS et al. Identification of ZO-1: A high molecular weight polypeptide associated with the tight junction (Zonula occludens) in a variety of epithelia. J Cell Biol 1986; 103:755-766.
13. Anderson JM, Stevenson BR, Jesaitis LA et al. Characterization of ZO-1, a protein component of the tight junction from mouse liver and Madin-Darby canine kidney cells. J Cell Biol 1988; 106:1141-1149.
14. Furuse M, Itoh M, Hirase T et al. Direct association of occludin with ZO-1and its possible involvement in the localization of occludin at tight junctions. J Cell Biol 1994; 127(6 Pt 1):1617-1626.
15. Jesaitis LA, Goodenough DA. Molecular characterization and tissue distribution of ZO-2, a tight junction protein homologous to ZO-1 and the Drosophila discs-large tumor suppressor protein. J Cell Biol 1994; 124:949-961.
16. Itoh M, Nagafuchi A, Yonemura S et al. The 220-kD protein colocalizing with cadherins in nonepithelial cells is identical to ZO-1, a tight junction-associated protein in epithelial cells: cDNA cloning and immunoelectron microscopy. J Cell Biol 1993; 121(3):491-502.
17. Anderson JM, Van Itallie CM. Tight junctions and the molecular basis for regulation of paracellular permeability. Am J Physiol 1995; 269(4 Pt 1):G467-475.
18. Haskins J, Gu L, Wittchen E et al. ZO-3, a novel member of the MAGUK protein family found at the tight junction, interacts with ZO-1 and occludin. J Cell Biol 1998; 141:199-208.
19. Citi S, Sabannay H, Jakes R et al. Cingulin, a new peripheral component of tight junctions. Nature 1988; 333(6170):272-276.
20. Zhong Y, Saitoh T, Minase T et al. Monoclonal antibody 7H6 reacts with a novel tight junction-associated protein distinct from ZO-1, cingulin and ZO-2. J Cell Biol 1993; 120(2):477-483.
21. Zahraoui A, Joberty G, Arpin M et al. A small rab GTPase is distributed in cytoplasmic vesicles in non polarized cells but colocalizes with the tight junction marker ZO-1 in polarized epithelial cells. J Cell Biol 1994; 124(1-2):101-115.
22. Denker BM, Saha C, Khawaja S et al. Involvement of a heterotrimeric G protein alpha subunit in tight junction biogenesis. J Biol Chem 1996; 271(42):25750-25753.
23. Dodane V, Kachar B. Identification of isoforms of G proteins and PKC that colocalize with tight junctions. J Membr Biol 1996; 149:199-209.
24. Madara JL. Intestinal absorptive cell tight junctions are linked to the cytoskeleton. Am J Physiol 1987; 253:C171-175.
25. Denker BM, Nigam SK. Molecular structure and assembly of the tight junction. Am J Physiol 1998; 274:F1-9.
26. Hecht G, Pestic L, Nikcevic G et al. Expression of the catalytic domain of myosin light chain kinase increases paracellular permeability. Am J Physiol 1996; 271:C1678-C1684.
27. Kitamura T, Brauneis U, Gatmaitan Z et al.. Extracellular ATP, intracellular calcium and canalicular contraction in rat hepatocyte doublets. Hepatology 1991; 14(4 Pt 1):640-647.

28. Mitic LL, Anderson JM. Molecular architecture of tight junctions. Annu Rev Physiol 1998; 60:121-142.
29. Wong V, Gumbiner BM. A synthetic peptide corresponding to the extracellular domain of occludin perturbs the tight junction permeability barrier. J Cell Biol 1996; 136:399-411.
30. Keon BH, Schafer S, Kuhn C et al. Symplekin, a novel type of tight junction plaque protein. J Cell Biol 1996; 134(4):1003-1018.
31. Gottardi CJ, Arpin M, Fanning AS et al. The junction-associated protein, zonula occludens-1, localizes to the nucleus before maturation and during the remodeling of cell-cell contacts. Proc Natl Acad Sci USA 1996; 93:10779-10784.
32. Balda MS, Matter K. Tight junctions. J Cell Sci 1998; 111:541-547.
33. Ridley AJ, Hall A. The small GTP-bonding protein rho regulates the assembly of focal adhesions and actin stress fibers in response to growth factors. Cell 1992; 70:389-399.
34. Nusrat A, Giry M, Turner JR et al. Rho protein regulates tight junctions and perijunctional actin organization in polarized epithelia. Proc Natl Acad Sci USA 1995; 92(23):10629-10633.
35. Hanks SK, Polte TR. Signaling through focal adhesion kinase. BioEssays 1997; 19:137-145.
36. Aungst B. Intestinal permeation enhancers. J Pharm Sci 2000; 89(4):429-442.
37. Donovan MD, Flynn GL, Amidon GL. The molecular weight dependence of nasal absorption: The effect of absorption enhancers. Pharm Res 1990; 7(8):808-815.
38. Thanou M, Florea BI, Langemeyer MWE et al. N-trimethylated chitosan chloride (TMC) improves the intestinal permeation of the peptide drug buserelin in vitro (Caco-2 cells) and in vivo (rats). Pharm Res 2000; 17(1):27-31.
39. Swenson ES, Milisen WB, Curatolo W. Intestinal permeability enhancement: Efficacy, acute local toxicity, and reversibility. Pharm Res 1994; 11(8):1132-1142.
40. Swenson ES, Curatolo WJ. Means to enhance penetration: (2) Intestinal permeability enhancement for proteins, peptides and other polar drugs: Mechanisms and potential toxicity. Adv Drug Deliv Rev 1992; 8:39-92.
41. LeCluyse EL, Sutton SC. In vitro models for selection of development candidates. Permeability studies to define mechanisms of absorption enhancement. Adv Drug Deliv Rev 1997; 23:163-183.
42. Swenson EC, Curatolo WJ. Intestinal permeability enhancement for proteins, peptides and other polar drugs: Mechanisms and potential toxicity. Adv Drug Deliv Rev 1992; 8(1):39-92.
43. Van Hoogdalem EJ, Wackwitz ATE, de Boer AG et al. Topical effects of absorption enhancing agents on the rectal mucosa of rats in vivo. J Pharm Sci 1990; 79(10):866-870.
44. Hovgaard L, Brondsted H, Nielsen HM. Drug delivery studies in Caco2 monolayers. II. Absorption enhancer effects of lysophosphatidylcholine. Int J Pharm 1995; 114:141-149.
45. Shimazaki T, Tomita M, Sadahiro S et al. Absorption-enhancing effects of sodium caprate and palmitoyl carnitine in rat and human colons. Dig Dis Sci 1998; 43(3):641-645.
46. Knipp GT, Ho NF, Barsuhn CL et al. Paracellular diffusion in Caco2 cell monolayers: Effect of perturbation on the transport of hydrophilic compounds that vary in charge and size. J Pharm Sci 1997; 86(10):1105-1110.
47. Anderberg EK, Lindmark T, Artursson P. Sodium caprate elicits dilatations in human intestinal tight junctions and enhances drug absorption by paracellular route. Pharm Res 1993; 10:857-864.
48. Tomita M, Hayashi M, Awazu S. Absorption-enhancing mechanism of EDTA, caprate and decanoylcarnitine in Caco2 cells. J Pharm Sci 1996; 85(6):608-611.
49. Lindmark T, Kimura Y, Artursson P. Absorption enhancement through intracellular regulation of tight junction permeability by medium chain fatty acids in Caco2 cells. J Pharmacol Exp Ther 1998; 284(1):362-369.
50. Gan LS, Thakker DR. applications of the Caco2 model in the design and development of orally active drugs: Elucidation of biological and physical barriers posed by the intestinal epithelium. Adv Drug Deliv Rev 1997; 23:80-100.
51. Liu DZ, LeCluyse EL, Thakker DR. Dodecylphosphocholine-mediated enhancement of paracellular permeability and cytotoxicity in Caco2 cell monolayers. J Pharm Sci 1999; 88(11):1161-1168.
52. Citi S. Protein kinase inhibitors prevent junction dissociation induced by low extracellular calcium in MDCK epithelial cells. J Cell Biol 1992; 117:169-178.
53. Hochman J, Artursson P. Mechanisms of absorption enhancement and tight junction regulation. J Contr Rel 1994; 29:253-267.
54. Sakai M, Imai T, Ohtake H et al. Cytotoxicity of absorption enhancers in Caco2 cell monolayers. J Pharm Pharmacol 1998; 50:1101-1108.
55. Gordon GS, Moses AC, Silver RD et al. Nasal absorption of insulin: Enhancement by hydrophobic bile salts. Proc Natl Acad Sci USA 1985; 82(21):7419-7423.

56. Pontiroli AE, Alberetto M, Pozza G. Metabolic effects of intranasally administered glucagon: Comparison with intramuscular and intravenous injection. Acta Diabetol Lat 1985; 22(2):103-110.
57. Haffejee N, Du Plessis J, Muller DG et al. Intranasal toxicity of selected absorption enhancers. Pharmazie 2001; 56(11):882-888.
58. Kirkpatrick FH, Gordesky SE. Marinetti GV. Differential solubilization of proteins, hospholipids, and cholesterol of erythrocyte membranes by detergents. Biochim Biophys Acta 1974; 345(2):154-161.
59. Kirkpatrick FH, Sandberg HE. Effect of anionic surfactants, nonionic surfactants and neutral salts on the conformation of spin-labeled erythrocyte membrane proteins. Biochim Biophys Acta 1973; 298(2):209-218.
60. Obata Y, Sesumi T, Takayama K et al. Evaluation of skin damage caused by percutaneous absorption enhancers using fractal analysis. J Pharm Sci 2000; 89(4):556-561.
61. Hochman JH, Fix J, LeCluyse EL. In vitro and in vivo analysis of the mechanism of absorption enhancement by palmitoyl carnitine. J Pharmacol Exp Ther 1994; 269(2):813-822.
62. Duizer E, van der Wulp C, Versantvoort C et al. Absorption enhancement, structural changes in tight junctions and cytotoxicity caused by palmitoyl carnitine in Caco2 cells and IEC-18 cells. J Pharmacol Exper Ther 1998; 287(1):395-287.
63. Lindmark T, Schipper N, Lazorova L et al. Absorption enhancement in intestinal epithelial Caco2 monolayers by sodium caprate: Assessment of molecular weight dependence and demonstration of transport routes. J Drug Target 1998; 5(3):215-223.
64. Tomita M, Hayashi M, Awazu S. Absorption-enhancing mechanism of sodium caprate and decanoyl carnitine in Caco2 cells. J Pharmacol Exp Ther 1995; 272:739-743.
65. Cho SY, Kim JS, Li H et al. Enhancement of paracellular transport of heparin disaccharide across Caco2 cell monolayers. Arch Pharm Res 2002; 25(1):86-92.
66. Sears CL, Kaper JB. Enteric bacterial toxins: Mechanisms of action and linkage to intestinal secretion. Microbiol Rev 1996; 60:167-215.
67. Baudry B, Fasano A, Ketley J et al. Cloning of a gene (zot) encoding a new toxin produced by Vibrio cholerae. Infect Immun 1992; 60(2):428-434.
68. Fasano A, Baudry B, Pumplin D et al. Vibrio cholerae produces a second enterotoxin, which affects intestinal tight junctions. Proc Natl Acad Sci USA 1991; 88:5242-5246.
69. Uzzau S, Fasano A. Cross-talk between enteric pathogens and the intestine. Cell Microbiol 2000; 2(2):83-89.
70. Uzzau S, Cappuccinelli P, Fasano A. Expression of Vibrio cholerae zonula occludens toxin and analysis of its subcellular localization. Microb Pathog 1999; 27:377-385.
71. Di Pierro M, Lu R, Uzzau S et al. Zonula occludens toxin structure-function analysis: Identification of the fragment biologically active on tight junctions and of the Zonulin receptor binding domain. J Biol Chem 2001; 276(22):19160-19165.
72. Fasano A, Uzzau S, Fiore C et al. The enterotoxic effect of Zonula occludens toxin on rabbit small intestine involves the paracellular pathway. Gastroentrology 1997; 112:839-846.
73. Wang W, Uzzau S, Goldblum S et al. Human Zonulin, a potential modulator of intestinal tight junctions. J Cell Sci 2000; 113:4435-4440.
74. Fasano A. Intestinal zonulin: Open sesame! Gut 2001; 49(2):159-162.
75. Fasano A. Toxins and the gut: Role in human disease. Gut 2002; 50(III):iii9-iii14.
76. Watts T, Not T, Berti I et al. Alterations in zonulin secretion parallels changes in intestinal permeability and precede the onset of diabetes in BB/Wor diabetic rats. J Invest Med 2000; 48:184.
77. Fasano A, Fiorentini C, Donelli G et al. Zonula Occludens Toxin modulates tight junctions through protein kinase C-dependent actin reorganization, in vitro. J Clin Invest 1995; 96(2):710-720.
78. Uzzau S, Lu R, Wang W et al. Purification and preliminary characterization of the zonula occludens toxin receptor from human (Caco2) and murine (IEC6) intestinal cell lines. FEMS Microbiol Lett 2001; 194:1-5.
79. Lu R, Wang W, Uzzau S et al. Affinity purification and partial characterization of the zonulin/zonula occludens toxin (Zot) receptor from human brain. J Neurochem 2000; 74(1):320-326.
80. Hidaka H, Fidge NH. Affinity purification of the hepatic high-density lipoprotein receptor identifies two acidic glycoproteins and enables further characterization of their binding properties. Biochem J 1992; 284:161-167.
81. Marinaro M, Di Tommaso A, Uzzau S et al. Zonula occludens toxin is a powerful mucosal adjuvant for intranasally delivered antigens. Infect Immun 1999; 67(3):1287-1291.
82. Fasano A. Cellular microbiology: Can we learn cell physiology from microorganisms? Am J Physiol 1999; 76:C765-C776.

83. Fasano A, Uzzau S. Modulation of intestinal tight junctions by Zonula Occludens toxin permits enteral administration of insulin and other macromolecules in an animal model. J Clin Invest 1997; 99(6):1158-1164.

84. Paganelli R, Levinsky RJ, Brostoff J et al. Immune complexes containing food proteins in normal and atopic subjects after oral challenge and effect of sodium cromoglycate on antigen absorption. Lancet 1979; 1(8129):1270-1272.

85. Bjarnason I, Peters TJ, Veall N. A persistent defect in intestinal permeability in coeliac disease demonstrated by a ^{51}Cr-Labelled EDTA absorption test. Lancet 1983; 1(8320):323-325.

86. Hollander D. Crohn's disease—a permeability disorder of the tight junction?. Gut 1988; 29(12):1621-1624.

87. Uzzau S, Fiore C, Margaretten K et al. Modulation of intestinal tight junctions: A novel mechanism of intestinal secretion. Gastroentrology 1996; 110:370a, (Abstr).

88. Watts TL, Alexander T, Hansen B et al. Utilizing the paracellular pathway: A novel approach for the delivery of oral insulin in diabetic rhesus macaques. J Invest Med 2000; 48:185.

89. Karyekar CS, Fasano A, Raje S et al. Zonula occludens toxin increases the permeability of molecular weight markers and chemotherapeutic agents across the bovine brain microvessel endothelial cells. J Pharm Sci 2003; 92(2):414-423.

90. Cox D, Gao H, Raje S et al. Enhancing the permeation of marker compounds and enaminone anticonvulsants across Caco2 monolayers by modulating tight junctions using zonula occludens toxin. Eur J Pharm Biopharm 2001; 52(2):145-150.

91. Cox D, Raje S, Gao H et al. Enhanced permeability of molecular weight markers and poorly bioavailable compounds across Caco2 cell monolayers using the absorption enhancer, zonula occludens toxin. Pharm Res 2002; 19(11):1680-1688.

92. Salama NN, Fasano A, Lu R et al. Effect of the biologically active fragment of Zonula Occludens Toxin, ΔG, on the intestinal paracellular transport and oral absorption of mannitol. Int J Pharm 2003; 251(1-2):113-121.

93. Salama NN, Fasano A, Thaker M et al. The effect of ΔG on the transport and oral absorption of macromolecules. J Pharm Sci 2004; 93(5):1310-1319.

94. Salama NN, Fasano A, Eddington ND. A significant increase in the transport and oral bioavailability of paracellular markers, antiviral and immunosuppressant drugs achieved by reversible tight junction opening with Delta G, a novel absorption enhancer. 18th Annual Meeting and Exposition of the American Association of Pharmaceutical Scientists. Salt Lake City, Utah: 2003:Oct 26-30 [Abstract].

95. Salama NN, Fasano A, Thakar M et al. The effect of ΔG on the oral bioavailability of low bioavailable therapeutic agents. J Pharmacol Exp Ther 2005; 312(1):199-205.

96. Rossi M, Maurano F, Luongo D et al. Zonula Occludens toxin (Zot) interfers with the induction of nasal tolerance to gliadin. Immunol Lett 2002; 81:217-221.

Index

Mammalian embryo 165, 170
Membrane-associated guanylate kinase
  (MAGUK) 8, 64, 65, 70, 76, 77, 79-81,
  83-85, 96, 102, 104, 110, 118, 122, 123,
  126, 166, 167, 184, 208
Metastasis 28, 116-120, 121, 122, 124, 157,
  211
Metazoan 1-3, 5, 6, 8-11, 13
Mitogen-activated protein kinase (MAPK) 27,
  109, 155, 157
Morphology 23, 27, 69, 164, 178, 181
MP20 33, 35-37
MUPP1 34, 35, 48, 49, 71, 103, 104, 117,
  118, 123, 124, 126, 182-184, 201
Mutant mice 39, 116, 121
Myelin 36, 40, 196-201, 203, 204
Myosin light chain (MLC) 136, 137, 141,
  186, 208
Myosin light chain kinase (MLCK) 136, 137,
  208

N

Na^+-glucose cotransport 137, 212
Na^+,K^+-ATPase 4-9, 12, 146, 155-158, 165
NACos 76, 77, 104, 146, 147, 157
Nuclear shuttling 80, 93
Null mice 19, 28, 34, 39, 200

O

OAP-1 202-204
OAP-1/Tspan-3 202-204
Occludin 10, 19-28, 34, 43, 44, 48, 49, 57-
  59, 64-68, 102, 103, 109, 110, 117, 118,
  120-122, 125, 136-140, 148-150,
  153-155, 164-166, 168, 169, 171, 175,
  178-182, 184, 199, 201, 206-208
Oligodendrocyte 39, 70, 102, 103, 196, 198,
  199, 201, 204
Oligodendrocyte-specific protein (OSP) 39,
  196-199, 201-204
Origin of metazoan 13
OSP/claudin-11 196, 198, 199, 201, 202-204
Ouabain 4, 9, 146, 147, 155-158

P

PALS1 70, 80, 118
PAR-3 48, 49, 70, 79, 103, 117, 118, 121,
  123, 136, 182-184, 201
PAR-6 49, 70, 79, 80, 81, 83, 85, 89, 103,
  118, 123, 136, 182-184, 201
Paracellin 38, 151
Paracellular permeability 23, 26, 43, 49, 50,
  68, 70, 110, 125, 126, 135, 137-141,
  149, 151, 154, 156, 170, 178, 180, 183,
  185, 208, 211
Paracellular transport 33, 40, 135-137, 207,
  209, 210
PATJ 48, 49, 66, 67, 70, 103, 118, 124, 126,
  182-184
PDZ 34, 35, 38, 39, 43-46, 48-50, 64, 65,
  67, 77, 79, 80, 84, 86, 92, 102, 104, 110,
  117, 118, 123-126, 146, 151, 167, 183,
  184, 187, 201
  PDZ binding motif 34, 35, 43-46, 48, 50,
    151
  PDZ domain 34, 35, 38, 39, 43, 45, 48,
    49, 64, 65, 67, 84, 102, 104, 110, 118,
    126, 183, 184, 201
Peripheral myelin protein 22 (PMP-22) 33,
  35-37, 39, 40, 196-198, 201
Permeability 2, 3, 23, 26, 34, 36, 38-40, 43,
  49, 50, 67-70, 110, 119, 120, 123, 125,
  126, 135-141, 146-149, 151, 153, 154,
  156, 164, 166, 170, 175, 177-181, 183,
  185-187, 207-213, 215
Pertussis toxin (PTX) 185
P-face 21, 23, 175-178, 180, 181
Phosphatidylinositol-3 kinase (PI3K) 103,
  109, 110, 117, 122, 123
Phosphorylation 4, 25-28, 50, 61, 105, 136,
  137, 139, 141, 146, 147, 151, 152, 155,
  168, 169, 180, 186, 208
PLP/DM20 196, 199-201, 204
Polarity 1-6, 8, 9, 28, 48, 49, 60, 65, 67, 68,
  70, 71, 77, 79, 103, 116-120, 122, 123,
  126, 136, 164, 165, 167, 178, 183, 184,
  187
Polychaetoid 102
Preimplantation 170
Protein kinase A (PKA) 27, 68, 140, 185, 206
Protein kinase C (PKC) 25-27, 50, 61, 118,
  123, 137, 139, 147, 154, 157, 169-171,
  182, 183, 185, 201, 204, 208, 212
Protein phosphatase 27, 118, 183
Proteolipid protein (PLP) 196, 198-201, 204
PSD-95 65, 70, 118
PTEN 103, 104, 110, 118, 122, 123, 154,
  182, 183